Astrobiology

Springer
Berlin
Heidelberg
New York
Barcelona
Hong Kong
London
Milan
Paris
Tokyo

Physics and Astronomy

ONLINE LIBRARY

http://www.springer.de/phys/

Gerda Horneck
Christa Baumstark-Khan (Eds.)

Astrobiology

The Quest
for the Conditions of Life

With 133 Figures
Including 39 Color Figures

Springer

Dr. Gerda Horneck
Dr. Christa Baumstark-Khan
DLR
Institut für Luft- und Raumfahrtmedizin
Linder Höhe
51147 Köln, Germany
e-mail: gerda.horneck@dlr.de
    christa.baumstark-khan@dlr.de

Cover picture: An artist's view of the DARWIN mission of the European Space Agency. DARWIN will search for the signature of habitable or life-bearing planets around nearby stars from 2012 onwards. See also the contribution by B.H. Foing in this volume. (Courtesy by ESA and ALCATEL)

Library of Congress Cataloging-in-Publication Data. Astrobiology : the quest for the conditions of life/ Gerda Horneck, Christa Baumstark-Khan, eds. p. cm. Includes bibliographical references and index. ISBN 3540421017 (alk. paper) 1. Life–Origin. 2. Exobiology. I. Horneck, G. (Gerda) II. Baumstark-Khan, C. (Christa) QH325.A78 2001 576.8'3–dc21 2001049849

ISBN-13:978-3-642-63957-9   e-ISBN-13:978-3-642-59381-9
DOI: 10.1007/ 978-3-642-59381-9

Springer-Verlag Berlin Heidelberg New York
a member of BertelsmannSpringer Science+Business Media GmbH

http://www.springer.de

© Springer-Verlag Berlin Heidelberg 2002
Softcover reprint of the hardcover 1st edition 2002

Typesetting: camera-ready copies by the editors
Cover design: Erich Kirchner, Heidelberg

Printed on acid-free paper   SPIN 10792081   54/3141/ba   5 4 3 2 1 0

Preface

This book guides the reader into the fascinating world of the newly emerging science of astrobiology. Its central focus is directed towards questions that have intrigued humans for a long time: Where do we come from? What is life? Are we alone in the Universe? They are jointly tackled by scientists converging from widely different fields, reaching from astrophysics to molecular biology and from planetology to ecology, among others. This spilling beyond the boundaries of classical sciences opens completely new opportunities for research, a state described by some contemporaries as the "Astrobiology Revolution of the Sciences". The book is written in such a way that on the one hand it provides the latest knowledge in this rapidly evolving field of astrobiology for the educated expert, while on the other hand describing most phenomena in a general and understandable way; it is thus also intended for interested laymen who are attracted to this new discipline.

In the first section on "Organic Material in Space and Habitable Zones" we invite the reader to explore the vast realms of the Universe for signatures of life beyond the Earth. In the interstellar medium, as well as in comets and meteorites, complex organics teem in huge reservoirs that eventually may provide the chemical ingredients for life. Astronomers are rapidly acquiring data on the existence of more and more planetary systems in our Galaxy, which supports the assumption that habitable zones are frequent and are not restricted to our own solar system. Within such a habitable zone, life may not be confined to its planet of origin: the impact scenario describes a natural mechanism of expulsion and transport of microbial communities through space.

In the following three sections, the physical and chemical conditions for life are discussed, the environmental requirements and boundaries for life on Earth, and the extraordinary capabilities of life to adapt to environmental extremes. These data are used to assess the habitability of other bodies within our solar system, especially Mars and Europa which are located within habitable zones. Among the wide field of environmental conditions life has to cope with, the most essential ones are discussed in detail. These are: water, one of the prerequisites for life; stresses associated with low levels of water; temperature extremes; electromagnetic fields and radiation, with an emphasis on environmental UV radiation and ionizing radiation; and gravity as a constant source of stress for life on planets.

Contemplating life, its origin, evolution and distribution, within the context of cosmic evolution, shows that it is the same principle that drives evolution towards increasing complexity, from the formation of the first elements to the self-organization of life and the appearance of consciousness. The section "Complexity and Life" presents examples of the emergence of complexity in astrobiological issues in general. A prominent example of this phenomenon is provided by molecular self-assembly in the context of the origin of life.

The book concludes with a description of ongoing or planned space missions from which we expect answers to the burning questions in astrobiology. Examples are astronomical missions which search for "biomarkers" in our Galaxy, planetary missions with targets of astrobiological interest, such as Mars, Saturn's moon Titan, the comet Wirtanen, and Jupiter's moon Europa, as well as experiments in Earth orbit on the likelihood of interplanetary transfer of life and the resistance of life to environmental extremes.

Most authors of this book were recruited from the participants of the 1st Symposium on Exo/Astrobiology in Germany, that took place on 22 March 2000 in Bremen, during the Annual Meeting of the Deutsche Physikalische Gesellschaft. We are grateful to the German Aerospace Center DLR for their support of this symposium. When, after the meeting, Christian Caron from the Astronomy and Space Sciences Editorial Department of Springer encouraged us to edit a book on Astrobiology, we realized that most of the areas that should be represented in one of the first books in astrobiology could be competently covered by the participants of the symposium. Where necessary, the authorship was complemented by colleagues and experts in astrobiology from other countries in Europe and from the US. Our special thanks go to Christian Caron, who invested so much interest in discussions on the structure and content of the book, in finding the right balance between exact and popular science and we thank him also for his patience during the finalization of the book. We appreciate the assistance of Lisa Steimel, our former secretary at DLR, for her invaluable help during the finalization of the manuscripts, including careful proofreading.

Köln, Germany   *Gerda Horneck*
August 2001   *Christa Baumstark-Khan*

Contents

Part III   Electromagnetic Fields, Radiation and Life

13   Martian Atmospheric Evolution:
   Implications of an Ancient Intrinsic Magnetic Field
Helmut Lammer, Willibald Stumptner and Gregorio J. Molina-Cuberos 203

14   The Ultraviolet Radiation Environment of Earth and Mars:
   Past and Present
Charles S. Cockell 219

15   Ultraviolet Radiation in Planetary Atmospheres
   and Biological Implications
Petra Rettberg and Lynn J. Rothschild 233

Part IV Gravity and Life

18 Graviperception and Graviresponse at the Cellular Level
Richard Bräucker, Augusto Cogoli and Ruth Hemmersbach 287

19 Gravistimulated Effects in Plants
Heide Schnabl..297

20 Gravitational Zoology: How Animals Use and Cope with Gravity
Ralf H. Anken and Hinrich Rahmann..315

Part V Complexity and Life

Introduction

Baruch S. Blumberg

A mission statement for astrobiology is: "Study the origin, evolution, distribution, and future of life on Earth and in the Universe." It addresses the profound questions "How did life originate?", "Are we alone in the Universe?", and "What is the future of life on Earth and in the Universe?" The wide range in the contents of this volume indicates the rich and exciting subject matter of this field. Much of this content can be grouped into three major components.

First, we know that life exists and, therefore, it must have originated; the first component is the study of how life began and evolved to its present form, with an emphasis on the early evolutionary processes.

Second, we don't know if life exists somewhere other than on Earth, and therefore the second component is to test the hypothesis that there is life elsewhere. To do this we must create models to test against the data that has been and will be collected at locations remote from Earth. The only observationally derived model for life that is available is life on Earth and, therefore, it is the model for the hypothesis-testing process. We fully expect that this model may be rejected to be replaced by another currently unknown model that is difficult to even imagine at our present level of knowledge. The major targets for the search for life in our solar system – Mars, Europa, Ganymede, etc. – are more similar to early Earth, i.e. the Earth that existed soon – say, within a billion years or so – after life began. Hence, the model for the search for life elsewhere is the life that existed early in the history of our home planet. The conditions at that time were extreme compared to the conditions in the regions that most animals currently occupy. As a consequence, there is a major program in astrobiology to study contemporary life in extreme environments: geothermal sites; regions with low and high temperatures; arid regions; regions with high or low pH, high or low salinity, low oxygen, and high or low pressures; and deep subsurface areas under the ground and under the bottom of the sea. The areas in and adjacent to impact craters are also of interest since impacts that could have affected life were more common in early Earth and on Mars and elsewhere in the solar system where life might now exist or have existed in the past.

There is also a model that unites the first two components. Life may have originated somewhere other than on Earth and may have been transported here in a comet or meteor, on "cosmic dust particles", or, possibly, as free-floating living material. An additional view is that the basic chemical components that make up life – prebiotic chemistry – arrived from elsewhere and that life was subsequently formed here.

The third component deals with the future of life on Earth and, if it exists, elsewhere in the Universe. It includes the study of changes in the Earth's environment and its inhabitants that may influence the existence of life, and of the conditions on Mars or elsewhere that could have destroyed or diminished life if it existed in the past. Independent of the existence of life away from Earth, this component relates to the broader question of how earthly life, including humans, can survive and explore away from the home planet, in near-Earth orbits and distant from Earth during interplanetary – particularly Martian – and deep-space travel. It connects at this level with the issue of astronaut and cosmonaut health and the basic science questions related to long trips through space and survival on distant planets and moons. The futuristic concept of terraforming, transforming places like Mars to be more like Earth, and colonization of distant planets could be included in this rubric.

In testing a hypothesis, for example that life exists elsewhere in the Universe, it is important to maintain an unbiased view on the outcome. If this is not done, there is a danger that data that is collected will be interpreted to support the favored result. It is easier to maintain an unbiased view if both outcomes, support or rejection, are interesting. Many people, including scientists, have an intuitive belief that the Universe is life-rich and that, in its immensity, there must be other inhabited places. However, when acting in the scientist mode, it is important to maintain objectivity in the evaluation of the data. The discovery of life elsewhere is obviously very interesting as it will open a vast area for exploration and research and radically change our perceptions of self and how we fit into the Cosmos. The rejection of the model will be difficult to do. More and more planets around stars other than our own Sun are being discovered and it is possible that life may exist on one or more of these. However, if the search is conducted for, say, 100 years, and life is not discovered, then it would decrease the probability that life exists elsewhere. We would have to consider that we are alone in this vast space. This also would have a profound effect on our perceptions of our role in the Universe and the future of humans in the Cosmos. It would reverse the concept of Copernican mediocrity, that we are the inhabitants of a relatively minor planet in a solar system that is one of billions situated somewhere away from the center of the Milky Way galaxy that itself is only one of billions in a universe that may not be unique. The realization of aloneness would profoundly change our attitudes.

Arising from these three components is an overriding question, "What is life?". Studying the origins of life on Earth and testing the hypothesis that life exists elsewhere will raise questions on its definition. For example, when does "prebiotic" chemistry become biology? Does the answer reside in the complexity of chemical interactions resulting in an emergent order that satisfies a definition of life? What are the differences between living bacteria themselves and the effects they have on rocks? This is illustrated in the ongoing investigations on the composition of the Mars meteorite ALH 84001. (The acronym and numbers indicate that the meteorite was the first found during the 1984 collecting season in the Alan Hills region of Antarctica.) Does it contain evidence of previous life? Are the very small ordered structures found in the rock the fossils of ancient Martian organisms or are they artifacts of geological or experimental origin? Does the magnetite in the meteorite originate from biological material or is it a non-organic structure? If it is biological, when did it stop being living matter and become mineral, i.e. what are the distinctions between life and non-life?

Astrobiology requires extensive international cooperation in order to be successful. A single nation cannot hope to alone accomplish this mission that is a quest of importance to all humankind. The United States National Aeronautics and Space Administration (NASA), to help advance the field, established the NASA Astrobiology Institute (NAI). It is structured to recognize and serve the specific requirements of astrobiology. I will describe the organization and early history of NAI to illustrate these characteristics. In the mid-1990s there was an increased interest in origins of life for the overall NASA program. It was fueled by several scientific reports, including the MARS satellite images that indicated that water might have been common early in the planet's history, the Jupiter Galileo probe, and, in particular, the discoveries related to ALH 84001. The NAI is a virtual institute, in the sense that several teams are located great distances from each other, but, despite this, they are expected to act as if they were members of a single organization. The NAI Mars Focus Group is helping in site selection of interest to astrobiology and is taking a major role in Mars Mission planning. The NAI Mission to Early Earth has assembled geologists, paleontologists, molecular biologists, and planetary scientists, to probe the nature of the earliest forms of life on Earth, as a model for the search for life on Mars and elsewhere in the Universe. The NAI Europa Focus Group is starting the astrobiological planning for a series of missions that are planned to take place during the next few decades. There are also Focus Groups, either in place or planned, for research on earthly meteorite impact sites, the study of Mars meteorites, mixed evolutionary genomics, the evolution of oxidation, and other topics. This has resulted in mutual cooperation between the teams and expanded the value of the NAI to each of the teams and to the field in general. Much of the work in astrobiology is focused on field trips, both on Earth and in the more elaborate and visionary field trips into space.

Concurrent with the activities of the NAI, European scientists interested in the various fields of astrobiology have established a European Exo/Astrobiology Network, that combines the astrobiology activities of different national groups. It had its inauguration assembly during the 1st European Workshop on Exo/Astrobiology, held from 21 to 23 May 2001 at ESRIN, Frascati, Italy. The current European member countries are Austria, Belgium, Denmark, France, Germany, Italy, Portugal, Spain, Sweden, Switzerland, The Netherlands, and the United Kingdom. The addition of other countries, such as Russia, is under consideration. Several European national groups, as well as the Australian astrobiology group, are in the process of entering into affiliations or associations with NAI. Hitherto, the Spanish Centro de Astrobiologia and the UK Astrobiology Forum have concluded agreements with NAI.

Although astrobiology stresses the importance of interdisciplinary research, it is also attempting to establish a new scientific field. There has been remarkable progress in this respect. Several journals dedicated to astrobiology and related fields are being established, there are astrobiology sections meeting in several of the related professional societies, graduate and undergraduate programs in the field are being established, and there is growing interest among younger scientists in entering the specialty. Astrobiology, and indeed most of space science, is a very long-term proposition. For example, under present conditions it will take 10 years or more for a round-trip mission to Europa, and the planning itself may take many years; a program to test a crucial hypothesis may take decades. A scientist in mid-career, planning a space astrobiology mission, may not be scientifically active long enough to allow the personal evaluation of the results of his or her experiment. It is necessary to think and act generationally, to be able to dedicate oneself to an activity that will be carried on and concluded by our scientific and, possibly, actual children. Space science is similar to the building of the medieval cathedrals that often took centuries, and required the passing on of skills from one generation of skilled workers to another.

There is a strong sense of mission in astrobiology. It will affect the currently planned missions of NASA, ESA and other space agencies and will lead to the design of missions that are generated mostly by the needs of the field. There are exciting prospects in the immediate future for addressing the questions on the origins and future of life on Earth and in the Universe, fundamental to human questing, that can be answered, in part, by the applications of the scientific process. The contents of this book will show how far the field has advanced and how it may proceed in the future.

Part I

Organic Material in Space and Habitable Zones

1   From Molecular Clouds to the Origin of Life

Pascale Ehrenfreund and Karl M. Menten

In our Milky Way and in external galaxies, the space between the stars is filled with an interstellar medium (ISM) consisting of gas and dust. The ISM can be divided in various different components with very different physical parameters, ranging from a very hot (10^6 K), dilute ($<10^{-2}$ particles cm^{-3}) component heated by supernova explosions, which fills more than half of its volume, to molecular clouds with temperatures from 10 to 100 K and molecular hydrogen (H_2) densities from a few hundred to 10^8 particles cm^{-3}, which make up less than 1% of the ISM's volume. Approximately 1% of its mass is contained in microscopic (micron-sized) interstellar dust.

While accounting for only a small fraction (~1%) of the Galaxy's mass, the ISM is nevertheless an important part of the Galactic ecosystem. Gravitational collapse of dense interstellar clouds leads to the formation of new stars, which produce heavier elements in their interiors by nucleosynthesis. The main reaction in stellar interiors is the nuclear fusion of H into He. In a later stage of stellar evolution C, N and O are formed. Further nucleosynthesis occurs in massive stars leading to elements as heavy as iron. At the end of their lifetime which is mainly determined by their initial mass, stars return material to the interstellar environment by mass outflows, forming expanding shells and envelopes, as well as by violent explosions. Thus, the ISM represents an environment, in which atoms, molecules, and solid matter undergo strong evolution and recycling.

Observations at radio, millimeter, sub-millimeter, and infrared wavelengths have led to the discovery of well over a hundred different molecules in interstellar clouds and circumstellar shells (see Table 1.1.). Many of these are organic species of considerable complexity, with $HC_{11}N$ [1] and diethyl ether [$(C_2H_5)_2O$] [2] being the largest detected so far. In all cases, accurate line frequencies predicted by laboratory spectroscopy ensure an unambiguous identification. These molecules, which have abundance ratios relative to molecular hydrogen of less, and in many cases much less than 10^{-4}, are important tracers of the physical conditions and chemistry of many different interstellar and circumstellar environments. Despite their relatively low abundances, the variety and complexity of organic compounds currently detected in space indicates an active chemistry and ubiquitous distribution [3, 4]. Extraterrestrial organics may have played a role in the origin and evolution of life [5-7].

The dense cold phases of the interstellar medium also host icy dust grains, whose molecular composition has recently been well constrained by observations from the Infrared Space Observatory (ISO) [8, 9]. These dust particles are important chemical catalysts and trigger molecular complexity in the interstellar gas and dust. In contrast to the case of gas phase molecules, the exact nature of interstellar solids cannot be unambiguously identified by astronomical observations but can be constrained by laboratory techniques. Our knowledge of the carbonaceous solid state and gas phase inventories of molecular clouds has recently been reviewed [4].

Numerous protoplanetary disks have been imaged with the Hubble Space Telescope (HST), indicating that the formation of extrasolar systems similar to our own solar system (Fig. 1.1) is a rather common process [10, 11, see Chapt. 2, Udry and Mayor]. During star formation, interstellar molecules and dust become the building blocks for protostellar disks, from which planets, comets, asteroids, and other macroscopic bodies form [12].

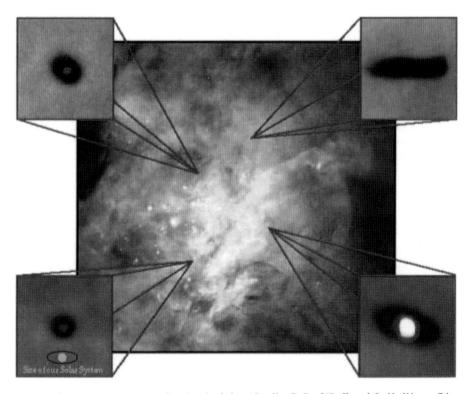

Fig. 1.1 Planetary systems now forming in Orion (Credit: C. R. O'Dell and S. K. Wong, Rice University, and NASA). The Orion nebula is a star-forming region located in the constellation Orion the Hunter about 1500 light-years away. The optically visible nebula is excited by one of the young massive stars that formed here about one million years ago together with thousands of lower mass stars. Many of the low mass stars are still surrounded by disks of placental cloud material of gas and dust that formed during the protostellar collapse. Using the Hubble Space Telescope, various of such protoplanetary disks have been detected in silhouette against the nebular emission background. The above mosaic shows several examples. In the bottom left insert the relative size of our own Solar System is shown for comparison. The discovery of protoplanetary disks around other stars provides strong evidence for the paradigm of solar system first proposed by Kant and Laplace.

Table. 2.1. Interstellar and circumstellar molecules as compiled per October 2000. Observations indicate the presence of molecules larger than 12 atoms, such as polycyclic aromatic hydrocarbons (PAHs), fullerenes and others in the interstellar medium.

Number of Atoms

2	3	4	5	6	7	8	9	10	11	12+
H_2	C_3	$c\text{-}C_3H$	C_5	C_5H	C_6H	CH_3C_3N	CH_3C_4N	CH_3C_5N?	HC_9N	C_6H_6
AlF	C_2H	$l\text{-}C_3H$	C_4H	$l\text{-}H_2C_4$	CH_2CHCN	$HCOOCH_3$	CH_3CH_2CN	$(CH_3)_2CO$		$HC_{11}N$
AlCl	C_2O	C_3N	C_4Si	C_2H_4	CH_3C_2H	CH_3COOH?	$(CH_3)_2O$	NH_2CH_2COOH?		PAHs
C_2	C_2S	C_3O	$l\text{-}C_3H_2$	CH_3CN	HC_5N	C_7H	CH_3CH_2OH			C_{60}^+
CH	CH_2	C_3S	$c\text{-}C_3H_2$	CH_3NC	$HCOCH_3$	H_2C_6	HC_7N			
CH^+	HCN	C_2H_2	CH_4	CH_3OH	NH_2CH_3		C_8H			
CN	HCO	CH_2D^+?	HC_3N	CH_3SH	$c\text{-}C_2H_4O$					
CO	HCO^+	HCCN	HC_2NC	HC_3NH^+						
CO^+	HCS^+	$HCNH^+$	HCOOH	HC_2CHO						
CP	HOC^+	HNCO	H_2CHN	NH_2CHO						
CSi	H_2O	HNCS	H_2C_2O	C_5N						
HCl	H_2S	H_2CO	H_2NCN							
KCl	HNC	H_2CN	HNC_3							
NH	HNO	H_2CS	SiH_4							
NO	MgCN	H_3O^+	H_2COH^+							
NS	MgNC	NH_3								
NaCl	N_2H^+	SiC_3								
OH	N_2O	CH_3								
PN	NaCN									
SO	OCS									
SO^+	SO_2									
SiN	$c\text{-}SiC_2$									
SiO	CO_2									
SiS	NH_2									
CS	H_3^+									
HF	H_2D^+									

Comets are agglomerates of frozen gases, ices, and rocky debris, and are likely to be the most primitive bodies in the solar system. Comets are formed in the outer solar system from remnant planetesimals that were not integrated into planets. Such comet nuclei were thrown into large high inclination orbits by perturbation of the major planets into the so-called Oort cloud at ~50 000 astronomical units (AU) from the sun. Comet nuclei that were formed near the plane of Neptune and beyond reside in the Edgewood-Kuiper belt [13]. When comets are perturbed and enter the solar system, solar radiation heats the icy surface and forms a gaseous cloud, the coma. During this sublimation process, "parent" volatiles are subsequently photolysed and produce radicals and ions, the so-called "daughter" molecules. Space missions to comet Halley and recent astronomical observations of two bright comets, Hyakutake and Hale-Bopp, allowed astronomers to make an inventory of cometary molecular species [14, 15]. Small bodies in the solar system, such as comets, asteroids, and their meteoritic fragments carry pristine material left over from the solar system formation process, thus sampling the molecular cloud material out of which the sun and planets formed.

1.1   The Search for Large Organic Molecules in Dense Clouds

To date, more than 120 molecules have been detected in the interstellar medium and the circumstellar environments of red giant stars (i.e. stars at the end of their life cycle, see Table 1.1.). Most of these have rotational spectra at radio-, millimeter-, and submillimeter wavelengths, which can be efficiently observed with modern telescopes and detector equipment. Many interstellar molecules only exist in dense clouds, where they are sufficiently shielded from UV irradiation. In particular, a number of prebiotic species such as H_2CO, HCN and NH_3 have been identified.

1.1.1   The Search for Amino Acids in the Interstellar Medium

For more than two decades, considerable effort has been devoted to various attempts of detecting the simplest amino acid, glycine (NH_2CH_2COOH) in interstellar molecular clouds. Glycine exists in a variety of conformations [16] and the extensive laboratory microwave spectroscopy of conformers I and II [17] allows meaningful astronomical searches. The very large partition function of a species as complex as glycine results in relatively weak lines which are difficult to detect in dense, warm molecular cloud cores due to confusion produced by a "forest" of weak lines from a large number of species that are also present in the region in question. Consequently, none of the searches conducted so far has resulted in an unambiguous detection; the best upper limits on the glycine-to-H_2 abundance ratio are of order 10^{-10} [18-25].

As discussed by Snyder [23], the use of high spatial resolution interferometric observations may bring some relief to the confusion problem as various classes of molecules (e.g., O-rich vs. N-rich species) have different spatial distributions and/or radial velocities [26].

The formation routes of amino acids in interstellar gas and dust are not yet established. Charnley [27] proposed that protonated glycine could be formed in hot cores by reaction of protonated aminomethanol and HCOOH.

$$NH_2CH_2OH^+ + HCOOH \rightarrow NH_2CH_2COOH^+ + H_2O \qquad (1.1)$$

Amino acids could also be formed in the solid phase by irradiation of interstellar grain mantles with ultraviolet photons [28]. However, amino acids are highly susceptible to UV photo-destruction, even under exposure to UV photons of relatively low energy [29]. Though low concentrations of amino acids may be detected by astronomical observations at radio wavelength in UV shielded environments, such as the hot cores and the inner cometary coma, all environments with an elevated UV flux should have merely traces of these compounds. This may explain the lack of detection of these compounds in space.

1.1.2   Organic Molecules in our Galactic Center

(Sub)millimeter line surveys of well known sources, such as the dense cores in the Orion and Sagittarius molecular clouds [30-32], show that molecules of considerable complexity can be found in these regions and others might yet have to be detected [25]. A large part of the molecular complexity found in these regions is due to gas-grain interactions. The sources mentioned above are examples of so-called "hot cores", dense (10^6 particles cm^{-3}) hot (~200 K) regions in the immediate vicinity of massive protostars. The high abundance of many molecular species found in hot cores cannot be explained by gas phase chemistry and one must invoke molecule formation on catalytic icy grain surfaces [3, 4] during the cold dark cloud phase of these cores. Heating by the newly formed protostar and/or energetic processes such as outflows producing shock waves lead to evaporation of the grain mantles into the gas, which is followed by gas-phase reactions. Pathways to the formation of larger molecules include alkyl cation transfer [25].

Figure 1.2 shows Berkeley-Illinois-Maryland Array (BIMA) observations near 3 mm wavelength of a methanol (CH_3OH) transition, the dust continuum emission, and the $12_{0,12}$-$11_{1,11}$ transition of methyl-ethyl-ether (MeOEt) toward the massive dense molecular cloud core Sagittarius B2(North) [Sgr B2(N)] near the Galactic center. These interferometer maps show a clear concentration of methanol around the Sgr B2(N) hot core (Fig. 1.2, top). The Sgr B2(N) regions is also known as the "Large Molecule Heimat" [23] due to the high concentrations and rich diversity of the complex organic molecules detected there. By contrast, few large organics are detected at Sgr B2(M) and this is generally believed to be because B2(M) is much more dynamically and chemically evolved; Sgr B2(N) appears to have only recently 'switched on' (less than about 10^4 years) and the contents of its evaporated mantles have not yet been destroyed in gas phase reactions. This situation is evident in Fig. 1.2 where MeOH and MeOEt are only present in the northern source. The map shown in Fig. 1.2 (bottom) shows that the MeOEt emission has a distribution that is very similar to that of the methanol, as one might expect if these molecules are chemically linked [25]. The BIMA data indicate [MeOEt/H_2] abundance ratios of 10^{-10}-10^{-9} in Sgr B2(N).

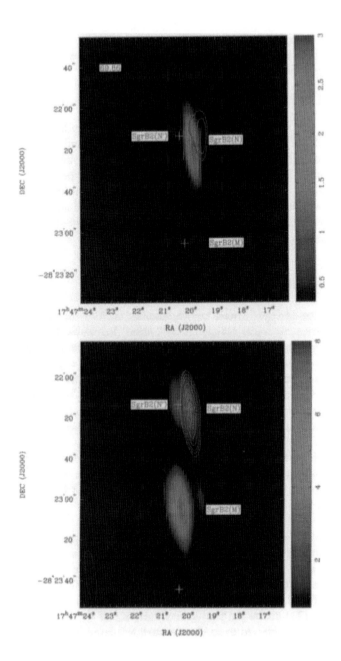

Fig. 1.2 BIMA maps of CH_3OH and MeOEt in Sgr B2(N) [25]. The contours roughly represent the strength of the molecular radiation. The size of the synthesized beam in each case is given by the ellipses in the lower-left corner. Top: CH_3OH $13_{-3}-14_{-2}$ at 84.4 GHz (contours) plus 3-mm continuum (color scale). For the 3 mm continuum the grey scale range is 0.51-6.5 Jy/beam. For MeOH, the sythesized beam was 21".91 x 4".74 and the contours are at 3, 4 and 5 sigma, with 1 rms=0.07 Jy/beam. Bottom: MeOEt $12_{0,12}-11_{1,11}$ at 79.6 GHz (contours) plus CH_3OH $13_{-3}-14_{-2}$ (color scale). The number on the upper-left corner is the LSR velocity. For MeOEt, the sythesized beam was 24".32 x 4".88 and the contours are at 3, 5, 7, 10, 20, 28 sigma, with 1 rms=0.15 Jy/beam. For MeOH, the gray scale range is 0.2-2.3 Jy/beam.

Radioastronomical observations are vital for the identification of large organic molecules in the interstellar medium and in cometary comae. Such observations help reconstructing the gas-grain chemical pathways in such regions.

1.2 Molecules in Protoplanetary Disks

In current scenarios of low mass star formation a protostar with an accretion disk and strong mass outflow is formed after gravitational collapse of a molecular cloud on a time-scale of 10^4-10^5 years [33, 34]. In its early evolutionary phases this protostar is still embedded in its placental cloud material. It then evolves into a T Tauri star and after ~10^6-10^7 years reaches the main sequence. During this phase, a planetary system may form. T Tauri stars are considered to resemble our sun when it was a few million years old. Studies of their surrounding gas and dust can therefore provide important clues on the early evolution of the solar nebula.

Infrared and millimeter surveys have shown that most T Tauri stars have circumstellar disks with masses of 10^{-3}-10^{-1} solar masses and sizes of 100-400 AU [35-37]. Such disks provide a reservoir of gas and dust for the formation of potential planetary systems [38]. For the purpose of discussing their chemistry, one may divide protoplanetary disks in a multi-layer structure consisting of a midplane, an intermediate region, and a surface region [39]. The densities and radiation fields in the surface region are similar to interstellar photodissociation regions. In the warm upper atmosphere of the disks exposed to both the central star and the interstellar radiation field, the molecules are rapidly destroyed by photodissociation [40]. The intermediate region has conditions similar to dense molecular clouds where ionmolecule and neutral-neutral reactions occur in addition to photochemistry. The midplane of the disk cannot be penetrated by UV photons at all and molecules freeze out onto dust grain surfaces.

Determining the molecular composition of such protostellar disks provides information about the protosolar nebula and the evolution of our solar system. Dutrey et al. [41] reported the detection of CN, HCN, HNC, CS, HCO^+, C_2H and H_2CO (ortho and para) in the protoplanetary disks of DM Tau and GG Tau. These systems are only 140 pc away and located in the outskirts of the Taurus molecular cloud complex. DM Tau is one of the oldest T Tauri stars in this region, whereas GG Tau is a young binary star. It was found that many of the above mentioned molecules are under-abundant with respect to standard dense clouds. The relatively large abundances of CN and C_2H indicate a rich photon-dominated chemistry [41]. It is worth pointing out that the small angular dimensions of the emission regions in question preclude detailed studies of the chemical and physical conditions on solar system size scales (i.e. less than 0.5 arcsec at the Taurus molecular cloud distance) with current millimeter-wavelength equipment. Future instruments such as the Atacama Large Millimeter Array (ALMA) combine the necessary spatial resolution and high sensitivity to allow such observations resolving size scales of a few astronomical units [42].

1.3   Diffuse Interstellar Clouds

Diffuse clouds are dominated by photochemistry. Such clouds have moderate extinctions (<1 mag) and densities of roughly 100-300 cm^{-3}. Interstellar extinction is measured in <u>mag</u>nitude (mag) at a given wavelength and reflects the starlight which is absorbed and scattered by dust grains and reaches the observer dimmed. The HST has provided many accurate ultraviolet measurements of metals like C, O, N, Mg, Si, S, Cr, Fe, Ge, Zn, and Kr, greatly advancing our knowledge on the metallicity of diffuse gas clouds [43]. In particular, measurements of the carbon abundance provide strong constraints on the budget available to form grains with carbonaceous mantles. Sightlines through diffuse clouds also allow measurements of extinction curves [44]. With many such measurements available, it has become clear that the long wavelength part of the extinction curve (>2500 Å) varies little from sightline to sightline, but that the converse is true for the short wavelength behavior, including the 2200 Å bump. These uncertainties very likely reflect changes in the composition and size distribution of (small) interstellar dust grains [45].

Organic molecules present in the diffuse medium can originate via gas phase reactions, either *in situ* [46] or by reactions in circumstellar envelopes followed by subsequent mixing into the diffuse medium, or by photoreactions of carbonaceous particles and sputtering by grain-grain collisions. Among the large organic molecules observed or suspected in diffuse clouds are <u>p</u>olycyclic <u>a</u>romatic <u>h</u>ydrocarbons (PAHs), fullerenes, carbon-chains, diamonds, amorphous carbon (hydrogenated and bare), and complex kerogen-type aromatic networks. The formation and distribution of large molecules in the gas and solid state is far from understood. In the envelopes of carbon-rich late-type stars, carbon is mostly locked in CO and C_2H_2. C_2H_2 molecules are precursors for soot formation where PAHs might act as intermediates [47]. The ubiquitous presence of aromatic structures in the ISM and in external galaxies has been well documented by numerous observations with the <u>I</u>nfrared <u>S</u>pace <u>O</u>bservatory, or ISO (see special volume 315 of Astronomy & Astrophysics, 1996). Possible evidence for carbon-chains and fullerenes arises from the characterization of the <u>D</u>iffuse <u>I</u>nterstellar <u>B</u>ands (DIBs) [48-51]. Diamonds were recently proposed to be the carriers of the 3.4 and 3.5 μm emission bands observed in planetary nebulae [52] and <u>h</u>ydrogenated <u>a</u>morphous <u>c</u>arbon (HAC) seems to be responsible for the 2200 Å bump in the interstellar extinction curve [53].

Carbonaceous dust in the interstellar medium may show considerable diversity and may include <u>a</u>morphous <u>c</u>arbon (AC), hydrogenated amorphous carbon (HAC), coal, soot, <u>q</u>uenched-<u>c</u>arbonaceous <u>c</u>ondensates (QCC), diamonds, and other compounds [54]. The coexistence of PAHs and fullerenes together with complex carbonaceous dust suggests a common link and an evolutionary cycle that is dominated by energetic processing [55, 56]. A detailed comparison of solid-state carbonaceous models of cosmic dust has been summarized in [57]. Figure 1.3 displays the chemical structure of some carbon compounds which are likely to be present in diffuse clouds and other space environments.

1.4   The Evolution of Organic Molecules During Solar System Formation

In comets and outer solar system asteroids, organic molecules formed in the presolar interstellar nebula may have survived solar system formation in relatively pristine form. Therefore, these small bodies carry important evidence on the formation of the solar system. A number of now extinct short-lived nuclides provide important information about the sources of solar system material and, moreover, are useful chronometers of chemical processes, which have occurred in the early solar nebula [58]. Primitive meteorites contain dust grains that predate the solar system. Isotopic analysis shows that such grains are formed in stellar atmospheres and thus represent samples of ancient stardust [59]. Isotopic analyses (e.g., of C, N, ^{26}Al, Sr, Zr, Mo etc.) of presolar grains allow reconstruction of the nucleosynthetic Processes, which occurred in the environment in which such particles were formed. Presolar material has been altered by chemical and physical processes in the solar nebula before becoming incorporated into small bodies. Dust and molecules have been affected by different processes dominating at various radial distances from the sun. It is assumed that the outer solar nebula was an environment of low temperature and pressure. Whereas heating and thermochemical reactions were important in the inner nebula, the outer solar nebula where comets formed, was mainly dominated by UV photochemistry and ion-molecule reactions [60].

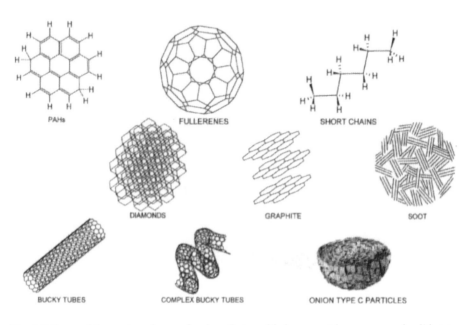

Fig. 1.3 Some of the various forms of carbon that are likely present in gaseous and solid state in the interstellar medium and in solar system material.

1.4.1   Comets

Small bodies of the solar system were formed in the region of the giant planets and beyond from remnant planetesimals which were not assembled into planets. Comets are amongst the most pristine objects and studies of their composition are thus of obvious interest for all models of the early solar system. The only way to measure the nuclear composition of a comet directly is via *in-situ* measurements by a space probe such as the Giotto mission to comet Halley [61]. Observations of the coma allow us in principle to deduce the molecular inventory of the nucleus, see Fig. 1.4. Remote sensing observations of comets Hyakutake and Hale-Bopp have revolutionized our understanding of the volatile chemical inventory of comets and the interstellar-comet connection. Many new cometary molecules were discovered by IR and radio observations [62-64, 14, 15]. Recent detections include SO, SO_2, HC_3N, NH_2CHO, HCOOH, and $HCOOCH_3$ [65]. Interesting to mention are also the upper limits for organic molecules such as glycine and ketene, which indicate that such molecules are not abundant in this type of comets [66]. With a few exceptions such as ethane (C_2H_6), N^+_2, CO^+, all molecules in the cometary coma are also observed in the interstellar medium.

Fig. 1.4 The spectacular appearance of comet West showing the ion and dust tails observed from the Observatoire de Haute Provence, France. Comets as bright as this appear only once or twice in a decade. Future space missions to rendezvous a comet will strongly improve our knowledge on the organic inventory, which may have seeded the early Earth.

Most of the current cometary inventory has been determined from remote sensing of the long period comets Halley, Hale-Bopp and Hyakutake. Much less is known about the composition of short-period comets. Interestingly, recent observations of such objects indicate significant chemical diversity in the giant planets region. Observations of comet 21P/Giacobini-Zinner show deficiencies of ethane and CO and comet Lee is depleted in CO compared to the long-period comets Hale-Bopp, and Hyakutake [67].

A large amount of information is expected to be obtained from space missions currently on their way or on the launching pad for a rendezvous with a comet such as STARDUST, CONTOUR and ROSETTA (see Chap. 24, Foing). On-board instrumentation on the ROSETTA spacecraft will measure the physical properties of comet Wirtanen and the chemical composition of its coma, but there will also be an attempt to land for the first time on a comet nucleus to perform *in situ* measurements. Such unprecedented encounters will strongly increase our knowledge on the chemistry and composition of comets.

1.4.2   Meteorites

Over a century ago, it was established that some meteorites contain carbonaceous material. These carbonaceous chondrites contain a few percent of carbon and some of them exhibit a large variety of organic compounds [68]. The best studied carbonaceous chondrite to date is the Murchison meteorite, of CM type, which fell in Australia on 28 September 1969. Subsequent analyses using a variety of methods have shown that the Murchison meteorite contains over 70 different amino acids, the majority of which have no known terrestrial occurrence [68]. It is generally thought that most meteorites, and in particular the CM type carbonaceous chondrites, originate from the asteroid belt. In fact, powdered samples of the Murchison meteorite heated up to 900 °C show strong similarities in their reflectance spectra to C and G type asteroids (C-type includes more than 75% of known asteroids; they are extremely dark, albedo 0.03), which points to an asteroidal origin [69]. However, based on mineralogical and chemical evidence, it has recently been suggested that both CI and CM meteorites could also be fragments of comets [70, 71]. The distinction between comets and asteroids is no longer clearly drawn, and several objects have currently a dual designation [72]. Understanding the link between small bodies such as comets, asteroids and their fragments enables us to reconstruct the processes occurring during planet formation [73].

1.5   Implications for the Origin of Life on Earth

Right after the formation of the Earth, about 4.5 x 10^9 years ago, the planet provided very hostile conditions for life to develop. Volcanic eruptions from the heated interior and external heavy bombardment by small bodies may have extinguished emerging life on a rapid time scale. The heavy bombardment phase, which has been scaled from the lunar record, ended about 3.8 x 10^9 ago. The first evidence for life follows 300 million years later and is provided by microfossils [74]. Ion probe measurements of the carbon isotope composition of carbonaceous inclusions (within apatite grains) from the oldest known sediments (banded-iron formation (BIF) from Isua) showed isotopically light carbon, indicative of biological activity even 300 million years earlier [75]. Those results provide evidence for the emergence of life on Earth ~3.8 x 10^9 years before present, just at the end of the late heavy bombardment phase. This leaves very little time for life to develop.

Today, research on the origin of life is an interdisciplinary field, which, apart from biologists, involves chemists, physicists, geologists and astronomers. Numerous theories for the origin of life exist which are based either on a terrestrial or an extraterrestrial origin [76]. Ideas for a terrestrial origin of life are focussed on the spontaneous formation of stable polymers out of monomers. It has been shown that amino acids spontaneously form polypeptides in aqueous solution under certain conditions and RNA oligomers spontaneously form on inorganic substances such as clay structures. For detailed information on the organic chemistry leading to higher complexity and life, we refer to the Chaps. 22, Heckl, and 23, Schidlowski.

An extraterrestrial origin of life on Earth via cosmic delivery of living organisms (panspermia), as proposed already in 1903 by Arrhenius, appears unlikely. Indeed, some organisms (and in particular their spores) are able to survive in extreme conditions of temperature and radiation on Earth and it has been argued that such species could survive interplanetary travel (see Chap. 4, Horneck et al.). However, recent studies have shown that the survival potential for living entities embedded in comets, asteroids, and cosmic dust impacting on the early Earth is negligible [77]. In contrast, the possible transport of extraterrestrial organic material via infalling comets and asteroids is a serious possibility [5, 6]. The very narrow window between the end of the heavy bombardment phase and the evidence for primitive organisms favors the idea that impacting prebiotic matter could have been the first step to life. Though it cannot be excluded that organics and living organisms have developed locally in protected areas on the Earth's surface or within the oceans, a substantial fraction of the Earth's prebiotic inventory of organic molecules and water may have been of extraterrestrial origin. Impact studies show that in particular small particles can be gently decelerated by the Earth's atmosphere and may have brought intact organics to the early Earth [78, 79].

The origin and development of life must be strongly dependent on the conditions of its host environment. One of the most important events in early Earth evolution is the formation of an atmosphere. The primitive atmosphere originating from the accumulation of gases released from the surface must be related in composition to volcanic emissions, whose composition, in turn, depends on the internal structure of the planet such as the oxidation state of the upper mantle [80]. Current evidence strongly favors the hot accretion model, in which the Earth essentially formed in a differentiated state [80]. In this case, non-reducing or mildly-reducing emissions, composed of CO_2 and N_2 (and only traces of other species), are predicted from the earliest times. Such an atmospheric composition is not favorable for the formation of abundant prebiotic molecules in contrast to the conditions for the Miller-Urey experiments [81]. These experiments showed that important biological molecules, including sugars and amino acids could be formed by spark discharge in an atmosphere of reducing conditions (containing CH_4, NH_3, H_2, H_2O). Exogenous delivery of organics may have been therefore more important than the endogenous production of organics on Earth. Even today, tons of organic material is brought to Earth via small particles, so called micrometeorites. More than 120 major craters found on Earth show the effects of violent impacts from space, see Fig. 1.5. For example, the Cretaceous-Tertiary (C/T) boundary sediments are sedimentary deposits that accrued from the end of the Cretaceous period to the beginning of the Tertiary period. They are distributed world wide and are recognized as a unique signature of a large asteroidal impact event near Chicxulub in Mexico [82].The discovery of high concentrations of extraterrestrial Ir in KTB sediments and Cr ratios are consistent with a chondritic type impactor [83].

The early atmosphere contained little or no free oxygen. The oxygen concentration increased markedly near 2.0×10^9 years ago due to photosynthetic activity of microorganisms. Greenhouse warming by small amounts of CH_4 in the atmosphere may have formed an organic haze layer, which cooled the climate and protected primitive life from UV irradiation in the period $3.5\text{-}2.5 \times 10^9$ years [84]. Recent analysis of precambrian sedimentary rocks have revealed a profound change in chemical reactions involving S and O in the atmosphere that occurred between 2.1×10^9 and 2.5×10^9 years ago [85]. During this exact period, the oxygen levels in the atmosphere strongly increased. The increase of oxygen in the atmosphere was a major step in the evolution of life on Earth (see Chap. 14, Cockell).

Fig. 1.5 49 000 years ago a large meteor created the Barringer Meteor Crater near Flagstaff, Arizona (credit D. Roddy, LPI). In 1920, it was the first feature on Earth to be recognized as an impact crater. Today, over 100 terrestrial impact craters have been identified.

1.6 Conclusions

The biogenic elements (H, C, N, O, S, P) and organic matter are some of the major constituents of the universe. Observations over the entire electromagnetic spectrum revealed a rich diversity of organic matter which is formed in interstellar and circumstellar regions. The route from a diffuse cloud to a self-gravitating molecular cloud core may take tens of millions of years, see Fig. 1.6. During this time, the interstellar gas undergoes strong chemical changes. To understand the process of star formation, one needs to comprehend the combined thermal and chemical balance of diffuse and dense interstellar gas clouds as they make their way from stellar winds to proto-stellar objects. The discoveries of proto-planetary disks around other stars suggest that the processes, which occurred in our solar nebula are common and the formation of solar systems like ours is not unique. The recent advances in the search for planets confirm this picture. The role of comets and other planetesimals in contributing organic matter to the primitive Earth and the prebiotic synthesis of biochemical compounds are major

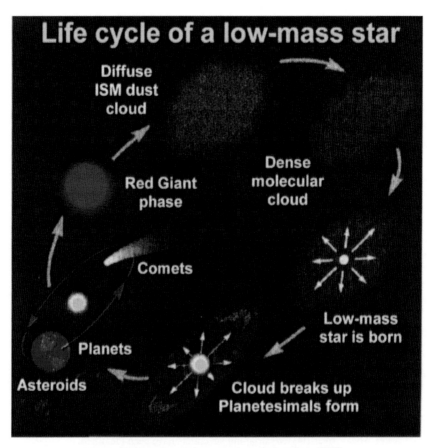

Fig. 1.6 The cycle of a low-mass star (taken from [86]).

questions, which remain to be answered in the future. Impacts may have led to extinction but may have also brought vital molecules to planet Earth, including organics and some H_2O. Valuable information may be obtained by future space missions investigating the existence of extinct and extant life on Mars, Europa and other bodies of the solar system and the search for new planets. Many of these questions will be answered by research groups, which are participating in the NASA Origins program. To provide answers to the questions how life originated is of vital importance in the frame of recent planetary detections and the possible emergence of life elsewhere.

Acknowledgement. The authors especially want to thank Malcolm Walmsley for a careful reading of the manuscript. This work was supported by the Netherlands Research School for Astronomy (NOVA). We thank Y. Kuan and S. Charnley for permission to show their BIMA maps of SgrB in advance of publication and Y. Pendleton for permission to reprint her figure from Sky & Telescope (1994).

1.7 References

1   M.B. Bell, P.A. Feldman, J.K.G. Watson, M.C. McCarthy, M.J. Travers et al., Astrophys. **518**, 740 (1999).

2   Y. Kuan, S.B. Charnley, T.L. Wilson, M. Ohishi, H.C. Huang, L. Snyder, BAAS **194**, 942 (1999).

3   E.F. van Dishoeck, G.A. Blake, Ann. Rev. Astron. Astrophys. **36**, 317 (1998).

4   P. Ehrenfreund, S.B. Charnley, 2000, Ann. Rev. Astron. Astrophys. **38**, 427 (2000).

5   J. Oro, Nature **190**, 389 (1961).

6   C. Chyba, P.J. Thomas, L. Brookshaw, C. Sagan, Science **249**, 366 (1990).

7   M. Bernstein, S.A. Sandford, L.J. Allamandola, J.S. Gillette, S.J. Clemett et al., Science **283**, 1135 (1999).

8   P. Ehrenfreund, W.A. Schutte, in: Y.C. Minh, E.F. van Dishoeck (Eds.) *Astrochemistry: From Molecular Clouds to Planetary Systems*, IAU Symposium 197, Sogwipo:Astron. Soc. Pac., 2000, pp. 135.

9   E. Gibb, D.C.B. Whittet, W.A. Schutte, J. Chiar, P. Ehrenfreund et al., Astrophys. J. **536**, 347 (2000).

10 M.J. McCaughrean, C.R. O'Dell, Astron. J. **111**, 1977 (1996).

11 D.L. Padgett, W. Brandner, K.R. Stapelfeldt, S.E. Strom, S. Terebey, D. Koerner, Astron. J. **117**, 1490 (1999).

12 V. Mannings, A. Boss, S. Russell (Eds.) *Protostars and Planets IV.*, Tucson: Univ. Ariz. Press, 2000.

13 D. Jewitt, J. Luu, C. Trujillo, Astron. J. **115**, 2125 (1998).

14 J. Crovisier, D. Bockelee-Morvan, Space Sci. Rev. **90**, 19 (1999).

15 W. Irvine, F. Schloerb, J. Crovisier, B. Fegley, M. Mumma, in: V. Mannings, A. Boss, S. Russell (Eds.) *Protostars and Planets IV.*, Tucson: Univ. Ariz. Press, 2000, p. 1159.

16 A.G. Csaszar, J. Am. Chem. Soc. **114**, 9568 (1992).

17 F. Lovas, Y. Kawashima, J.U. Grabow, R.D. Suenram, G.T. Fraser, E. Hirota, Astrophys. J. **455**, L201 (1995).

18 R.D. Brown et al., MNRAS **186**, 5 (1979).

19 J.M. Hollis, L.E. Snyder, R.D. Suenram, F.J. Lovas, Astrophys. J. **241**, 1001 (1980).

20   L.E. Snyder, Y. Kuan, Y. Miao, F.J. Lovas, in: S. Shostak (Ed.) *Progress in the Search for Extraterrestrial Life*, ASP Conf. Series 74 , San Francisco, 1995, pp. 106.

21   I.I. Berulis, G. Winnewisser, V.V. Krasnov, R.L. Sorochenko, Sov. Astron. Lett. **11**, 251 (1985).

22   F. Combes, N. Q-Rieu, G. Wlodarczak, Astron. Astrophys. **308**, 618 (1996).

23   L. Snyder, Orig. Life Evol. Biosphere **27**, 115 (1997).

24   C. Ceccarelli, L. Loinard, A. Castets, A. Faure, B. Lefloch, Astron. Astrophys. **362**, 1122 (2000).

25   S.B. Charnley, P. Ehrenfreund, Y. Kuan, Spectrochimica Acta (2001), in press.

26   D.M. Mehringer, L. E. Snyder, Y. Miao, F.J. Lovas, Astrophys. J. **480**, L71 (1997).

27   S.B. Charnley, in: F. Giovannelli (Ed.) *The Bridge Between the Big Bang and Biology*, Rome:Consiglio Nazionale delle Ricerche, 2000, in press.

28   M. Bernstein et al., Astrophys. J. (2001) in preparation.

29   P. Ehrenfreund, M. Bernstein, J. Dworkin, S. Sandford, L. Allamandola, Astrophys. J. **550**, L95 (2001).

30   P. Schilke, T.D. Groesbeck, G.A. Blake, T.G. Phillips, Astrophys. J. Suppl. **108**, 301 (1997).

31   P. Schilke, D.J. Benford, T.R. Hunter, D.C. Lis, T.G. Phillips, Astrophys. J. Suppl. **132**, 281 (2001).

32   A. Nummelin, P. Bergman, A. Hjalmarson, P. Friberg, W.M. Irvine, T.J. Millar, M. Ohishi, S. Saito, Astrophys. J. **117**, 427 (1998).

33   C.J. Lada, in: C.J. Lada, N.D. Kylafis (Eds.) *The Physics of Star Formation and Early Stellar Evolution.*, Kluwer, Dordrecht, 1991, pp. 329.

34   F.H. Shu, in: C.J. Lada, N.D. Kylafis (Eds.) *The Physics of Star Formation and Early Stellar Evolution*, Kluwer, Dordrecht, 1991, pp. 365.

35   S.V.W. Beckwith, A.I. Sargent, Nature **383**, 139 (1996).

36   Dutrey, S. Guilloteau, G. Duvert, L. Prato, M. Simon, K. Schuster, F. Menard, Astron. Astrophys. **309**, 493 (1996).

37   L.G. Mundy, L.W. Looney, W. J. Welch, in: V. Mannings, A. Boss, S. Russell (Eds.) *Protostars and Planets IV*, Tucson: Univ. Ariz. Press, 2000, p. 355.

38   F. Shu, J. Najita, D. Galli, E. Ostriker, in: *Protostars and Planets III.*, Tucson: Univ. Ariz. Press, 1993, p. 3.

39   Y. Aikawa, E. Herbst, Astron. Astrophys. **351**, 233 (1999).

40   K. Willacy, W.D. Langer, Astrophys. J. **544**, 903 (2000).

41   A. Dutrey, S. Guilloteau, M. Guelin, Astron. Astrophys **317**, L55 (1997).

42   K.M. Menten, in: J. Bergeron, A. Renzini (Eds.) *From Extrasolar Planets to Cosmology: The VLT Opening Symposium*, ESO Astrophysics Symposia, Springer, Berlin, 2000, pp. 78.

43   J.A. Cardelli, J.S. Mathis, D.C. Ebbets, B.D. Savage, Astrophys. J. **402**, L17 (1993).

44   P. Jenniskens, J.M. Greenberg, Astron. Astrophys. **274**, 439 (1993).

45   B. Draine, in: *The Evolution of the Interstellar Medium*, Astron. Soc. Pac., 1990, p. 193.

46   Bettens, E. Herbst, Astrophys. J. **468**, 686 (1996).

47   M. Frenklach, E.D. Feigelson, Astron. Astrophys. **341**, 372 (1989).

48   P. Freivogel, J. Fulara, J.P. Maier, Astrophys. J. **431**, L151 (1994).

49   M. Tulej, D.A. Kirkwood, M. Pachkov, J.P. Maier, Astrophys. J. **506**, L69 (1998).

50   B.H. Foing, P. Ehrenfreund, Nature **369**, 296 (1994).

51   B.H. Foing, P. Ehrenfreund, Astron. Astrophys. **317**, L59 (1997).

52   O. Guillois, G. Ledoux, C. Reynaud, Astrophys. J. **521**, L133 (1999).

53   V. Mennella, L. Colangeli, E. Bussoletti, P. Palumbo, A. Rotundi, Astrophys. J. **507**, L177 (1998).

54   T. Henning, F. Salama, Science **282**, 2204 (1998).

55   P. Jenniskens, G.A. Baratta, A. Kouchi, M.S. De Groot, J.M. Greenberg, G. Strazzulla, Astron. Astrophys. **273**, 583 (1993).

56   Scott, W.W. Duley, G.P. Pinho, Astrophys. J. **489**, L193 (1997).

57   R. Papoular, J. Conard, O. Guillois, I. Nenner, C. Reynaud, J. Rouzaud, Astron. Astrophys. **315**, 222 (1996).

58   G.J. Wasserburg, M. Busso, R. Gallino, Astrophys. J. **466**, L109 (1996).

59   E. Zinner, S. Amari, in: T. Le Bertre, A. Lebre, C. Waelkens (Eds.) *Asymptotic Giant Branch Stars*, IAU Symposium 191, Astron. Society of the Pacific, 1999, pp. 59.

60   B. Fegley, Space Sci. Rev. **90**, 239 (1999).

61   H.U. Keller, W.A. Delamere, H.F. Huebner, H.J. Reitsema, H.U. Schmidt et al., Astron. Astrophys. **187**, 807 (1987).

62   D.C. Lis, J. Keene, K. Young, T.G. Phillips et al., Icarus **130**, 355 (1997).

63   M. Mumma, in: Y. Pendleton, A. Tielens (Eds.) *From Stardust to Planetesimals*, Provo. Utah. Astron. Soc. Pac. 122, 1997, p. 369.

64   D. Bockelee-Morvan, H. Rickman, Earth, Moon and Planets **79**, 55 (1999).

65   D. Bockelee-Morvan, D.C. Lis, J.E. Wink, D. Despois, J. Crovisier et al., Astron. Astrophys. **353**, 1101 (2000).

66   J. Crovisier, D, Bockelee-Morvan, P. Colom, N. Biver, D. Despois, D. Lis, D.J. Benford, D. Mehringer, DPS **31**, 3202 (1999).

67   M. Mumma, M.A. DiSanti, N. Dello-Russo, K. Magee-Sauer, T.W. Rettig, Astrophys. J. **531**, L155 (2000).

68   J.R. Cronin, S. Chang, in: J.M. Greenberg, C.X. Mendoza- Gomez, V. Pirronello (Eds.) *The Chemistry of Life's Origins*, Kluwer, Dordrecht,1993, pp. 209.

69   T. Hiroi, C.M. Pieters, M.E. Zolensky, M.E. Lipschutz, Science **261**, 1016 (1993).

70   H. Campins, T.D. Swindle, Science **33**, 1201 (1998).

71   L. Lodders, R. Osborne, Space Sci. Rev. **90**, 289 (1999).

72   D. Yeomans, Nature **404**, 829 (2000).

73   D. Cruikshank, in: Y. Pendleton, A. Tielens (Eds.) *From Stardust to Planetesimals*, Provo:Utah. Astron. Soc. Pac. 122, 1997, pp. 315.

74   J.W. Schopf, Science **260**, 640 (1993).

75   S.J. Mojzsis, G. Arrhenius, K. McKeegan, T. Harrison, A. Nutman, C. Friend, Nature **384**, 55 (1996).

76   C.P. McKay, in: D.C.B. Whittet (Ed.) *Planetary and Interstellar Processes Relevant to the Origins of Life*, Kluwer Academic Publishers, 1997, pp. 263.

77   NRC Report *Task Group: Evaluating the biological potential in samples returned from planetary satellites and small solar system bodies*, National Academy of Sciences, 1998.

78   E. Anders, Nature **342**, 255 (1989).

79   C. Chyba, C. Sagan, Nature **355**, 125 (1992).

80   J. F. Kasting, Science **259**, 920 (1993).

81   S.L. Miller, Biochim.Biophys. Acta **23**, 480 (1957).

82   F.T. Kyte, Nature **396**, 237 (1998).

83   A. Shukolykov, G. Lugmair, Science **282**, 927 (1998).

84   A.A. Pavlov, J.F. Kastings, L.L. Brown, J. of Geophys. Res. **105**, 11981 (2000).

85   J. Farquhar, H. Bao, M. Thiemens, Science **289**, 756 (2000).

86   Y. Pendleton, D. Cruikshank, Sky & Telescope **87/3**, 36 (1994).

2   The Diversity of Extrasolar Planets
Around Solar Type Stars

Stéphane Udry and Michel Mayor

An important step in the search for an environment favorable to the development of exobiological life was accomplished in 1995 with the discovery by Mayor and Queloz [1] of the first extrasolar planet orbiting a star similar to the Sun, 51 Pegasi. As the proximity of the planet to the star and the large luminosity contrast between the star and the planet prevent the planet to be seen directly, its presence was inferred from the induced modulation of the observed stellar radial velocity. This method provides interesting information on the orbital characteristics of the system, although giving only access to the minimum mass of the planetary companion (projection effect due to the non-alignment of the orbital plane with the line of sight). Nevertheless, the radial-velocity technique, whose precision allows giant-planet detections, has proven since then to be very efficient. During the past 6 years an impressive series of new enthusiastic results in the domain were announced. New candidates are regularly pointed out by the teams of «planet hunters» monitoring high-precision radial velocities. Also a number of various approaches to the field are explored aiming to better understand and constrain planetary formation.

The new detected candidates present a large variety of characteristics, often unexpected from the observation of our own Solar System: some of the giant planets are found very close to their parent stars (a few stellar radii), some are on very elongated orbits, some of the planets are also very massive. In fact, new observational results often tend to set new questions rather than bring definitive answers to the large variety of extrasolar system properties. However, the regular increase of candidate discoveries already allows us to point out some preliminary trends that should help us to better understand the formation of extrasolar planets.

The goal of this presentation is to give to the readers a synthetic, up-to-date view of the field, keeping in mind that it is evolving very rapidly. The main milestones of exoplanet discoveries will be recalled, with a special emphasize on the most recent announcements of the Geneva group. The global properties of the extrasolar planet sample as a whole will be discussed as well. The second part of this contribution will be devoted to a global description of the extrasolar planet properties in terms of mass function and orbital element distributions, emphasizing their differences from the equivalent distributions for stellar companions to solar type stars. Giant planets and stellar binaries have different formation mechanisms whose fossil traces should be revealed by a comparison of their orbital characteristics. Chemical properties of stars with planets will then be discussed in comparison with stars without known planets. Finally, future ambitious programs aiming to search for terrestrial planets and signature of life in their atmospheres will be mentioned.

2.1   Detections: Milestones and Recent Announcements

It is not possible in a few lines to give an exhaustive report of the extremely rich ensemble of results obtained during the past years in the domain of extrasolar planets. We will simply recall the main prominent discoveries of the past 6 years, with a special mention to the *latest news*.

1995: A pioneer Canadian team around G. Walker published their negative results from a high-precision radial-velocity systematic search for giant planets, carried out over more than 10 years, in a sample of 21 solar type stars [2].

October 1995: Mayor and Queloz [1] announced the detection of the first extrasolar planet orbiting a solar type star, 51 Pegasi. The discovery came from a systematic radial-velocity monitoring of 142 dwarfs of the solar neighborhood, started with ELODIE at the Haute-Provence Observatory 1.5 years earlier. The very *exotic* properties of the planet ($P = 4.23$d, $a = 0.05$AU, $T_{eq} \approx 1300$K) raised interesting questioning on the standard views of planetary formation.

1996: At Lick, G. Marcy and P. Butler, following a sample of 120 stars, rapidly announced 5 more candidates among which 3 presented properties similar to 51 Peg b [3].

1998: Summer 1998, 8 planets were known [4]. At the end of the same year, 16 planets were known. The acceleration of the discoveries was due to the growing time base of the observations and to an enlargement of the monitored samples.

April 1999: Twenty planets had been detected, among them the first multi-planet system, υ And (Andromeda) [5]. Two years after the discovery, the proposed 3-planet model always fits the observations very well [6]. The 3 planets have periods of 4.6, 241 and 1308 days, and minimum masses of 0.68, 2.05 and 4.29 M_{Jup}, respectively.

November 1999: Announcement of the transit of a planet in front of the stellar disk of HD 209458. From the photometric data [7, 8] the real mass, radius and mean density of the planet have been determined [9]. Information on the system geometry was also inferred from the spectroscopic observations of the transit [10].

Spring 2000: Discoveries of several planetary candidates with minimum masses below the mass of Saturn [11, 12].

August 2000: Nine new planetary candidates were announced at the IAU Symp. 202 in Manchester (IAU General Assembly), among them the second and third known multi-planet systems: HD 83443 harboring 2 subsaturn-mass planets close to their parent star [13] and HD 168443 a system with a massive inner planet and a still more massive very low-mass companion further out [14]. HD 83443 seems to be a resonant system with a ratio of 10 between the periods.

January 2001: A new stunning 2-planet resonant system is proposed to model the radial velocities of Gl 876 [15], for which the single-planet model left large residuals around the Keplerian solution [16, 17]. The periods of the 2 planets are found to be close to a ratio of 2.

March 2001 : Very recently, 11 new planetary candidates including 2 new multi-planet systems have been announced by the Geneva group and international collaborators [18]. Several of the candidates present interesting properties. They will be briefly described in the following part as well as the two new multi-planet systems, HD 74156 and HD 82943.

In summary, beginning of May 2001, 63 objects are known with minimum masses below 10 Jupiter masses (or 67 with $m_2\sin(i) < 17\,M_{\mathrm{Jup}}$; Table 2.1). Among them, 6 multi-planet systems have been detected. Taking into account that the large on-going planet-search programs are still incomplete and inhomogeneously followed, we can estimate the fraction of the sample stars with giant planets to be larger than 5%. A more definitive estimate will become available with the completion of the large surveys of statistically well-defined samples as e.g., the CORALIE planet-search program in the Southern hemisphere [19].

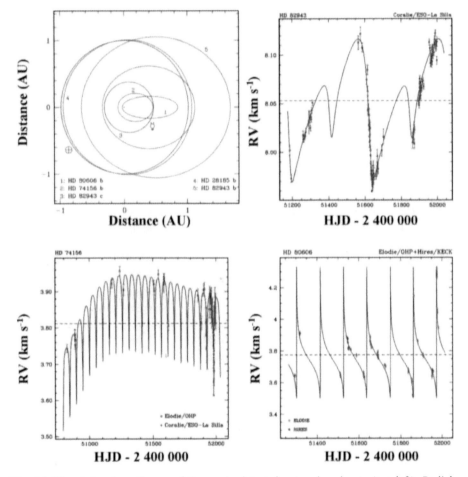

Fig. 2.1 Visual drawing of some of the newly detected extrasolar planets (top left). Radial-velocity measurements and best Keplerian solutions for HD 80606 (bottom right) and the 2-planet systems HD 82943 (top right) and HD 74156 (bottom left); HJD=Heliocentric Julian Date; RV=Radial Velocity.

Table. 2.1. Main orbital characteristics of the exoplanets and very low-mass brown dwarfs. The list is sorted by increasing minimum masses

Object	P [1] (days)	e [2]	$m_2\sin(i)$ [3] ($10^{-3} M_{Sun}$)	A [4] (AU)	References
HD 83443 c	29.83	0.42	0.15	0.17	[13]
HD 16141	75.82	0.28	0.215	0.35	[11]
HD 168746	6.407	0.0	0.24	0.066	[12,68]
HD 46375	3.024	0.02	0.25	0.041	[11]
HD 108147	11.05	0.57	0.33	0.1	[12,68]
HD 83443 b	2.985	0.08	0.36	0.0375	[13]
HD 75289	3.475	0.06	0.42	0.04	[19]
51 Peg	4.230	0.00	0.45	0.05	[1]
HD 6434	22.09	0.29	0.46	0.15	[12,69]
BD -10:3166	3.487	0.05	0.47	0.046	[70]
Gl 876 c	30.12	0.27	0.53	0.13	[15]
HD 187123	3.097	0.01	0.57	0.042	[71]
HD 209458	3.524	0.0	0.65	0.047	[8,9]
υ And b	4.617	0.02	0.68	0.059	[5,6]
HD 192263	24.13	0.0	0.72	0.15	[72,75]
HD 179949	3.092	0.0	0.76	0.045	[74]
HD 38529	14.31	0.27	0.77	0.129	[25]
ε Eridani	2518	0.6	0.80	3.4	[73]
HD 82943c	221.8	0.55	0.84	0.73	[18]
HD 121504	64.62	0.13	0.85	0.32	[12,69]
55 Cnc	14.66	0.03	0.88	0.12	[3]
HD 130322	10.72	0.04	1.0	0.088	[19]
HD 37124	154.8	0.31	1.0	0.55	[75]
HD 52265	119.2	0.35	1.03	0.5	[70,76]
ρ CrB	39.64	0.07	1.1	0.23	[77]
HD 177830	391.6	0.41	1.2	1.1	[75]
HD 210277	435.6	0.34	1.25	1.09	[78]
HD 217107	7.110	0.14	1.28	0.071	[79]
HD 27442	426.0	0.02	1.37	1.18	[80]
HD 74156 b	51.61	0.65	1.49	0.276	[18]
HD 134987	259.6	0.24	1.5	0.81	[75]
HD 82943 b	443.7	0.39	1.56	1.16	[18]
16 Cyg B	804.4	0.67	1.67	1.61	[81]
Gl 876 b	61.02	0.1	1.8	0.2	[15,16,17]
HD 19994	454.2	0.2	1.8	1.23	[12,69]
HD 160691	763.0	0.62	1.88	1.65	[80]
υ And c	241.2	0.24	2.1	0.83	[5,6]
HD 8574	228.8	0.4	2.13	0.76	[18]
ι Hor	311.3	0.22	2.25	0.93	[82]
47 UMa	1084	0.13	2.45	2.1	[83]

Table. 2.1. Main orbital characteristics of the exoplanets and very low-mass brown dwarfs. The list is sorted by increasing minimum masses (continued)

Object	P [1] (days)	e [2]	$m_2\sin(i)$ [3] $(10^{-3} M_{Sun})$	A [4] (AU)	References
HD 12661	264.0	0.33	2.8	0.789	[25]
HD 169830	229.9	0.35	2.9	0.82	[76]
HD 80606	111.8	0.927	3.7	0.47	[18,23]
GJ 3021	133.7	0.51	3.35	0.494	[76]
HD 195019	18.30	0.01	3.43	0.135	[79]
HD 213240	759.0	0.31	3.53	1.6	[18, 22]
Gl 86	15.83	0.04	3.6	0.11	[84]
HD 92788	340.8	0.36	3.6	0.97	[12, 69, 25]
τ Boo	3.313	0.02	4.1	0.047	[3]
υ And d	1308	0.31	4.25	2.55	[5,6]
HD 50554	1279	0.42	4.68	2.38	[18]
HD 190228	1161	0.5	4.8	2.3	[85]
HD 222582	575.9	0.71	5.2	1.35	[75]
14 Her	1650	0.37	5.4	2.84	[14]
HD 28185	385.0	0.06	5.6	1.007	[18, 22]
HD 178911 B	71.5	0.145	6.18	0.326	[18]
HD 10697	1072	0.12	6.3	2.12	[75]
HD 106252	1500	0.54	6.5	2.61	[18]
HD 168443 b	58.12	0.53	6.9	0.29	[14, 86]
HD 74156 c	2300	0.4	7.16	3.47	[18]
HD 89744	265.0	0.7	7.2	0.91	[87]
70 Vir	116.7	0.4	7.4	0.48	[88]
HD 141937	658.8	0.4	9.26	1.49	[18]
HD 114762	84.03	0.35	11.0	0.37	[89]
HD 162020	8.428	0.28	13.7	0.074	[12,90]
HD 202206	256.4	0.43	14.3	0.76	[12,90]
HD 168443 c	1667	0.27	14.4	2.67	[14,86]

[1] P=Period; [2] e=orbital eccentricity; [3] m_2=mass of the planet, $m_2\sin(i)$=minimum mass of the planet; [4] a=semi-major axis

2.2   Very Recent ELODIE and CORALIE Detections

ELODIE and CORALIE are two high-resolution spectrographs designed for high-precision radial-velocity measurements [20]. They are installed on the 193-cm telescope at the Haute-Provence Observatory (France) and on the 1.2-m Leonard Euler Swiss telescope at La Silla Observatory (ESO, Chile), respectively. The ELODIE planet-search sample is magnitude-limited and consists of ~350 F-K dwarfs of the solar vicinity. It is an extension of the original *Mayor-Queloz* sample out of which 51 Peg b was discovered [1]. The CORALIE planet-search sample consists of more than 1650 F8 to M0 solar type stars selected according to distance in order to obtain a statistically well-defined volume-limited set of dwarfs of the solar neighborhood [19]. In addition to the planet search, this sample will allow us to collect information on spectroscopic binaries and give us a synthetic view on companions to solar type stars from $q=m_2/m_1=1$ down to $q \leq 0.001$.

On April 4th 2001 [18], we announced the latest results related to these two programs, namely the detection of 11 new planetary candidates, among them two multi-planet systems. Their main orbital properties are given in Table 2.1. Some have special characteristics:

- The 2-planet system around the star HD 82943 [18] (Fig. 2.1, top right) is resonant in the same way as Gl 876 [15]: the period of the inner planet (221.8 days) is almost exactly half of that of the outer one (443.7 days). Future observations should confirm the 1:2 ratio between the periods. Gravitational interactions between the two planets are expected to be strong. Consequently, a two-Keplerian model will rapidly diverge from the real temporal evolution of the system, requiring new developments for investigating those systems [21].

- The 2-planet system HD 74156 [18] is reminiscent from HD 168443 [14] with a Jupiter-size planet on a 51.6-day period orbit and a second, heavier companion further out. The outer period is estimated to be around 2300 days from the minimization of the residuals around the solution (Fig. 2.1, bottom left). The long period is not completely covered yet and could be somehow larger than the estimated value. In such a case however, the inferred minimum mass of 7.4 M_{Jup} for the outer planet will not change much as its dependency with the orbital period is very weak. Such systems set the question of the possible formation of *super-massive* planets in protoplanetary disks.

- The orbit of HD 28185 b [18, 22] is very nearly circular and its period of 385 days completely locates the planet in the "habitable zone" of the central star, at a distance of 1.007 AU, almost exactly the Earth-Sun separation. The giant, gaseous nature of the planet (minimum mass of 5.6 M_{Jup}) seems not favorable for the development of life on its surface. However, moons potentially orbiting the planet may well harbor more bio-friendly environments. The hypothesis of natural satellites around giant extrasolar planets is not so speculative when observing our own Solar System.

- HD 80606 b [18, 23] is the planet with the most elongated orbit detected so far ($e = 0.927$; Fig 2.1, bottom right). This star was followed in the framework of an ELODIE/HIRES-Keck collaboration involving the Geneva, Grenoble, Haute-Provence Observatories, the CfA and the Tel Aviv University. Along its orbit, the planet explores a region between 0.034 AU (a few stellar radii) and 0.905 AU from the star. The origin of the elongated shapes of most of the long-period exoplanetary orbits is far from being fully understood yet (see below). A number of them, including the present candidate, could owe their large eccentricities from the dynamical perturbation of an additional stellar or planetary companion. HD 80606 belongs to a visual double system with a similar star, HD 80607, 1800 AU apart.

With these 11 new discoveries the CORALIE/ELODIE programs have contributed to the detections of 32 among the 63 known planetary candidates with minimum

masses below 10 Jupiter masses (or 36 among the 67 objects with $m_2\sin(i) < 17\ M_{Jup}$).
Furthermore, among the new detections we have 2 new planetary systems bringing to
6 the number of known multi-planet systems, 4 of which (with HD 83443 [13] and
HD 168443 [14]) owe their detection to CORALIE/ ELODIE measurements. This
demonstrates the outstanding role that small-size telescopes can still play in modern
astrophysics.

2.3   Observed Properties of Extrasolar Planets

The observed properties of planetary candidates are diverse and often surprising because different from the characteristics of the giant planets of our Solar System. A
visual idea of the global properties of the sample is given in Fig. 2.2 displaying the
eccentricities of the planet orbits as a function of the star-planet separations. The sym-
bol sizes scale with the planet minimum masses inferred from the Keplerian orbital
solutions.

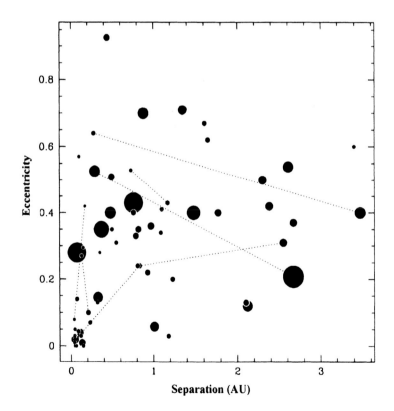

Fig. 2.2 *Separation–eccentricity* diagram for the sample of known extrasolar planets and very
low-mass brown dwarfs. The dot size is related the the minimum mass of the objects. Components of multi-planet system are connected by dotted lines.

The figure is interesting. It clearly shows several of the unexpected properties of extrasolar planets. From the observation of our Solar System we were expecting Jupiter-like planets on quasi-circular orbits with periods of the order of several years. We have actually found planets with minimum masses up to ten times the mass of Jupiter, often on elongated orbits and with periods that can be very short, of the order of a few days. These different aspects will be discussed in the following parts. A more extended review of the extrasolar planet properties can be found e.g., in the PPIV contribution of Marcy, Cochran and Mayor [24].

In figure 2.2 the components of multi-planet systems are connected by dotted lines. As for single-detected planets, they present a large variety of characteristics: well hierarchized systems, resonant planets, very light or super-massive components. Except for HD 83443, there seems to be a trend for the outer planets to be more massive than the inner ones. This is most probably due to an observational bias, inherent to the radial-velocity technique more sensitive to closer and heavier planets. Multi-planet systems are still sparsely detected, but their number will grow rapidly with the increasing time-base and precision of the surveys. Already in a sample of 12 stars with planets, followed over more than 2 years at Lick, half of the stars present a drift of their systemic velocity, indicating the presence of an additional planetary or stellar companion [25].

2.4   Hot Jupiters

Thirteen among the extrasolar planets detected to date reside in close orbits, with $a < 0.09$ AU. Such small orbits were not predicted by the *standard theory* (e.g., [26-28]). The surprising small orbits stand in apparent contrast to the prediction that the giant planets formed first from ice grains, which exist only beyond ~3-5 AU. Such grain growth provides the supposed requisite solid core around which gas could rapidly accrete, over the lifetime of the protoplanetary disk ($\sim 10^7$ years). In the inner regions where volatiles elements are swept out by the radiation of the newborn star, only "light" objects (≤ 1 M_{Earth}) can form from the remaining material (dust, silicates, etc.).

According to this picture, giant planets should form in the outer regions of the protostellar nebula and then move towards the system center where they are actually observed. Such a migration was already predicted in the early eighties from numerical simulations of gravitational disk-planet interactions [29-31]. The model has been improved now with the newly-formed core rapidly moving inwards and the planet growing by accretion during the migration process [32]. The migration time scale is found to be very short, of the order of 10^6 years. The difficulty is then to stop the migrating planet before it falls into the central star. Several processes may be invoked (see [33] for a review):

- A central magnetospheric cavity around the star, free of material, extending out to the stellar corotation, at the edge of the disk (magnetic coupling between the star and the disk) [34]: the migration naturally stops at the disk edge.
- Consecutive formation of giant planets that migrate and fall into the star. When the disk disappears some of the planets are still there.

- Exchange of angular momentum between the star and the planet by tidal interaction [35] or mass transfer through Roche lobe overflow [36] when the planet
comes close to the star.
- Planet evaporation. It takes place when the thermal velocity of the gas (mainly
Hydrogen) becomes important with regards to the escape velocity from the
planet, typically for V_{esc}/V_{therm} smaller than about 5 [37]. This ratio depends
mainly on the planet mass and the lighter candidates come close to the evaporation limit [33]. In the future, if some gaseous giant planets on short-period
orbits are detected with masses well below 0.2 M_{Jup}, they will be subjected to
evaporation-related evolution. We can speculate that these objects will have
Hydrogen-depleted atmospheres.

An observational evidence for a stop of the planet migration at a "well-defined"
distance from the star is given by the sharp cut-off at the lower end of the cumulative
distribution of exoplanet periods (see below, Fig. 2.5 left).

Other scenarios, not involving planet migration, are also proposed for explaining
hot Jupiters:

- *Jumping Jupiters*: Several giant planets, formed simultaneously in the protoplanetary disk, perturb each other [38, 39]. These gravitational, chaotic interactions modify their orbital characteristics. Some of the planets may be ejected
from the system, some are brought close to the central star. It seems however
difficult with this process to bring planets on short-period circular orbits. On
the other hand, some close-in planets are found on elongated orbits.
- *In situ formation*: Planets form where they are observed, much more rapidly
than generally assumed, in a standard way [40, 41] or from gravitational instabilities in the disk leading to a fast collapse [42]. The perturbation of a close
stellar companion could favor such instabilities.

2.4.1   Hot Jupiters: Direct Detections

It is possible to take advantage of the proximity of hot Jupiters to their parent stars to
directly "see" their presence. Several methods have been proposed.

Transit. Short-period giant planets have a significant probability for their orbital
plane to be sufficiently close to the line of sight to observe an eclipse of the star by the
planet (transit). The probability is about 10% for a 4-day period orbit. The induced
decrease of the star luminosity is estimated to be around 1% for a Jupiter-type planet
(from the surface ratio of the disks of the two bodies). The modeling of the observed
light curve provides then the physical parameters of the planet (giving the needed
parameters for the parent star): orbital inclination, real mass, radius, mean density.

The prove of the giant gaseous nature of hot Jupiters was brought by the photometric observation of a transit of a planet in front of the disk of HD 209458 [7, 8] (Fig.
2.3, top right) at the time predicted by radial-velocity measurements [9, 8] (Fig. 2.3,
top left). It has provided us for the first time with physical characteristics of an exoplanet, R_{pl} = 1.4 R_{Jup}, M_{pl} = 0.69 M_{Jup}, ρ = 0.31 gcm^{-3}, in complete agreement with
theoretical predictions [43, 44]. Observations with better facilities have then drasti-
cally improved these early results (see e.g., the impressive transit luminosity curve
obtained with the Hubble Space Telescope [45]).

In the same time, the transit was also observed spectroscopically with ELODIE [10] (Fig. 2.3, bottom right). In the same way as spots, the shadow on the stellar disk of a transiting planet induces deformations of the spectral lines by hiding part of the light, coming from the approaching or receding sides of the rotating star. This effect influences the measured radial velocities in the form of an anomaly of the orbital curve (Fig. 2.3, bottom left) whose shape depends on the geometry of the transit and on stellar parameters (e.g., stellar rotation). The main results can be summarized as follow: 1) the positive sign of the anomaly curve during the ingress indicates that the stellar rotation is in the same direction as the planet motion; 2) a small angle of $\sim 3.9°$ is estimated between the rotation and orbital axes; 3) the method provides an independent estimate of the stellar rotation: $v\sin i = 3.75\,\mathrm{kms}^{-1}$.

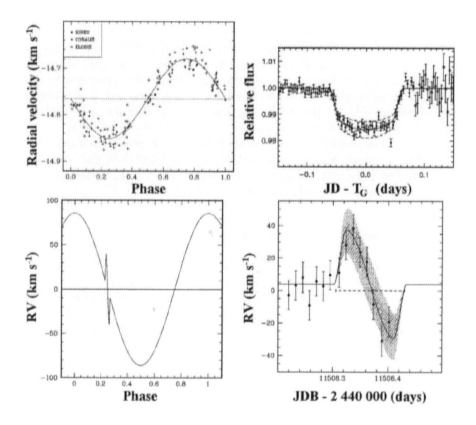

Fig. 2.3 Transit of HD 209458 b: *Top.* First photometric observations on the 9th and 16th of September 1999 (right, [7]), at the phase predicted by the radial-velocity measurements (left, [9]). *Bottom.* Anomaly of the orbital radial-velocity curve (left) observed with ELODIE (right, [10]); JD-T_G=time difference to the center to the transit; JDB=Julian Date.

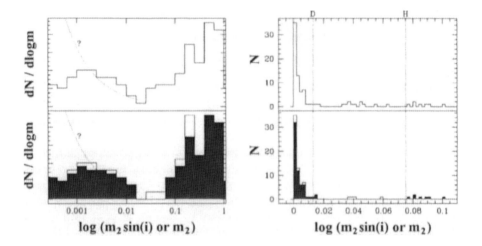

Fig. 2.4 Mass function of companions to solar type stars, including our latest detections, in log (left) and linear (right) scales. The dotted vertical lines indicate the H- and D-burning limits. Top: m₂sin(i). Bottom: composite histograms of m2 (open part) and m₂sin(i) (if sini not known; black part).

Reflected light. Hot Jupiters present a large reflecting surface to the light coming from the close star. In the visible, the reflected light represents about 10^{-5} of the total observed flux. Despite this very unfavorable flux ratio, the large Doppler shift of the reflected light could allow us to separate the latter from the stellar light in very high signal-to-noise spectra. This technique is even more promising in the IR where the intrinsic emission of the planet adds up to the reflected light. In the visible, the technique was intensively applied to τ Boo, but no conclusive results have been obtained yet [46].

Spectral signatures of the planet atmosphere. Because of the hot temperature, some elements of the planet atmosphere evaporate and are blown away by the high-speed solar wind. Doppler shifts of the spectral lines of those elements will then be very different for phases "off" (wind perpendicular to the line of sight) and "in" (wind in the observer direction) "transit" [47, 48]. The difficulty resides in finding the atmospheric lines which, on the other hand, provide information on the chemical composition of the planetary atmosphere.

2.5 The Mass Function of Substellar Companions

Radial velocities give only access to the minimum mass $m_2\sin(i)$ of low-mass companions. On the long run, most of the orbital planes of known brown-dwarf and planetary companions will be determined by precise astrometric measurements (often combined with spectroscopic data) obtained either with ground-based interferometers (VLTI, KeckI) or astrometric space missions (FAME, SIM, GAIA). A significant number of orbital planes of very low-mass companions have, however, already been determined thanks to several different techniques: Hipparcos astrometric data for the heavier candidates [49, 50], transit observations [7-9] or synchronization considerations for short-period systems. For these companions we have an estimate of their real mass $m2$. The histogram of minimum masses of companion to solar type stars, including our latest planet detections, is illustrated in Fig. 2.4 in logarithmic (left) and linear (right) scales. The top panels present the $m_2\sin(i)$ distributions whereas the bottom diagrams show composite histograms of $m2$ (open) or $m_2\sin(i)$ (black, when $\sin i$ is not known).

Han et al. [51] have recently suggested that most of the exoplanet candidates discovered so far have masses well above the planetary limit, in the brown-dwarf or even in the stellar domain, the orbits being seen nearly pole-on. They reached this conclusion by trying to extract the astrometric orbit (hence the orbital inclination) from the Hipparcos intermediate data. As already cautioned by Halbwachs et al. [49], this approach is doomed to fail for systems with apparent separations that are below the Hipparcos sensitivity, because due to measurements errors a positive motion of the star on the sky is always observed, even for "single" stars. This is the case for most of the extrasolar planets. Moreover, the Han et al. results have been shown [52, 53] to be statistically incompatible with the hypothesis of a random distribution of orbital inclinations, expected for volume-limited samples as the CORALIE one [19].

In figure 2.4, the observed gap in the mass distribution between giant planets and stellar secondaries strongly suggests the existence of two distinct companion populations for solar type stars. The huge planetary peak is not the tail of the binary distribution. The sharp drop of the mass function observed around 8 M_{Jup} (Fig. 2.4, right) strongly suggests a maximum mass for giant planets close to 10 M_{Jup}. In particular, the shape of the mass function does not suggest a relation between the D-burning limit (13.6 M_{Jup}) and the maximum mass for giant planets (The discovery of *free-floating brown dwarfs* in σ-Orionis with masses probably below 10 M_{Jup} [54] further refutes the D-burning limit as a good indicator of the *brown dwarf-planet* transition.).

Despite the huge observational bias against the detection of small-mass companions, we observe an increasing number of low-mass planets. The planetary mass function even increases towards the lower masses. In the same time, the easier-detected brown dwarfs are rare.

These conclusions completely hold when a deconvolution scheme is applied to the $m_2\sin(i)$ distribution in order to derive a statistical planetary mass function [55].

2.6   Orbital Element Distributions: Traces of Planet Formation

From the observation of a very clear gap between their mass distributions, giant planets and stellar binaries are believed to have different formation mechanisms whose fossil traces should be revealed by a comparison of their orbital properties.

2.6.1   The Distribution of Periods

Due to the still strong observational bias affecting the detection of long-period planets, a significant comparison of the period distributions of planetary candidates ($m_2\sin(i) \leq 10 \, M_{Jup}$) and spectroscopic binaries ($m_2\sin(i) \geq 20 \, M_{Jup}$) is only possible for relatively short-period systems, for which the detection bias is vanishing. On the domain of very-short periods ($P \leq 10d$), the distribution of giant-planet periods is steeply rising for decreasing periods down to about 3 days where a cut-off is observed. The latter clearly appears in Fig. 2.5 (left) comparing the cumulative distributions of periods smaller than 10 days for giant planets and spectroscopic binaries. Several reasons related to the possible end of the migration process close to the central star may be invoked to set this limit like e.g., magnetospheric central cavity of the accretion disk, tidal interaction, Roche lobe overflow or evaporation (see above or [33]).

2.6.2   The Distribution of Eccentricities

The comparison of orbital eccentricities of spectroscopic binaries for G, K and M primaries [56-58] with the equivalent distribution for giant planets is remarkable. In Fig. 2.6 we have plotted the orbital eccentricities as a function of log P both for double stars and giant planets. At the first glance we do not observe significant differences of orbital elements between planets and spectroscopic binaries. If the formation mechanisms for planets and double stars are different, why do we observe so similar (e, log P) distributions?

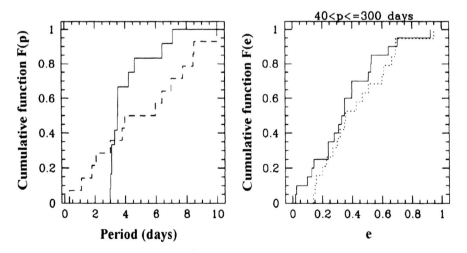

Fig. 2.5 Cumulative distributions for planetary (solid line) and stellar companions (dashed line) to solar type stars of: periods smaller than 10 days (left) and eccentricities (e) in the 40- to 300-day period range (right).

Nevertheless, some small differences may be emphasized. For planets with very short periods, most of the orbits are quasi-circular as for double stars. Planets having suffered a strong orbital migration are probably circularized by the tidal interaction with the accretion disk [29, 31]. However, we observe several quasi-circular planetary orbits with periods larger than the stellar circularization limit ($P \leq 10$ days). In this range of periods, planetary systems have smaller eccentricities than stellar binaries indicating different formations or evolutions. On another hand, a few shorter-period planets present eccentric orbits. Such configurations could be explained by the presence of an additional companion or by gravitational interaction between giant planets formed in the outer regions of the system [38, 39].

Most of the orbits of double stars with long periods are fairly eccentric. Quasi-circular long-period orbits are very rare. A similar situation exists for giant exoplanets (except for HD 27442 and HD 28185), although they are still strongly observationally biased in that period range. For periods larger than about 40 days, the comparison of double star and planet eccentricities is really surprising. With the limited-size samples presently available, we cannot see any significant distinctions between both populations. This is shown in Fig. 2.5 (right), which gives the corresponding cumulative

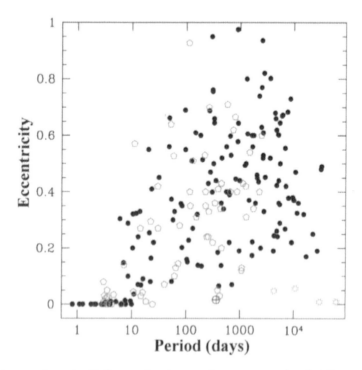

Fig. 2.6 Comparative (e,log$P$) diagram for planetary (open pentagons) and stellar companions to G+K+M solar type stars of the field (filled circles). The Earth position is indicated and starred symbols represent giant planets of the Solar System.

function of eccentricities for periods between 40 and 300 days. If the orbital eccentricity of binaries finds its origin in the disruption of small N-body systems, the generated distribution of eccentricities could be close to the distribution generated by the gravitational interactions between giant planets.

The origin of the eccentricity of extrasolar giant planets has been searched in the gravitational interaction between multiple giant planets [38, 39] or between the planets and the planetesimals in the early stages of the system formation [59]. Among the giant-planet candidates several eccentric orbits show a drift of their mean velocity, indicating the presence of a long-period companion (stellar or planetary) whose gravitational perturbation can also be suspected to be responsible for the observed (high) planetary eccentricity as e.g., for the planet orbiting 16 Cyg B [60].

In conclusion, the differences observed between planetary systems and double stars in the shape of the secondary mass functions and in the period and eccentricity distributions argue for the two populations being formed by distinct processes. More observations, however, are still needed to bring clear constraints on the possible formation and evolution scenarios.

2.7 Metal Enrichment of Stars Bearing Planets

Shortly after the discovery of the first extrasolar planets, it was pointed out that stars with planets were in average more metal rich compared to the common dwarfs of the solar neighborhood [61, 62]. With the growing number of detected candidates, this early trend is confirmed [63] and strengthened by an homogeneous metallicity determination for a set of planet-hosting and comparison stars [64] (Fig. 2.7, left). The abundance ratios of "non-Iron" chemical elements in the stellar atmosphere ([Li/H], [C/H] and [N/H]) are found to be comparable for stars with and without planets [63, 64].

The observed relation between the presence of a planet and the chemical anomaly in the stellar atmosphere is of great interest for constraining planet-formation scenarios. The main explanation for metal enrichment of stars with planets rests on the idea that an environment rich in heavy elements favors giant-planet formation. This point of view is supported by the negative result of an intensive photometric search for transiting planets in 47 Tuc [65], a metal deficient globular cluster ($<$[Fe/H]$>= -0.7$). No transit was detected whereas several dozens were expected. It should be noted however, that in the case of 47 Tuc, the absence of planet could be due to the cluster high stellar density in the monitored region. Close stellar neighbors could prevent the formation of the protoplanetary disk.

Another possible explanation relates to the hypothesis that during the period of planet formation, one or several migrating planets could fall into the central star. In such a case, the contamination should be more effective for stars with small convective zones (massive dwarf stars), what is not observed [64]. However, for HD 82943, there are spectroscopic indications of a planet engulfment by a star with two known extrasolar planets [66], suggesting that contamination could nevertheless be acting during the planet formation phase and produce stellar metal enrichment.

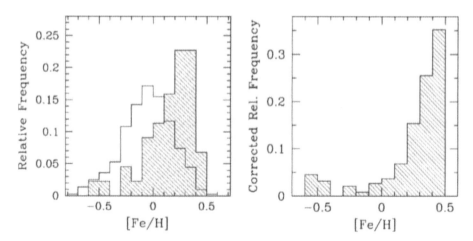

Fig. 2.7 Left: Comparative distributions of metallicities for stars with planets (dashed histogram) and comparison stars (open histogram). Metallicities are homogeneously derived from high-resolution spectroscopy. Right: Corrected relative frequency of planets per metallicity interval (taken from [64]).

Normalizing the [Fe/H] histogram of star hosting planets by the number of stars of the solar vicinity in given metallicity intervals, we obtain the "corrected" frequency of stars with planets per metallicity interval [64] (Fig. 2.7, right). The distribution is steeply rising for increasing metallicities and it turns out that a very large fraction of rich stars harbor giant planets. In the diagram, the Sun already resides in the "low-frequency" tail of the metallicity distribution. This could be taken as an indication that planetary systems resembling more to our own should be searched for around stars with moderate metallicity, for which the building (and perhaps the migration) of massive giant planet could be more difficult. Such systems would appear as more interesting for the search of exobiological life.

For our planet-search surveys, we take care of not introducing a bias in favor of metal-rich stars in our observational strategy, in order not to bias our interpretation of the results.

2.8 Summary and Future Perspectives

In about five and a half years, our understanding of planetary formation had to integrate several new peculiar characteristics brought by an increasing number of extrasolar planet discoveries. We can summarize the prominent results as follows:

- More than 5% of dwarf stars of the solar neighborhood harbor giant planets, about 1/6 of which are very close to their parent stars (hot Jupiters).
- Giant planets are detected around stars with masses ranging from 0.3 to 1.4 M_{Sun}.
- The orbital characteristics of extrasolar planets are diverse. Periods range from

2.985 days to several years and the orbits can be very elongated, up to $e = 0.927$.

- From the mass distribution of substellar companions to solar type stars, the upper mass limit of extrasolar planet is estimated around 10 M_{Jup}.
- The mass distribution of extrasolar planets increases for decreasing planet masses, despite the strong observational bias against very light companions. In particular, several planets have been detected with subsaturn minimum masses [11, 12].
- The real mass, radius and mean density have been determined for the giant planet orbiting the star HD 209458 by the observation of the photometric transit of the planet in front of the stellar disk [7-9]. Information on the geometry of the system has been obtained by the observation of the spectroscopic transit of the planet [10].
- Six multi-planet systems have been described. One is hierarchically organized (υ And [5, 6]), two have massive outer components (HD 168443 [14], HD 74156 [18]) and three are resonant systems (HD 83443 [13], Gl 876 [15], HD 82943 [18]).
- Giant planets are preferentially found around metal-rich stars [61-64]. No other chemical anomaly is pointed out for elements non related to Iron.
- The observed differences between the mass functions and the orbital element distributions of double stars and planetary candidates strongly suggest two different formation and evolution histories for the two populations.

What are now the expected progresses in the domain of extrasolar planet search, on short and longer time scales?

Radial-velocity programs. The efforts invested by the "planet-hunter" teams go in two complementary directions. On the one hand, the samples are enlarged to improve the available statistics and bring stronger constraints to the theoretical approaches of the domain. Between 2000 and 3000 stars are monitored by the different groups. They should provide several tens of additional extrasolar planet detections in the coming months/years. On the other hand, the precision is improved to get a faster access to lighter planets and multi-planet systems. The latter will probably play a preponderant role in our understanding of planetary formation. In this context, newly or soon available ESO instruments (UVES/VLT or HARPS/3.6-m [67]) open new possibilities for European astronomers.

Astrometry. Complementarity to the radial-velocity measurements, precise astrometric data (measure of the motion of the star on the sky due to the planet perturbation) of stars with planets will rapidly provide real masses for the known candidates. The needed precision will be achieved by interferometric techniques on large ground-based telescopes (VLTI, KeckI) or space missions (SIM, FAME). Contrarily to the radial-velocity technique mainly sensitive to short-period orbits, astrometry is more efficient for longer periods. Moreover, it will directly provide us with the real planet masses.

The complementarity between astrometric and radial-velocity measurements is illustrated in Fig. 2.8, displaying the mass of detected companions to solar type stars in function of the component separations. Precision limits of the mentioned methods are clearly indicated. The efficiency of astrometry for extrasolar planet search depends on the precision achieved for the measurements of star positions. With a precision of 10-50 µas, the VLTI will be sensitive to most of the known planets. It should also allow the detection of a "real" Jupiter up to a distance of about 200 pc from our Sun. With an expected precision of 4 to 1 µas, SIM should be able to point out terrestrial planets around stars in our close vicinity (a few parsecs).

In addition to interferometric facilities, satellites especially designed for the precise measurements of stellar positions, parallaxes and proper motions are being studied (e.g., GAIA; see Chap. 24, Foing). Such satellites are expected to significantly contribute to the progress in the field of extrasolar planets, especially for stars non-accessible to the radial-velocity technique like TTauri, A or B stars.

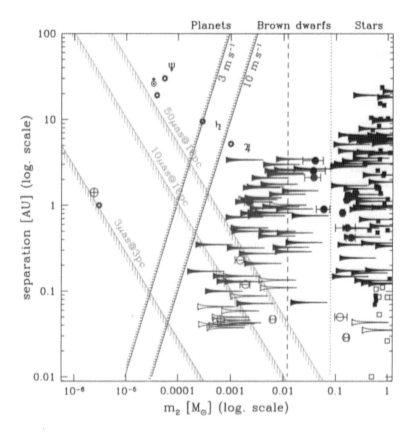

Fig. 2.8 *Mass-separation* diagram of companions to solar type stars (planets, brown dwarfs and stellar binaries). The elongated symbols represent the sin$i$ probability in logarithmic scale for candidates with only minimum mass determinations. Open symbols are used for low-eccentricity orbits. Solar-System planets are located by starred symbols. Finally, the precision limits of the radial-velocity technique and interferometric astrometry (VLTI: 10 and 50 µas for a star at 10 pc; and SIM: 3 µas for a star at 3 pc), are indicated by the labeled inclined shaded lines.

Photometric transits. Because of their simplicity and the importance of the obtained results, photometric-transit programs from the ground develop rapidly. Instruments with large field of view promise to be very efficient as the expected number of detections is statistically proportional to the number of monitored stars. From the ground, the achieved photometric precision easily allows for the detection of transiting giant planets. The detection of Earth-like planets, however, is only possible from space (induced luminosity variation of about 0.01%). Several space missions aiming to detect terrestrial planets are foreseen (COROT, Eddington, Kepler). Eddington, for example, is expected to find around 2000 terrestrial planets, among them a few tens in the habitable zone of the star.

On the long term: the search for life. Terrestrial planets detected by space interferometric and astrometric programs or by photometric transit searches will provide ideal targets for more ambitious projects, aiming to find bio-tracers in the atmosphere of those planets. Two similar projects are being presently studied: Darwin/ESA and TPF/NASA (Terrestrial Planet Finder). They basically consist in a battery of IR telescopes in space whose light will be combined in a clever way (nulling interferometry technique) to remove the light coming from the target star and thus reveal the planetary companion. Traces of Carbone dioxide in the observed low-resolution spectra of the planet will indicate the presence of an atmosphere, that should be "habitable" if water vapor is found and even "inhabited", at least by primitive forms of life, if Ozone is present.

So, in about 20 years, we should be able to scientifically give an answer to a fundamental philosophical question, recurrent throughout our history, on the origin, and unicity of life in the Universe. Today, the first element to the answer has already been brought by the discovery of planets around stars similar to our Sun.

2.9 References

1   M. Mayor, D. Queloz, Nature 378, 355 (1995).
2   G.A.H. Walker, A.R. Walker, A. Irwin, A.M. Larson, S.L. Yang, D.C. Richardson, Icarus 116, 359 (1995).
3   P. Butler, G. Marcy, E. Williams, H. Hauser, P. Shirts, ApJ Letters 474, L115 (1997).
4   G. Marcy, P. Butler, ARA&A 36, 57 (1998).
5   P. Butler, G. Marcy, D. Fischer, T. Brown, A. Contos, S. Korzennik, P. Nisenson, R. Noyes, ApJ **526**, 916 (1999).
6   G. Marcy, P. Butler, D. Fischer et al., in : A. Penny et al. (Eds.) *Planetary Systems in the Universe*, IAU Symp.202, ASP Conf. Ser., 2000, in press.
7   D. Charbonneau, T. Brown, D. Latham, M. Mayor, ApJ **529**, L45 (2000).
8   G. Henry, G. Marcy, P. Butler, S. Vogt, ApJ **529**, L41 (2000).
9   T. Mazeh, D. Naef, G. Torres et al., ApJ **532**, L55 (2000).
10 D. Queloz, A. Eggenberger, M. Mayor, C. Perrier, J.-L. Beuzit, D. Naef, J.-P. Sivan, S. Udry, A&A **359**, L13 (2000).
11 G. Marcy, P. Butler, S. Vogt, ApJ **536**, L43 (2000).
12 ESO Press Release, May 4th 2000, http://www.eso.org/outreach/press-rel/pr-2000/pr-13-00.html

13   M. Mayor, D. Naef, F. Pepe, D. Queloz, N.C. Santos, S. Udry, M. Burnet, in : A. Penny, P. Artymowicz, A.-M. Lagrange, S. Russel (Eds.) *Planetary Systems in the Universe*, IAU Symp.202, ASP Conf. Ser., 2000, in press.

14   S. Udry, M. Mayor, D. Queloz, in : A. Penny, P. Artymowicz, A.-M. Lagrange, S. Russel (Eds.) *Planetary Systems in the Universe*, IAU Symp.202, ASP Conf. Ser., 2000, in press.

15   G. Marcy, P. Butler, D. Fischer, S. Vogt, J. Lissauer, E. Rivera, ApJ (2001) in press.

16   X. Delfosse, T. Forveille, M. Mayor, C. Perrier, D. Naef, D. Queloz, A&A **338**, L67 (1998).

17   G. Marcy, P. Butler, S. Vogt, D. Fischer, J. Lissauer, ApJ **505**, L147 (1998).

18   ESO Press Release, April 4th 2001, http://www.eso.org/outreach/press-rel/pr-2001/pr-07-01.html

19   S. Udry, M. Mayor, D. Naef, F. Pepe, D. Queloz, N.C. Santos, M. Burnet, B. Confino, C. Melo, A&A **356**, 590 (2000).

20   A. Baranne, D. Queloz, M. Mayor et al., A&AS **119**, 373 (1996).

21   G. Laughlin, J.E. Chambers, ApJ (2001) in press.

22   N.C. Santos, M. Mayor, D. Naef, F. Pepe, S. Udry, M. Burnet, D. Queloz, in preparation.

23   D. Naef, D. Latham, M. Mayor et al., A&A (2001) in press.

24   G. Marcy, W. Cochran, M. Mayor, in : V. Mannings, A. Boss, S. Russel (Eds.) *Protostars and Planets IV*, University of Arizona Press, Tucson, 2000, p. 1285.

25   D. Fischer, G. Marcy, P. Butler, S. Vogt, S. Frink, K. Apps, ApJ (2001) in press.

26   J.B. Pollack, P. Bodenheimer, in : S. Atreya, J.K. Pollack, M.S. Matthews (Eds.) *Origin and Evolution of Planetary and Satellite Atmospheres*, ,University of Arizona Press,1989, p. 564.

27   A. Boss, Science **267**, 360 (1995).

28   J. Lissauer, Icarus **114**, 217 (1995).

29   P. Goldreich, S. Tremaine, ApJ **241**, 425 (1980).

30   D. Lin, J. Papaloizou, ApJ **309**, 846 (1986).

31   W.R. Ward, ApJ **482**, L211 (1997).

32   P. Bodenheimer, O. Hubickyj, J. Lissauer, Icarus **143**, 2 (2000).

33   M. Mayor, S. Udry, in : F. Garzon, C. Eiroa, D. de Winter, T.J. Mahoney (Eds.) *Disks, Planetesimals and Planets*, , ASP Conf. Ser., 2000, in press.

34   F. Shu, J. Najita, E. Ostriker et al., ApJ **429**, 781 (1994).

35   D. Lin, P. Bodenheimer, D.C. Richardson, Nature **380**, 606 (1996).

36   D. Trilling, W. Benz, T. Guillot, J. Lunine, W. Hubard, A. Burrows, ApJ **500**, 428 (1998).

37   J.S. Lewis, R.G. Prinn, *Planets and their Atmospheres. Origin and Evolution*, Inter. Geophys. Ser. **33** (1984).

38   S.J. Weidenschilling, F. Marzari, Nature **384**, 619 (1996).

39   D. Lin, S. Ida, ApJ **477**, 781 (1997).

40   G. Wuchterl, in : F. Paresce (Ed.) *Science with the VLT Interferometer*, ESO Astrophys. Symp., Springer, 1997, p. 64.

41   G. Wuchterl, BAAS **31**, 36.07 (1999).

42   A. Boss, in : A. Penny, P. Artymowicz, A.-M. Lagrange, S. Russel (Eds.) *Planetary Systems in the Universe*, IAU Symp.202, ASP Conf. Ser., 2000, in press.

43   D. Saumon, W.B. Hubbard, A. Burrows, T. Guillot, J. Lunine, G. Chabrier, ApJ **460**, 993 (1996).

44   T. Guillot, Science **286**, 72 (1999).

45 T. Brown, D. Charbonneau et al., in : A. Penny, P. Artymowicz, A.-M. Lagrange, S. Russel (Eds.) *Planetary Systems in the Universe*, IAU Symp.202, ASP Conf. Ser., 2000, in press.

46 A. Cameron, K. Horne, A. Penny, D. James, in : A. Penny, P. Artymowicz, A.-M. Lagrange, S. Russel (Eds.) *Planetary Systems in the Universe*, IAU Symp.202, ASP Conf. Ser., 2000, in press.

47 A. Coustenis, J. Schneider, R. Wittemberg et al., in : R. Rebolo, E. Martin, M.R. Zapatero Osorio (Eds.) *Brown Dwarfs and Extrasolar Planets*, ASP Conf. Ser. **134**, 1998, p. 296.

48 H. Rauer, D. Bockelée-Morvan, A. Coustenis, T. Guillot, J. Schneider, A&A **355**, 573 (2000).

49 J.-L. Halbwachs, F. Arenou, M. Mayor, S. Udry, D. Queloz, A&A **355**, 581 (2000).

50 S. Zucker, T. Mazeh, ApJ **531**, L67 (2000).

51 I. Han, D. Black, G. Gatewood, ApJ **548**, L57 (2001).

52 D. Pourbaix, A&A (2001) in press.

53 D. Pourbaix, F. Arenou, A&A (2001) in press.

54 M.-R. Zapatero Osorio, V. Béjar, E. Martin et al., Science **290**, 103 (2000).

55 A. Jorissen, M. Mayor, S. Udry, A&A (2001) submitted.

56 A. Duquennoy, M. Mayor, A&A **248**, 485 (1991).

57 M. Mayor, S. Udry, J.-L. Halbwachs, F. Arenou, in: H. Zinnecker, R. Mathieu (Eds.) *The Formation of Binary Stars*, IAU Symp.200, ASP Conf. Ser., 2000, in press.

58 S. Udry, M. Mayor, X. Delfosse, T. Forveill, C. Perrier, in: B. Reipurth, H. Zinnecker (Eds.) *Birth and Evolution of Binary Stars*, IAU Symp.200P, 2000.

59 H.F. Levison, J. Lissauer, M.J. Duncan, AJ **116**, 1998 (1998).

60 T. Mazeh, Y. Krymolowski, G. Rosenfeld, ApJ **477**, L103 (1997).

61 G. Gonzalez, MNRAS **285**, 403 (1997).

62 G. Gonzalez, A&A **334**, 221 (1998).

63 N.C. Santos, G. Israelian, M. Mayor, A&A **363**, 228 (2000).

64 N.C. Santos, G. Israelian, M. Mayor, A&A (2001) in press.

65 R. Gilliland, T. Brown, P. Guhathakurta et al., ApJ **545**, L47 (2000).

66 G. Israelian, N. Santos, M. Mayor, R. Rebolo, Nature **411**, 163 (2001).

67 F. Pepe, M. Mayor, B. Delabre, D. Kohler, D. Lacroix, D. Queloz, S. Udry, W. Benz, J.-L. Bertaux, J.-P. Sivan, in: *Astronomical telescopes and instrumentations 2000*, Garching, SPIE Conf., 2000, in press.

68 F. Pepe, M. Mayor, D. Naef, D. Queloz, N.C. Santos, S. Udry, M. Burnet, in preparation.

69 D. Queloz, M. Mayor, D. Naef, F. Pepe, N.C. Santos, S. Udry, M. Burnet, in: A. Penny, P. Artymowicz, A.-M. Lagrange, S. Russel (Eds.) *Planetary Systems in the Universe*, IAU Symp.202, ASP Conf. Ser., 2000, in press.

70 P. Butler, S. Vogt, G. Marcy, D. Fisher, G. Henry, K. Apps, ApJ **545**, 504 (2000).

71 P. Butler, G. Marcy, S. Vogt, K. Apps, PASP **110**, 1389 (1998).

72 N.C. Santos, M. Mayor, D. Naef, F. Pepe, D. Queloz, S. Udry, M. Burnet, Y. Revaz, A&A **356**, 590 (2000).

73 A. Hatzes, W. Cochran, B. McArthur et al., ApJL (2001) in press.

74 C. Tinney, P. Butler, G. Marcy et al., ApJ **551**, L507 (2001)

75 S. Vogt, G. Marcy, P. Butler, K. Apps, ApJ **536**, 902 (2000).

76 D. Naef, M. Mayor, F. Pepe, D. Queloz, N.C. Santos, S. Udry, M. Burnet, A&A (2001) in press.

77 R. Noyes, S. Jha, S. Korzennik et al., ApJ **483**, L111 (1997).

78   G. Marcy, P. Butler, S. Vogt, D. Fischer, M. Liu, ApJ **520**, 239 (1998).

79   D. Fischer, G. Marcy, P. Butler, S. Vogt, K. Apps, PASP **111**, 50 (1998).

80   P. Butler, C. Tinney, G. Marcy et al., ApJ (2001) in press.

81   W. Cochran, A. Hatzes, P. Butler, G. Marcy, ApJ **483**, 457 (1997)

82   M. Kürster, M. Endl, S. Els et al., A&A **353**, L33 (2000).

83   P. Butler, G. Marcy, ApJ **464**, L153 (1996)

84   D. Queloz, M. Mayor, L. Weber, A. Blecha, M. Burnet, B. Confino, D. Naef, F. Pepe, N.C. Santos, S. Udry, A&A **354**, 99 (2000).

85   J.-P. Sivan, M. Mayor, D. Naef, D. Queloz, S. Udry, C. Perrier, J.-L. Beuzit, in: A. Penny, P. Artymowicz, A.-M. Lagrange, S. Russel (Eds.) *Planetary Systems in the Universe*, IAU Symp.202, ASP Conf. Ser., 2000, in press.

86   G. Marcy, P. Butler, S. Vogt, ApJ, submitted.

87   S. Korzennik, T. Brown, D. Fischer, P. Nisenson, R. Noyes, ApJL (2001) in press.

88   G. Marcy, P. Butler, ApJ **464**, L147 (1996).

89   D. Latham, R. Stefanik, T. Mazeh, M. Mayor, G. Burki, Nature **339**, 38 (1989).

90   S. Udry, M. Mayor, D. Naef, F. Pepe, D. Queloz, N.C. Santos, M. Burnet, in preparation.

3 Habitable Zones in Extrasolar Planetary Systems

Siegfried Franck, Werner von Bloh, Christine Bounama, Matthias Steffen,
Detlef Schönberner and Hans-Joachim Schellnhuber

If we ask the question about the possible existence of the life outside the Earth, we first have to determine the habitable zone (HZ) for our solar system. The HZ of distances between a main sequence star and an Earth-like planet is roughly defined as the range of mean orbital radii which imply moderate planetary surface temperatures suitable for the development and subsistence of carbon-based life. The latter precondition is usually taken as the requirement that liquid water is permanently available at the planet's surface. The HZ concept was introduced by Huang [1, 2] and extended by Dole [3] and Shklovskii and Sagan [4].

For our purposes, an Earth-like planet is one similar in mass and composition to Earth. It's mass has to be sufficient to maintain plate tectonics in order for the global carbon cycle to operate and stabilize the surface temperature.

It is generally accepted that the Earth's climate is mainly determined by the atmospheric CO_2 level. On geological time scales, i.e. over hundreds and thousands of million years, the Earth's climate is stabilized against increasing insolation by a negative feedback provided by the global carbon cycle: higher surface temperatures increase the precipitation and so increase the weathering rates resulting in decreasing atmospheric CO_2 content and decreasing greenhouse effect. In the case of lower surface temperatures, the negative feedback loop acts analogously.

We know that at present only our Earth has liquid water at its surface. It is well-known that Venus is much too hot for the existence of liquid water. At the Venusian orbit the insolation is too strong that the above described negative feedback breaks down: on Venus the atmosphere became so full of water vapor that no infrared radiation from the surface was able to escape to space. The resulting higher surface temperatures forced the vaporization of water to the atmosphere. This positive feedback effect is called "runaway greenhouse". On the other hand, the negative feedback loop stabilizing Earth's climate may also fail, if we would shift the planet too much away from the Sun. At such distances CO_2 condenses to form CO_2 clouds that increase the planetary albedo, i.e. the reflection of solar radiation, and cause lower surface temperatures. If the planet's surface would be covered with snow and ice, the albedo would increase further. This positive feedback loop is called "runaway glaciation".

Concerning Mars, we presently know of no life, but there is an ongoing discussion about the possibility that life might have been there in the past. The present Martian surface temperature is so low that CO_2 condenses and the polar ice caps contain a mixture of CO_2 ice and water ice (see Chap. 6, Jaumann et al.). However, the climate on Mars may not always have been so inhabitable. Early in its history, the climate is thought to have been more suitable for the existence of liquid water at or near the surface. The evidence comes from the interpretation of images that show the geology of the surface features (see e.g., [5]). According to our investigation of the HZ for the solar system [6], the Martian orbit position was within the HZ up to about 500 million years ago.

Jovian-type planets do not have a solid or liquid surface, covered by an atmosphere, near which organisms may exist. Therefore, usually they are considered as inhabitable. But there is the possibility that moons of giant planets are within a habitable zone. The best candidate for producing a habitable environment is Europa, the second Galilean satellite of Jupiter, with a mean density of about 3 g × cm^{-3} and therefore mostly composed of rock, but there are also enough volatiles. Due to low surface temperatures and additional internal heat sources (tidal heating), only a subsurface ocean could exist. A good analogue for possible life underneath the ice of Europa can be found in Lake Vostok, Antarctica. There, microorganisms exist at ice depth of 3 km. NASA plans to investigate the possible subsurface ocean of Europa with the help of a spacecraft. Another interesting object is Saturn's moon Titan with a methane-rich atmosphere, in which photochemical reactions may create organic molecules. Titan's atmosphere will also be studied by the mission CASSINI (see Chap. 24, Foing). A detailed investigation about habitability of moons around giant planets is given by Williams et al. [7].

The same type of stability calculations described above for the solar system with the Sun as the central star can also be performed for stars other than our Sun. Such investigations are of special importance, because we now have novel techniques for the detection of extrasolar planetary systems (see also Chap. 2, Udry and Mayor). The expected basic results for the HZ around other central stars are relatively simple: to have a surface temperature in the range similar to the Earth's, a planet orbiting a central star with lower mass would have to be closer to the star than 1 Astronomical Unit (AU, i.e. mean distance between Earth and Sun), whereas a planet orbiting a brighter star that has more mass than our Sun, would have to be farther than 1 AU from the star. But the problem is a little bit more complicated: we also have to take into account the different times that stars spend on the so-called main sequence. The main sequence is a band running from the upper left to the lower right on a plot of luminosity versus effective radiating temperature. Such a plot is called Hertzsprung-Russell diagram. Stars on the main sequence receive their energy mainly from hydrogen burning, i.e. the fusion of hydrogen to helium. In our investigation of extrasolar planetary systems [8], we have used a parameterization for the luminosity of main sequence central stars in the mass range between 20% and 250% solar masses.

Dole's [3] estimations of the HZ have been based on an optical thin atmosphere and a fixed albedo model. He finds 0.725 AU for the inner boundary and 1.24 AU for the outer boundary of the HZ, respectively. Hart [9, 10] calculated the hypothetical evolution of the terrestrial atmosphere over geologic time for different orbital radii. He found that the HZ, i.e. the "ecological niche" between runaway greenhouse and runaway glaciation processes is amazingly narrow for stars like our Sun: It is delimited from below by $R_{inner} = 0.958$ AU and from above by $R_{outer} = 1.004$ AU, where $R_{inner\,r}$ and R_{outer} are the inner and outer limits to the mean orbital radius, respectively.

A main disadvantage of those calculations was the neglect of the negative feedback between atmospheric CO_2 partial pressure and mean global surface temperature via the carbonate-silicate cycle as discovered by Walker et al. [11]. The inclusion of that feedback by Kasting et al. [12] produced the interesting result of an almost constant inner boundary of the HZ but a remarkable extension of its outer boundary. In subsequent years, the HZ approach experienced a number of refinements and the extension to other classes of main sequence stars [13-15]. An overview is provided by Doyle [16]. Recent studies conducted by our group (see particularly, [17, 8]) have generated a rather comprehensive characterization of habitability, based on the possibility of photosynthetic biomass production under large-scale geodynamic conditions. Thus, not only the availability of liquid water on a planetary surface is taken into account but also the suitability of CO_2 partial pressure. Our definition of habitability is described in detail in the next part, especially by Eq. (3.5).

At the present time, the determination of habitable zones in extrasolar planetary systems is of special interest, because in the last few years up to ~50 objects have been identified with the help of novel techniques [18]. Unfortunately, most of the discovered planets are giants on orbits surprisingly close to the central star. Nevertheless, there is hope to find also Earth-like planets with the help of those astronomical observing programs, launched in the early 1990s that rely on planet detection in the Milky Way via gravitational microlensing observation and other techniques [19]. The most important programs are Massive Compact Halo Objects (MACHO), Probing Lensing Anomalies Network (PLANET), Experience pour la Recherche d'Objects Sombres (EROS), and Optical Gravitational Lensing Experiments (OGLE).

Gravitational microlensing events occur, when a faint or dark star passes the line of sight of a more distant, brighter star. The light rays emanating from the latter are bent by the gravitational field of the closer, fainter star. This results in a discernible magnification of the image of the brighter object. A planet orbiting the faint star can cause a minor extra peak in the magnification record.

Summarizing we can state that the HZ is the range of orbital distances from a star, in which a planet can maintain liquid water and biological productivity on its surface. The HZ can be calculated with the help of climatological approaches or within the framework of Earth system science. According to our model, the HZ for the present solar system extends between about 0.95 AU and about 1.2 AU [6] and was broader in the past. For extrasolar systems we can postulate a distinct HZ for young central stars in the mass range between about 0.4 and 2 solar masses.

The next two parts describe model calculations for the Sun and for other single main sequence stars, respectively. In the final we give our main conclusions and point out several areas for future work.

3.1 Models for Calculating the HZ in the Solar System

Since the early work of Hart [9, 10], there have been many improvements in climatic constraints on the inner and outer boundaries of the HZ. The most comprehensive work in this field is the paper by Kasting et al. [13]. The authors define the boundaries of the HZ via so-called critical solar fluxes. For the inner radius of the HZ they give three different estimations based on the following assumptions:

1. loss of planetary water by a moist greenhouse [20],
2. loss of planetary water by a runaway greenhouse,
3. observation that there was no liquid water on Venus' surface at least for the last 1 Ga.

The outer radius of the HZ is also calculated by three different approaches:

1. estimation based on observations and arguments that early Mars had a warm and wet climate (see also the recent papers [5, 21]),
2. maximum possible CO_2 greenhouse heating (but see also [22]),
3. first condensation limit of CO_2 clouds that increase the planetary albedo.

Assuming the possibility of a "cold start", i.e. an originally ice covered planet that was initially beyond the outer HZ boundary enters the HZ due to HZ boundary shifts with time, Kasting et al. [13] found the following values for the present HZ in the solar system:

1. most conservative case: 0.95 AU ... 1.37 AU,
2. least conservative case: 0.75 AU ... 1.90 AU,
3. intermediate case (favored): 0.84 AU ... 1.77 AU.

The modeling approach by Franck et al. [17, 8] is based on the ideas introduced by Caldeira and Kasting [23]. Therefore, a careful simulation of the coupling between increasing solar luminosity, the silicate rock weathering as parameterized by the mean rate F_{wr}, and the global energy balance forms the cornerstone of the investigation. As a direct product, the partial pressure of atmospheric carbon dioxide P_{atm} and the biological productivity Π can be estimated as a function of time t throughout planetary past and future. In the following we give the crucial elements of the causal web employed.

The global energy balance of the planet's climate is usually expressed with the help of the Arrhenius-equation [24].

$$(1-a)\,S(t) = 4\sigma\,T_{bbr}^4 \tag{3.1}$$

where a is the planetary albedo, σ is the Stefan-Boltzmann constant, and T_{bbr} is the effective black-body radiation temperature. The time dependence of the solar constant S is fitted with the help of a formula given by Gough [25]. The surface temperature of the planet T_s is related to T_{bbr} by the greenhouse warming factor:

$$T_S = T_{bbr} + \Delta T. \tag{3.2}$$

Usually ΔT is parameterized as a function of T_s and P_{atm} [23, 26].

The total process of weathering embraces first the reaction of silicate minerals with carbon dioxide, second the transport of weathering products, and third the deposition of carbonate minerals in sediments. The basic assumptions and limitations of this approach are given in Franck et al. [26]. The weathering rate F_{wr} is a key function in our model. For any given weathering rate the surface temperature T_s and the carbon dioxide concentration in the soil P_{soil} can be calculated self-consistently [23, 26, 17].

The main role of the biosphere in the context of our model is to increase P_{soil} in relation to the atmospheric carbon dioxide partial pressure and proportional to the biologic productivity, Π. Π is considered to be a function of temperature and carbon dioxide partial pressure in the atmosphere only.

$$\frac{\Pi}{\Pi_{max}} = \left(1 - \left(\frac{T_s - 25\,°C}{25\,°C}\right)^2\right)\left(\frac{P_{atm} - P_{min}}{P_{1/2} + (P_{atm} - P_{min})}\right) \tag{3.3}$$

Π_{max} is the maximum productivity and is assumed to be twice the present value [27]. $P_{1/2} + P_{min}$ is the value, at which the pressure dependent factor is equal to 1/2 and $P_{min} = 10$ ppm the minimum value for photosynthesis. For fixed P_{atm}, Eq. (3.3) produces maximum productivity at the optimum temperature ($T_s = 25\,°C$) and zero productivity outside the temperature tolerance interval (0 - 50 °C).

First we have solved the system of equations under the assumption that the weathering rate F_{wr} is always equal to the present value $F_{wr,0}$. This is clearly a rather rough approximation. We call this approach the geostatic model (GSM).

Franck et al. [17] have introduced the geodynamic model (GDM). In this case a balance between the carbon dioxide sink in the atmosphere-ocean system and the metamorphic (plate-tectonic) sources is expressed with the help of dimension less quantities [28]:

$$f_{wr} \cdot f_A = f_{sr}, \tag{3.4}$$

where $f_{wr} \equiv F_{wr} / F_{wr,0}$ is the weathering rate normalized by the present value, $f_A \equiv A_c / A_{c,0}$, is the continental area normalized by the present value, and $f_{sr} \equiv S / S_0$ the spreading rate normalized by the present value. With the help of Eq. (3.4) we can calculate the weathering rate from geodynamical theory [17]. Our model is sketched in Fig. 3.1.

Based on our calculation scheme, we define the HZ as the range of all orbital distances, where the biological productivity is greater than zero:

$$HZ := \{R | \Pi (P_{atm}(R,t), T_s(R,t)) > 0\} \tag{3.5}$$

In our calculation with the help of the model shown in Fig. 3.1 [17] we found the following values for the present HZ: $R_{inner} = 0.97$ AU and $R_{outer} = 1.39$ AU.

3.2 HZ Around Other Main Sequence Stars

The same type of HZ calculations, both on the base of climatic constraints and on the base of Earth system modeling as well, can be performed for stars with masses different from the solar mass.

Kasting et al. [13] restricted themselves to stellar lifetimes greater than 2 Ga which corresponds to masses less than 1.5 M_s (1 M_s = one solar mass). At the low-mass end they restricted themselves to masses greater than 0.5 M_s, because stars with masses ≤ 0.5 M_s show negligible evolution. Stellar luminosities and temperatures were taken directly from Iben [29, 30], climatic constraints correspond to their so-called intermediate case. As expected, stellar HZ's for more massive stars are rather short, because they have to be truncated at the end of the main sequence. HZ's for low mass stars are essentially the same over time. In Fig. 3.2 we show the so-called zero age main sequence HZ from Kasting et al. [13] as function of stellar mass.

In our calculation of HZ in extrasolar planetary systems [8] we used the luminosity evolution of central stars on the main sequence in the mass range between 0.8 and 2.5 M_s. The results have been obtained by polynomial fitting of detailed stellar evolution models like the one presented by Schaller et al. [31]. The temperature tolerance window for the biological productivity given in Eq. (3.3) was extended to the range between 0 °C and 100 °C in order to incorporate thermophiles [32]. Furthermore, for this study a linear continental growth model was applied. To present the results of our modeling approach, we have delineated the HZ for an Earth-like extrasolar planet at a given but arbitrary distance R in the stellar mass-time plane (Fig. 3.3).

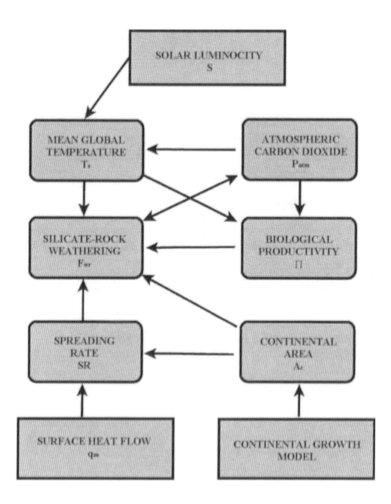

Fig. 3.1 Sketch of our Earth system model. Arrows may describe both negative and positive feedbacks.

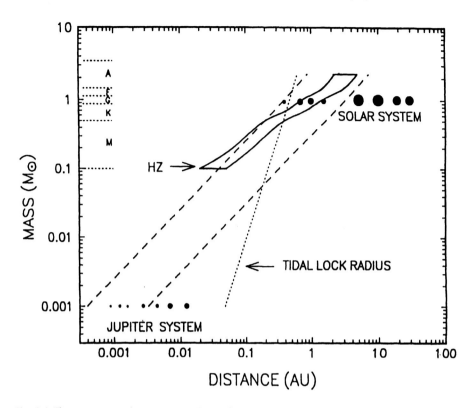

Fig. 3.2 The zero age main sequence HZ as a function of central star mass (in solar masses M⊙) for the intermediate case of climatic constraints. The long-dashed lines delineate the probable terrestrial planet accretion zone. The dotted line is the orbital distance, for which an Earth-like planet in a circular orbit would be locked into synchronous rotation (tidal locking). (Figure of Kasting et al. [13], by kind permission of Dr. J. F. Kasting and Academic Press.)

In Fig. 3.3 the HZ is limited by the following effects:

1. Stellar lifetime on the main sequence decreases strongly with mass. Using simple scaling laws [33], we estimated the central hydrogen burning period and got τ_H <0.8 Ga for M >2.2 M_s. Therefore, there is no point in considering central stars with masses larger than 2.2 M_s, because an Earth-like planet may need ~0.8 Ga of habitable conditions for the development of life [9, 10]. Quite recently, smaller numbers for the time span required for the emergence of life have been discussed, for instance 0.5 Ga [34]. If we perform calculations with τ_H <0.5 Ga, we obtain qualitatively similar results, but the upper bound of central star masses is shifted to 2.6 M_s.

2. When a star leaves the main sequence to turn into a red giant, there clearly remains no HZ for an Earth-like planet. This limitation is relevant for stellar masses in the range between 1.1 and 2.2 M_s.

3. In the stellar mass range between 0.6 and 1.1 M_s the maximum life span of the biosphere is determined exclusively by planetary geodynamics, which is inde-

pendent (in a first approximation, but see limiting effect 4) from R. So we obtain the limitation $t < t_{max}$.

4.   There have been discussions about the habitability of tidally locked planets. We take this complication into account and indicate the domain, where an Earth-like planet on a circular orbit experiences tidal locking. That domain consists of the set of (M,t) couples which generate an outer HZ boundary below the tidal-locking radius. This limitation is relevant for $M < 0.6$ M_s. As an illustration we depict the HZ for $R = 2$ AU in Fig. 3.3.

3.3   Conclusions

The question, whether an Earth-like planet discovered outside the solar system may accommodate life, can be answered with the help of Figs. 3.2 and 3.3 if the mass and age of the central star as well as the planet's orbit are known. This is only the present state of the art in theoretical modeling of HZ. There are of course a lot prerequisites for such calculations that have been summarized recently by Lissauer [35].

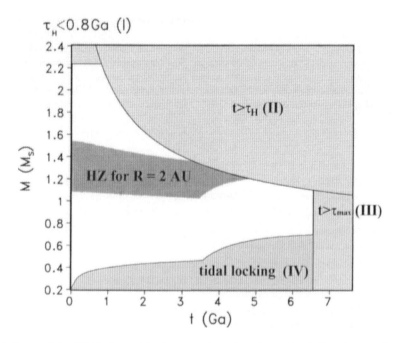

Fig. 3.3 Shape of the HZ (dark grey) in the mass-time plane for an Earth-like planet with photosynthesis at distance $R = 2$ AU from the central star. The potential overall domain for accommodating the HZ of planets at some arbitrary distance is limited by a number of R-independent factors that are explained in the text. (Figure slightly changed from Franck et al. [8], copyright by the American Geophysical Union).

A main assumption is that extraterrestrial life would be carbon-based and needs liquid water. In this way, the carbon-based life is connected with the global carbon cycle between the components of the whole system. The carbon cycle may operate only on a geologically active planet with liquid water. The present understanding of plate tectonics, however, is not sufficient to enable us to predict whether a given planet would exhibit such a phenomenon or not. First theoretical steps to tackle this problem were made by Solomatov and Moresi [36].

Furthermore, planetary orbits are generally chaotic. For stable orbits over a long period of time a minimum separation is required in a system of Earth-like planets [37]. Fortunately, this separation is comparable to the width of the HZ.

To detect Earth-sized extrasolar planets, NASA and ESA are both designing space missions for the second decade of this century. Examples for such project are "Terrestrial Planet Finder" (TPF), i.e. an infrared interferometer operating in an orbit to detect Earth-like extrasolar planets and investigate their atmospheres (http:// astrobiology.arc.nasa.gov), and DARWIN, i.e. an infrared space interferometer to search for signs of life on any Earth-like planet found (http://ast.star.rl.ac.uk/darwin, see Chap. 24, Foing).

Acknowledgements. This work was supported by the German Science Foundation (DFG, grant IIC5-Fr910/9-3) and by the Ministry of Science, Research, and Culture of the State of Brandenburg/Germany (HSP III 1.6; 04/035).

3.4   References

1   S.S. Huang, Am. Sci. **47**, 397 (1959).

2   S.S. Huang, Sci. Am. **202**(4), 55 (1960).

3   S.H. Dole (Ed.) *Habitable Planets for Man*, Blaisdell, New York, 1964, 158 pp.

4   I.S. Shklovskii, C. Sagan (Eds.) *Intelligent Life in the Universe*, Holden-Day, San Francisco, 1966, 509 pp.

5   M.P. Golombek, Science **283**, 1470 (1999).

6   S. Franck, A. Block, W. von Bloh, C. Bounama, H.J. Schellnhuber, Y. Svrezhev, Planet. Space Sci. **48**, 1099 (2000).

7   D.M. Williams, J.F. Kasting, R.A. Wade, Nature **385**, 234 (1997).

8   S. Franck, A. Block, W. von Bloh, C. Bounama, M. Steffen, D. Schönberner, H.J. Schellnhuber, JGR-Planets **105**, No. E1, 1651 (2000).

9   M.H. Hart, Icarus **33**, 23 (1978).

10 M.H. Hart, Icarus **37**, 351 (1979).

11 J.C.G. Walker, P.B. Hays, J.F. Kasting, J. Geophys, Res. **86**, 9776 (1981).

12 J.F. Kasting, Icarus **74**, 472 (1988).

13 J.F. Kasting, D.P. Whitmire, R.T. Reynolds, Icarus **101**, 108 (1993).

14 J.F. Kasting, Origins of Life **27**, 291 (1997).

15 D.M. Williams, *The stability of habitable planetary environments*, A Thesis in Astronomy and Astrophysics, Pennsylvania State University, 1998, 140 pp.

16 L.R. Doyle (Ed.), *Circumstellar habitable zones*, Proc. First International Conference, Travis House Publications, Menlo Park, California, 1996, 525 pp.

17 S. Franck, A. Block, W. von Bloh, C. Bounama, H.J. Schellnhuber, Y. Svirezhev, Tellus **52B**, No.1, 94 (2000).

18 J. Schneider, *Extrasolar Planets and Exobiology*, http:// www. obspm.fr/encycl/ encycl.html, 2000.

19 D.P. Bennett, S.H. Rhie, Astrophys. J. **472**, 660 (1996).

20 J.F. Kasting, O.B. Toon, J.B. Pollack, Sci. Am. **256**, 90 (1988).

21 M.C. Malin, K.S. Edgett, Science **288**, 2330 (2000).

22 F. Forget, R.T. Pierrehumbert, Science **278**, 1273 (1997).

23 K. Caldeira, J.F. Kasting, Nature **360**, 721 (1992).

24 S.A. Arrhenius, Philos. Mag. **41**, 237 (1896).

25 D.O. Gough, Sol. Phys. **74**, 21 (1981).

26 S. Franck, K. Kossacki, C. Bounama, Chem. Geol. **159**, 305 (1999).

27 T. Volk, Am. J. Sci. **287**, 763 (1987).

28 J.F. Kasting, Am. J. Sci. **284**, 1175 (1984).

29 I. Iben, Annu. Rev. Astron. Astrophys. **5**, 571 (1967).

30 I. Iben, Astrophys. J. **147**, 624 (1967).

31 G. Schaller, D. Schaerer, G. Meynet, A. Meader, Astron. Astrophys. Suppl. Ser. **96-2**, 269 (1992).

32 D. Schwartzman, M. McMenamin, T. Volk, Bio Sci. **43**, 390 (1993).

33 R. Kippenhahn, A. Weigert (Eds.) *Stellar Structure and Evolution*, Springer-Verlag, Berlin, 1990, 468 pp.

34 B. Jakosky (Ed.) *The Search for Life on Other Planets*, Cambridge University Press, Cambridge, 1998, 326 pp.

35 J.J. Lissauer, Nature **402**, Supp., C11 (1999).

36 V.S. Solomatov, L.N. Moresi, Geophys. Res. Lett. **24**, 1907 (1997).

37 J.J. Lissauer, Rev. Mod. Phys. **71**, 835 (1999).

4 Viable Transfer of Microorganisms in the Solar System and Beyond

Gerda Horneck, Curt Mileikowsky, H. Jay Melosh, John W. Wilson,
Francis A. Cucinotta and Brett Gladman

It is now generally accepted that at the beginning of our solar system a considerable amount of organic molecules and water were imported to the early Earth as well as to the other terrestrial planets via asteroids and comets [1-3]. The period of heavy bombardment lasted until approximately 3.8 billion years (Ga) ago. These impactors would on the one hand have been delivering the volatiles as precursors of life, on the other hand, if sufficiently large and fast, would have eroded the atmosphere and perhaps sterilized the Earth and/or Mars, if life existed there [4, 5].

Impactors of sizes larger than 1 km lead to the ejection of a considerable amount of soil and rocks that are thrown up at high velocities, some fraction reaching escape velocity [6]. These ejecta leave the planet and orbit around the Sun, usually for time scales of a few hundred thousand or several million years until they either impact another celestial body or are expelled out of the solar system [7]. Meteorites of lunar and some of Martian origin detected within in the last decades are witnesses of these processes [8, 9]. The question arises whether such rock or soil ejecta could also be the vehicle for life to leave its planet of origin, or, in other words, whether spreading of life in the solar system via natural transfer of viable microbes is a feasible process.

4.1 Scenario of Interplanetary Transfer of Life Within the Solar System

The supposition that life can be naturally transferred from one planet to another or even between solar systems goes back to the last century [10] and was formulated as hypothesis of Panspermia by S. Arrhenius in 1903 [11]. It postulates that microscopic forms of life, for example spores, can be dispersed in space by the radiation pressure from the Sun thereby seeding life from one planet to another, or one solar system to another, respectively. This hypothesis has been subjected to several criticisms with arguments, such as it cannot be experimentally tested and spores will not survive long-time exposure to the hostile environment of space, especially vacuum and radiation (reviewed in [12]). It has also been pointed out that Panspermia only shunts aside the question of the origin of life to another celestial body (see Chap. 1, Ehrenfreund and Menten).

However, a variety of recent discoveries have shed new light on the likelihood of viable transfer in space such as (i) the detection of meteorites, some of lunar and some of Martian origin [9]; (ii) the detection of organics and the still highly debated supposition of microbial fossils in one of the Martian meteorites [13]; (iii) the probability of small particles of diameters between 0.5 μm and 1 cm [14], or even boulder-sized rocks reaching escape velocities by the impact of large comets or asteroids on a planet, e.g., on Earth [6] or Mars [15, 16]; (iv) the ability of bacterial spores to survive to a certain extent the shock waves of such a simulated impact [17]; (v) the high UV-resistance of microorganisms at the low temperatures of deep space, tested at temperatures down to 10 K [18]; (vi) the reported survival of bacterial spores over millions of years, if enclosed in amber or salt stocks [19, 20], or in space over periods extending up to 6 years [21]; (vii) the paleogeochemical evidence of a very early appearance of life on Earth in the form of metabolically advanced microbial prokaryotic ecosystems leaving not more than approximately 0.4 Ga for the evolution of life from the simple precursor molecules to the level of a prokaryotic, photoautotrophic cell [22]; (viii) the biochemical evidence of a common ancestor for all life forms on Earth [23].

Viable transfer from one planet to another requires that life, probably of microbial nature, survives the following three steps: (i) the escape process, i.e. ejection into space, e.g., caused by a large impact on the parent planet; (ii) the journey through space, i.e. time scales in space comparable with those experienced by the Martian meteorites (approximately 1-15 Ma); and (iii) the landing process, i.e. non-destructive deposition of the biological material on another planet (Fig. 4.1).

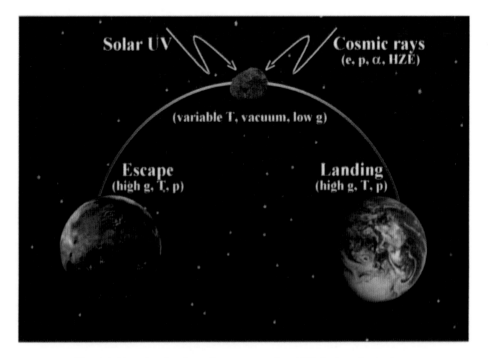

Fig. 4.1 Scenario of an interplanetary transfer of life in the solar system

4.2   Survival of the Escape Process

The most plausible process capable of ejecting microbe-bearing surface material from a planet or moon into space is the hypervelocity impact of a large object such as an asteroid or comet, under strong or moderate shock metamorphism of the ejected rock fragments. The peak shock pressure estimates for the presently studied 15 Martian meteorites range from about 20 GPa to about 45 GPa and estimates of associated post-shock temperature range from about 100 °C at 20 GPa to about 600 °C at 45 GPa [24]. Although these impacts are very energetic, a certain fraction of ejecta is not heated above 100 °C. These low temperature fragments are ejected from the so-called spall zone, i.e. the surface layer of the target where the resulting shock is considerably reduced by superimposition of the reflected shock wave on the direct one [25]. Estimates suggest that within the last 4 Ga, more than 10^9 fragments of a diameter of ≥ 2 m and temperatures ≤ 100 °C were ejected from Mars of which about 5% arrived on Earth after a journey in space of ≤ 8 Ma (Fig. 4.2). The corresponding numbers for a transfer from Earth to Mars are about 10^8 fragments ejected from the Earth with about 0.1% arriving on Mars within 8 Ma [26]. During the preceding period of "heavy bombardment" even 10 times higher numbers are estimated. Hence, the 15 Martian meteorites, so far detected on Earth, represent probably only an infinitesimal fraction of those imported from Mars within Earth's history.

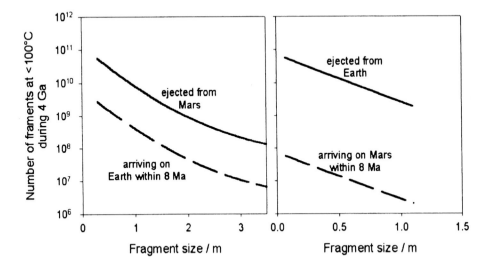

Fig. 4.2 Number of fragments ejected within the last 4 Ga at temperatures ≤ 100 °C from Mars or from the Earth and arriving within 8 Ma on Earth or on Mars (data based on calculations in [26]).

In experiments simulating impacts comparable to those experienced by the Martian meteorites, the survival of microbes was tested after subjecting spores of *Bacillus subtilis* to accelerations, jerks or shock waves. Accelerations as they occur during planetary ejections are apparently not a barrier to interplanetary transfer of life. Bacteria are routinely treated with even higher values of acceleration in normal microbiological separation techniques, although with much slower rise times. Ballistic experiments provide rise times equivalent to those estimated for an object receiving escape velocity during an impact. It has been shown that bacterial spores as well as cells of *Deinococcus radiodurans* survived gun shots with accelerations up to 4500 km s^{-2} ($460\,000 \times g$) with a rise time of <1 ms, which are equivalent or even higher than the acceleration and jerk values calculated for an object ejected from Mars [27, 28].

It is important to note that during a meteorite impact, the total duration of the pressure pulse is very short, i.e. on the order of 1 ms [26]. At impacts forming craters of a diameter of >10 km (as possibly required for the Martian meteorites studied) the pressure pulse can last up to 20 ms. In shock recovery experiments with an explosive set-up, the survival of spores of *B. subtilis* was studied after a shock treatment in the pressure range, which some Martian meteorites have experienced [17]. It was found that the a substantial fraction of spores (up to 10^{-4}) was able to survive a peak shock pressure of 32 ± 1 GPa and a post-shock temperature of about 250 °C. These data support the hypothesis that bacterial spores may survive an impact-induced escape process in a scenario of interplanetary transfer of life. Assuming a mean spore density of 10^8 spores/g in e.g., desert soil or rock, a 1 kg rock would accommodate approximately 10^{11} spores, of which up to 10^7 could survive even extremely high shock pressures occurring during a meteorite impact. At more moderate shock waves as they would occur in the spall zone of an impact, even substantial higher survival rates are expected than in these simulation studies at shock pressures of 32 GPa. To prove this, further survival studies using such moderate shock waves are required.

4.3   Survival of the Interplanetary Transfer Phase

4.3.1   Space Environment of Interest

Once rocks have been ejected from the surface of their home planet, microbial passengers have to cope with an entirely new set of problems affecting their survival, namely exposure to the space environment (reviewed in [29-32]). This environment is characterized by a high vacuum, an intense radiation of galactic and solar origin and extreme temperatures (Table 4.1.).

In interplanetary space, pressures down to 10^{-14} Pa prevail. Within the vicinity of a body, the pressure may significantly increase due to outgassing. In a low Earth orbit, pressure reaches values of 10^{-6} to 10^{-4} Pa. The major constituents of this environment are molecular oxygen and nitrogen as well as highly reactive oxygen and nitrogen atoms.

Table 4.1. The environment in Earth orbit and of interplanetary space (modified from [34])

Space parameter	Earth orbit (≤500km)	Interplanetary space
Space vacuum		
Pressure (Pa)*)	10^{-6}-10^{-4}	10^{-14}
Residual gas (part/cm⁻³)	10^5 H, 2×10^6 He, 10^5 N, 3×10^7 O	1 H
Solar electromagnetic radiation		
Irradiance (W/m²)	1360	Different values[a]
Spectral range (nm)	Continuum from 2×10^{-12} to 10^2 m	Continuum from 2×10^{-12} to 10^2 m
Cosmic ionizing radiation		
Dose (Gy/a) *)	0.1-3000[b]	≤ 0.25 [c]
Temperature (K)	100-400[a]	> 4[a]
Microgravity (g)	10^{-3}-10^{-6}	< 10^{-6}

*) 1 Pa = 10^{-5} bar, 1 Gy = 100 rad; [a] depending on orientation and distance to Sun; [b] depending on altitude and shielding, highest values at high altitudes and in the radiation belts; [c] depending on shielding.

The radiation environment of our solar system is governed by components of galactic and solar origin. The galactic cosmic radiation entering our solar system is composed of protons (85%), electrons, α-particles (14%) and heavy ions (1%) of charge Z>2, the so-called HZE particles (high charge Z and high energy E). The solar particle radiation, emitted in solar wind and during solar particle events, is composed of 90-95% protons, 5-10% α-particles and a relatively small number of heavier ions. In interplanetary space, the annual radiation dose amounts to ≤0.25 Gy/a, depending on mass shielding with the highest dose at 30 g/cm² shielding due to built up secondary radiation. In the vicinity of the Earth, the radiation dose can increase due to the radiation belts where protons and electrons are trapped by the geomagnetic field.

The spectrum of solar electromagnetic radiation spans several orders of magnitude, from short wavelength X-rays to radio frequencies. At the distance of the Earth from the Sun (1 AU), solar irradiance amounts to 1360 W m⁻², the solar constant. Of this radiation, 45% is attributed to the infrared fraction, 48% to the visible fraction and only 7% to the ultraviolet range. The extraterrestrial solar spectral UV irradiance has been measured during several space missions, such as Spacelab 1 and EURECA [33].

The temperature of a body in space which is determined by the absorption and emission of energy, depends on its position with respect to the Sun and other orbiting bodies, and also on its surface, size, mass, and albedo (reflectivity). In Earth orbit, the energy sources include solar radiation (1360 W m⁻²), the Earth's albedo (480 W m⁻²) and terrestrial radiation (230 W m⁻²). When orbiting a planet, an object can be shaded from the Sun as it passes on the planet's night side. Therefore, the temperature of a body in space can reach both extremely high and low values. In experiments in Earth orbit, temperatures between 240 K and 320 K were measured.

4.3.2   Approaches to Studying the Biological Effects of Space

In order to study the survival of resistant microbial forms in the upper atmosphere or in free space, microbial samples have been exposed *in situ* by use of balloons, rockets or space crafts and their responses were investigated after recovery (reviewed in [29, 31]). For this purpose, several facilities were developed such as the Exposure Device on Gemini, MEED (microbal ecology exposure device) on Apollo, ES029 on Spacelab 1, ERA (exobiology radiative Assembly) on EURECA, UV-RAD on Spacelab D2, BIOPAN on FOTON, and EXPOSE for the International Space Station (ISS) [30, 34] (see also Fig. 4.6). These investigations were supported by studies in the laboratory, in which certain parameters of space (high and ultrahigh vacuum, extreme temperature, UV-radiation of different wavelengths, ionizing radiation) were simulated. The microbial responses (physiological, genetic and biochemical changes) to selected factors applied separately or in combination were determined.

Many spore-forming bacteria are found in terrestrial soils and their spores have been recognized as the hardiest known forms of life on Earth. The developmental pathway from a vegetatively growing bacterial cell to a spore, i.e. the dormant state, is triggered by depletion of nutrients in the bacterial cell's environment [35]. In the dormant stage, spores undergo no detectable metabolism and exhibit a high degree of resistance to inactivation by various physical insults such as cycles of extreme heat and cold, extreme desiccation including vacuum, UV and ionizing radiation, as well as oxidizing agents or corrosive chemicals (recently reviewed by Nicholson et al. [32]). The high resistance of *Bacillus* endospores is mainly due to two factors: (i) a dehydrated, highly mineralized core enclosed in a thick protective envelop, the cortex and the spore coat layers (Fig. 4.3), and (ii) the saturation of their DNA with small, acid-

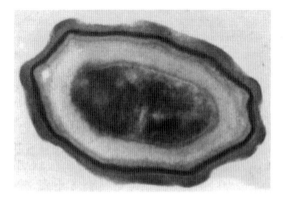

Fig. 4.3 Electonmicrograph of a spore of *B. subtilis* with the inner core containing the DNA surrounded by protective layers, the long axis of the spore is 1.2 μm, the core area 0.25 μm² (courtesy of S. Pankratz).

soluble proteins whose binding greatly alters the chemical and enzymatic reactivity of the DNA [36]. In the presence of appropriate nutrients, spores respond rapidly by germination and outgrowth, resuming vegetative growth. Hence, spore formation represents a strategy, by which a bacterium escapes temporally and/or spatially from unfavorable conditions: spores exhibit incredible longevity and can be relocated e.g., by wind and water, to remote areas. Among the bacterial spores, the endospores of the genus *Bacillus* are the best investigated ones [32].

In addition, a variety of microorganisms exist that are adapted to grow or survive in extreme conditions of our biosphere. Some of them may be suitable candidates for studies on microorganisms in space. Examples are endo- or epilithic communities, consisting of cyanobacteria, algae, fungi and/or lichens, which represent a simple microbial ecosystem living inside or on rocks [37, 38], or osmophilic microbial assemblages that are trapped in evaporite deposits [39, 40], or extremely radiation-resistant microorganisms, like bacteria of the species *D. radiodurans* as the most radiation-resistant bacteria known to exist on Earth today. Although *D. radiodurans* is non-sporulating, it can go into a kind of dormancy under certain environmental adverse conditions such as lack of food, desiccation or low temperatures [41, 42].

4.3.3   Biological Effects of the Vacuum of Space

Because of its extremely dehydrating effect, space vacuum has been considered to be one of the factors that may prevent interplanetary transfer of life [43]. However, space experiments have shown that up to 70% of bacterial and fungal spores survive short-term (e.g., 10 days) exposure to space vacuum, even without any protection [29]. The chances of survival in space are increased, if the spores are embedded in chemical protectants such as sugars, or salt crystals, or if they are exposed in thick layers. For example, 30% of *B. subtilis* spores survived nearly 6 years of exposure to space vacuum, if embedded in salt crystals, whereas approximately 70% survived in the presence of glucose [21] (Table 4.2.). Sugars and polyalcohols stabilize the structure of

Table 4.2. Survival of spores of *B. subtilis* after exposure to space vacuum (10^{-6} to 10^{-4} Pa) during different space missions.

Mission	Duration of vacuum exposure	Survival fraction at end of exposure in thin layers (%)		Survival fraction at end of exposure in thick layers and presence of protective sugars (%)		References
		in space	ground control	in space	ground control	
SL 1	10 d	69.3 ± 15.8	85.3 ± 2.6	n.d.	n.d.	[29]
EURECA	327 d	32.1 ± 16.3	32.7 ± 5.6	45.5 ± 0.01	62.7 ± 8.2	[44]
LDEF	2 107 d	1.4 ± 0.8	5.4 ± 2.9	67.2 ± 10.2	77.0 ± 6.0	[21,29]

n.d. = not determined

the cellular macromolecules during vacuum-induced dehydration, leading to increased rates of survival.

To determine the protective effects of different meteorite materials, "artificial meteorites" were constructed by embedding *B. subtilis* spores in clay, meteorite dust or simulated Martian soil [44] and exposing them to the space environment. Crystalline salt provided sufficient protection for osmophilic microbes in the vegetative state to survive at least 2 weeks in space [40]. For example, a species of the cyanobacterium *Synechococcus* that inhabits gypsum-halite crystals was capable of nitrogen and carbon fixation and about 5% of a species of the extreme halophile *Haloarcula* survived after exposure to the space environment for 2 weeks in connection with a FOTON space flight.

The mechanisms of damage due to vacuum exposure are based on the extreme desiccation. If not protected by internal or external substances, cells in a vacuum experience dramatic changes in lipids, carbohydrates, proteins and nucleic acids. Upon desiccation the lipid membranes of cells undergo dramatic phase changes from planar bilayers to cylindrical bilayers [45]. The carbohydrates, proteins and nucleic acids undergo so-called Maillard reactions, i.e. amino-carbonyl-reactions, to give products that become cross-linked eventually leading to irreversible polymerization of the biomolecules [45]. Concomitant with these structural changes are functional changes, including altered selective membrane permeability, inhibited or altered enzyme activity, decreased energy production, alteration of genetic information, etc.

Vacuum-induced damage to the DNA is especially dramatic, because it may be lethal or mutagenic. The mutagenic potential of space vacuum was first demonstrated during the Spacelab I mission in spores of *B. subtilis*, which showed an up to tenfold increased mutation rate over the spontaneous rate [46]. This vacuum-induced mutagenicity is accompanied by a unique molecular signature of tandem-double base changes at restricted sites in the DNA [47]. In addition, DNA strand breaks have been observed to be induced by exposure to space vacuum [48, 49]. Such damage would accumulate during long-term exposure to space vacuum, because DNA repair is not active during this state of anhydrobiosis. Survival ultimately depends on the efficiency of the repair systems after germination.

4.3.4   Biological Effects of Galactic Cosmic Radiation

If not sufficiently shielded by meteorite material, microbes may be affected by the ionizing components of radiation in space. Especially the heavy primaries of galactic cosmic radiation, the so-called HZE particles, are the most biologically effective species (reviewed in [50, 51]). Because of their low flux (they contribute to approximately 1% of the flux of particulate radiation in space), methods have been developed to localize precisely the trajectory of an HZE particle relative to the biological object and to correlate the physical data of the particle relative to the observed biological effects along its path. In the Biostack method visual track detectors are sandwiched between layers of biological objects in a resting state, e.g., *B. subtilis* spores [52] (Fig. 4.4). This method allows (i) to localize each HZE particle's trajectory in relation to the biological specimens; (ii) to investigate the responses of each biological individual

hit separately, in regard to its radiation effects; (iii) to measure the impact parameter b
(i.e. the distance between the particle track and the sensitive target); (iv) to determine
the physical parameters [charge (Z), energy (E) and linear energy transfer (LET)]; and
finally (v) to correlate the biological effects with each HZE particle parameters.

The small size of bacterial spores (their cytoplasmic core has a geometrical cross
section of 0.2 to 0.3 μm^2) requires to develop special techniques in connection with
the Biostack method. In this case, spores in monolayers are directly mounted on the
track detector. After exposure, the track detector with the spores is etched under microscopical control on the spore-free side only and spores located around a track of an
HZE particle (at a radius of ≤5 µm) are removed by micromanipulation and incubated
each individually in special incubation chambers [53] (Fig. 4.4). Using this microscopical method, the accuracy in determining the impact parameter was ≥0.2 µm,
depending on the dip angle of the trajectory. Figure 4.4 shows the frequency of inactivated spores as a function of the impact parameter b. Spores within b ≤0.2 µm were
inactivated by 73%. The frequency of inactivated spores dropped abruptly at b >0.2
µm. However, 15-30% of spores located within 0.2 < b <3.8 µm were still inactivated.
A statistical analysis showed that all data at b ≤3.8 µm are significantly different from
the control value (at b >10 µm) (95% confidence) [54].

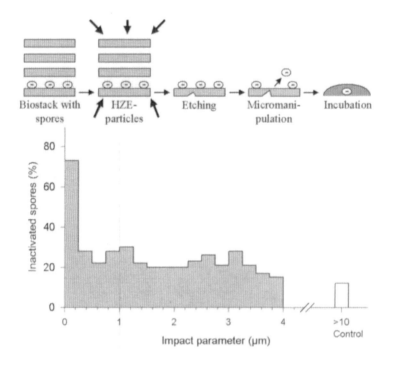

Fig. 4.4 Biostack method to localize the effect of single particles (HZE particles) of cosmic
radiation (biological layers are sandwiched between track detectors) and results on the inactivation probability of spores of *B. subtilis* as a function of their distance from the particles
trajectory, the impact parameter (data from space experiments [54]).

As shown above, *B. subtilis* spores can survive even a central hit of an HZE particle of cosmic radiation. Such HZE particles of cosmic radiation are conjectured to set the ultimate limit on the survival of spores in space, because they penetrate even thick shielding. However, since the flux of HZE particles is relatively low, it may last up to several hundred thousand to one million of years in space until a spore might be hit by an HZE particle (e.g., iron of LET >100 keV/µm).

With increasing shielding thickness, e.g., by the outer layers of the meteorite, the dose rate caused by cosmic radiation goes through a maximum, because the heavy ions interact with the shielding material and create secondary radiation. Based on experimental data from accelerator experiments with *B. subtilis* spores [55] and biophysical models, relating the structure of HZE tracks to the probability of inactivating the spores [56], an estimated density of the meteorite of 3 g/cm³ (taken from data on Martian meteorites) and a NASA model on an HZE transport code for cosmic radiation [57], the dose rates and probabilities for inactivating spores have been calculated behind different shielding thicknesses: the physical dose rates reach a maximum behind a shielding layer of about 10 cm (30 g/cm² shield thickness); behind 30 cm (90 g/cm²) the value is approximately the same as obtained without any shielding and only for higher shielding thicknesses the dose rate reduces significantly (Fig. 4.5) [26]. The calculations also show that even after 25 Ma in space, a substantial fraction of a spore population (10^{-6}) would survive the exposure to cosmic radiation if shielded by 2 to 3 m of meteorite material. The calculations are based on the assumption that the rock may accommodate about 10^8 spores/g, of which at least 100 spores/g would survive. The same surviving fraction would be reached after about 600 000 years without

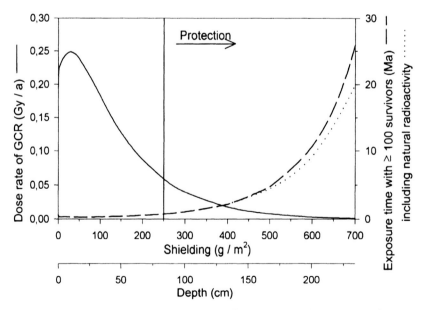

Fig. 4.5 Shielding of spores of *B. subtilis* against galactic cosmic radiation (GCR) by meteorite material and survival times (≥ 10^{-6} survivors) at different depths of the meteorite due to GCR (dashed line) or to GCR plus natural radioactivity of 0.8 mGy/a (dotted line) (data from [26]).

any shielding, after about 300 000 years behind 10 cm of shielding (maximum dose rate) and after about 1 Ma behind 1 m of shielding.

4.3.5   Biological Effects of Extraterrestrial Solar UV Radiation

Solar UV radiation has been found to be the most deleterious factor of space, as tested with dried preparations of viruses, and of bacterial and fungal spores [29]. The full spectrum of extraterrestrial UV radiation kills unprotected spores of *B. subtilis* within seconds [2] and is about thousand times more efficient than the UV radiation at the Earth's surface due to the effective protection of our biosphere by the stratospheric ozone layer [58, 59] (see also Chap. 14, Cockell; Chap. 15, Rettberg and Rothschild). The reason for this high biological efficiency of extraterrestrial UV is the highly energetic UV-C (190-280 nm) and vacuum UV region (<190 nm) that is directly absorbed by the genetic material of the cells, the DNA [44].

This high biological efficiency of extraterrestrial UV is even increased, if bacterial spores are simultaneously exposed to solar UV-radiation and space vacuum [29, 46]. Upon dehydration (e.g., in space vacuum) DNA undergoes substantial conformational changes. This conversion in the physical structure leads to an altered DNA photochemistry. The following photoproducts are generated within the DNA of *B. subtilis* spores exposed to UV radiation in vacuum: (i) two thymine decomposition products, namely the *cis-syn* and *trans-syn* cyclobutadithymine (Thy<>Thy); (ii) 5,6-dihydro-5(α-thyminyl) thymine (TDHT); (iii) DNA protein cross-linking [46, 60]. From the efficiency of repair processes (photoenzymatic repair, spore photoproduct specific repair) it is concluded that photoproducts other than *cis-syn* Thy<>Thy and TDHT seem to be responsible for the UV supersensitivity of spores, if irradiated under vacuum conditions [29].

However, one has to bear in mind that a few micrometers of meteorite material may be sufficient to protect the microorganisms enclosed against UV radiation. Therefore, microorganisms travelling through space inside a meteorite are probably not under a serious threat of being killed by solar UV radiation.

4.3.6   Bacterial Survival During Long-Term Dormancy

According to the Martian meteorites, so far detected, travelling times in space may span over several millions of years. Because of the enormous number of launched ejecta (Fig. 4.2), much shorter transfer times should be feasible. Simulations of Mars meteorite transfer show that some ejecta may arrive in tens of thousands of years and even time spans of less than a century are possible [61]. These relatively fast arriving meteorites are the most interesting ones with regard to viable transfer. The extreme environment of space obviously does not support active metabolism and growth of the microorganisms enclosed; however, a variety of organisms exist that are adapted to survive in extreme conditions when in the dormant state [62]. The question arises whether microorganisms in the dormant state, e.g., as bacterial spores, could survive the extreme conditions of space over extended periods of time.

Besides cosmic radiation and vacuum, threats to the stability of the DNA of spores inside meteorites may arise from various sources such as natural radioactivity of the meteorite, hydrolysis, chemicals, and temperature extremes. The natural radioactivity of the known Martian meteorites reaches values up to 0.8 mGy/a. It has been calculated by Mileikowsky et al. [26] that for smaller Martian ejecta (<2-3 m in diameter) the effects of cosmic radiation dominate over those of natural radioactivity. With increasing depth, however, in larger ejecta, the galactic cosmic radiation (GCR) dose rate decreases and finally the effects of natural radioactivity become more important than those of cosmic radiation (Fig. 4.5). The effects of natural radioactivity may be more serious for terrestrial ejecta, which on the average are about 5 times higher in natural radioactivity than Martian rocks [26].

So far, most data dealing with the instability and decay of DNA by hydrolysis, chemicals or temperature were obtained with moist biological systems at atmospheric pressure [63]. However, bacterial endospores are especially resistant to such stresses. Their cytoplasm is partially dehydrated and mineralized, causing enzymes to become inactive and the DNA stabilized [32]. Recent reports suggest that spores of the genus *Bacillus* can remain intact over millions of years, if preserved in amber [19] or in brine inclusions in salt crystals [20]. In these two latter studies, which report revitalization of bacterial spores in 25-40 Ma old amber [19] or even in 250 Ma old buried salt crystals [20], thorough sterilization procedures of the sample surface were applied to avoid contamination from contemporary microbes. However, the phylogenetic analysis revealed their very close relationship to contemporary species, which initiated a controversial discussion on the real age of the spores. Hence, several open questions have still to be solved, before assuming that bacterial spores are nearly immortal.

4.3.7   Combined Effects of the Complex Matrix of Space Parameters

During the Long Duration Exposure Facility (LDEF) mission, for the first time, spores of *B. subtilis* were exposed to the full environment of space, i.e. space vacuum, solar UV-radiation and most of the components of cosmic radiation, for an extended period of time, namely nearly 6 years, and their survival was determined after retrieval. The spores were exposed in multilayers and predried in the presence of glucose as chemical protectant. After retrieval, the spore samples had turned from white into yellow, a phenomenon, which is probably due to photochemical processes of the outer layers. However, in each sample thousands of spores survived the space journey, from an initial sample size of 10^8 spores [21]. One possible explanation is that all spores in the upper layers were completely inactivated by the high flux of solar UV-radiation, thereby forming a protective crust, which considerably attenuated the solar UV-radiation for the spores located beneath this layer. Therefore, the survivors probably originated from the innermost part of the samples. Of spores, covered by an aluminum foil ,which protected them against UV radiation, up to 70% survived the 6 years lasting space journey (Table 4.2.). These results are the first experimental proof that a very high percentage of spores survive at least 6 years in space, if efficiently shielded against solar UV radiation. The shielding could be achieved by the outer layers of rocks as used by those microbial communities that inhabit rocks. However, calculations show that most ejecta would require thousands or millions of years before reaching another planet [6, 7, 14]. Therefore, 6 years seem to be by far too short for an interplanetary transfer of life.

4.4   Survival of the Landing Process

When captured by a planet with an atmosphere, most meteorites are subjected to very high temperatures during landing. However, because the fall through the atmosphere takes a few seconds only, the outermost layers form a kind of heat shield and the heat does not reach the inner parts of the meteorite. During entry, the fate of the meteorite strongly depends on its size: large meteorites may brake into pieces, however, these may be still large enough to remain cool inside until hitting the surface of the planet; medium sized meteorites may obtain a melted crust, whereas the inner part still remains cool; micrometeorites of a few μm in size may tumble through the atmosphere without being heated at all above 100 °C. Therefore, it is quite possible that a substantial number of microbes can survive the landing process on a planet. However, no experiments have been done so far to investigate the effects of the landing process experimentally. Recently, the European Space Agency (ESA) has developed a facility, called STONE which is attached to the heat shield of a FOTON satellite to test mineral degradations during landing [64]. This facility might be an ideal tool to study the effects of landing of bacterial spores embedded in an artificial meteorite.

4.5   Conclusions: On the Likelihood of Interplanetary Transfer of Life as a Mode of Distribution of Life Throughout the Solar System

Although it is difficult at present to prove definitely that life has been transported through our solar system, model calculations and experiments at simulation facilities and in space allow to estimate the chances of resistant microbial forms to survive the different steps of such a scenario. Experiments in space which were performed on free platforms since the Apollo era and more sophisticated on the external platform of Spacelab and on free flying satellites such as LDEF, EURECA, and FOTON, have given some insight into responses of bacterial spores and other extremophile microorganisms to the parameters of space. The most interesting results are summarized in the following: (i) extraterrestrial solar UV radiation is thousand times more efficient than UV at the surface of the Earth and kills within a few seconds 99% of *B. subtilis* spores; (ii) space vacuum increases the UV sensitivity of the spores; (iii) although spores survive extended periods of time in space vacuum (up to 6 years), genetic changes occur such as increased mutation rates; (iv) after 6 years in space, up to 70% of bacterial spores survive, if protected against solar UV radiation and dehydration; (v) spores could escape a hit of a cosmic HZE particle (e.g., iron ion) for up to 1 Ma. Calculations using radiative transfer models for cosmic rays and biological data from accelerator experiments have shown that a meteorite layer of 1 m or more effectively protects bacterial spores against galactic cosmic radiation.

The data obtained so far on the responses of resistant microorganisms to the environment of space support the supposition that space – although it is very hostile to terrestrial life – is not a barrier for cross-fertilization in the solar system. Ejection of microbes inside rocks and their transport through the solar system is a feasible process. If protected against solar UV and galactic cosmic radiation, spores may survive inside meteorites over extended periods of time.

However, several aspects justify continued research, e.g., on the survival rate during the ejection and landing process, on the shelf-life of spores in vacuum, as well as on the best mechanisms to effectively protect the microorganisms enclosed against the three steps of interplanetary transfer. Relevant studies to tackle these questions will be performed in future studies in space and at ground based facilities [34, 65, 66]. For future research on bacterial spores and other microorganisms in space, ESA is developing the EXPOSE facility that is to be accomodated outside of the ISS for 1.5 years (Fig. 4.6). EXPOSE will support long-term *in situ* studies of microbes in artificial meteorites, as well as of microbial communities from special ecological niches such as endolithic and endoevaporitic ecosystems [34]. These experiments on the Responses of Organisms to the Space Environment (ROSE) include the study of photobiological processes in simulated radiation climates of planets (e.g., early Earth, early and present Mars, and the role of the ozone layer in protecting the biosphere from harmful

Fig. 4.6 EXPOSE facility to be mounted on the truss structure of the ISS to study the sensitivity of organics and microorganisms to space environment.

Table 4.3. Experiments of the ROSE consortium to study the responses of organisms to space environment on the EXPOSE facility of the ISS [34]

Code	Objective	Assay system
ENDO	Impact of ozone depletion on microbial primary producers from sites under the "ozone hole"	1. Endolithic microbial communities 2. Mats of cyanobacteria 3. Mats of algae
OSMO	Protection in evaporites against solar UV and anhydrobiosis	1. Synechococcus in and beneath gypsum-halite 2. Haloarcula (pigmented and non-pigmented) in and beneath NaCl 3. Haloarcula DNA in KCl
SPORES	Protection of spores by meteorite material against space (UV, vacuum, and ionizing radiation)	1. *B. subtilis* spores 2. Fungal spores 3. Lycopod spores All in or beneath meteorite material
PHOTO	Main photoproducts in dry DNA and DNA from dry spores	1. DNA 2. Bacterial spores
PUR	Sensitivity of the biologically effective UV radiation to ozone	1. T7 bacteriophage 2. Phage-DNA 3. Uracil
SUBTIL	Mutational spectra induced by space vacuum and solar UV	1. *B. subtilis* spores 2. Plasmid DNA

UV-B radiation), as well as studies of the probabilities and limitations for life to be distributed beyond its planet of origin (Table 4.3.). Here-to-fore, the results from the EXPOSE experiments will eventually provide clues to a better understanding of the processes regulating the interactions of life with its environment.

4.6 Outlook: On the Likelihood of Transport of Viable Microorganisms Between Solar Systems

Today more than 50 extrasolar planets have been detected, all of the size of Saturn or Jupiter (see Chap. 2, Udry and Mayor). The present detection methods – Doppler effect, astrometry and transit photometry - do not possess sufficient sensitivity to discover Earth-like planets nor giant planets, orbiting at distances larger than 3 AU from their sun. For that we have to wait for future planned space missions, e.g., Darwin of ESA (see Chap. 24, Foing) which are especially designed to detect Earth-like planets by searching for signatures indicative of a biosphere on a planet, such as atmospheric oxygen as a suggested biomarker for photosynthetic activity. Nevertheless, the information on the existence of extrasolar planets has corroborated the interest in the question whether life forms can be transported between the planets or moons of different solar systems.

In a follow-on study, Mileikowsky et al. [67] have estimated the likelihood of viable transfer of life from planets or moons of extrasolar planetary systems to the Earth where the source planet is located in either the general galactic star field or in a possible temporary "sibling" cluster born together with the Sun. To leave the planet, an impact ejection mechanism has been assumed similar as described in the studies of viable transfer within our solar system [26]. Because the information on the existence and frequency of Earth-like planets is still missing – and probably for many years to come -, they have based their calculations on specific estimations of the unknown circumstances as described below.

Major parameters of the calculations are (i) the star density and (ii) the relative speed between the emitting and receiving planetary system. In the case of the general galactic field, the present numerical density of stars in the solar neighborhood was used, namely 1 star per 10 pc^3 = 0.1 pc^{-3} (1pc = parsec = 3.26 light years). This is based on the generally held opinion of astronomers that the general star field during the youth of our solar system was probably quite similar to today's. The distribution of speeds of the stars in our galaxy is a very good Maxwellian approximation with about 20 km/s as the most common speed and 0.5 km/s being orders of magnitude less frequent. In the case of a possible cluster of "sibling" stars born together with our Sun, the cluster would last only limited time after its formation from the molecular cloud, before the stars disperse in all directions into the general galactic star field. However, the distances between the stars within the cluster would be much smaller than in the general galactic star field. Furthermore, the relative speed between the members of a cluster is less than 1 km/s which is much lower than the most frequent one between stars in the general galactic star field. To calculate the number n of hits on Earth by ejecta from extrasolar systems, the following formula was used [67]:

$$n \approx 2 \cdot 10^{-15} \; T \; R \; t \; \frac{N}{\sigma} \left(1 - e^{-\frac{v_0^2}{2\sigma^2}} \right)$$

(4.1)

With $N = N_d \times f_{ps} \times f_{TP} \times f_{HZM} \times f_i$,
N_d = numerical density of stars in pc^{-3}; f_{ps} = fraction of stars with planetary systems; f_{TP} = fraction of planetary systems with terrestrial-like planets; f_{HZM} = fraction of terrestrial-like planets orbiting in zones habitable for microbes; f_i = fraction of other factors; T = time period studied comprising escape from a planet, expulsion into interstellar space, capture by our solar system, orbiting in our solar system, capture by the Earth, and hitting the Earth (Ma); t = time period for microbial survival inside ejecta in space (Ma); R = rate of expulsion from extrasolar systems into interstellar space (Ma^{-1}); σ = dispersion in the Maxwellian distribution of the velocities of the stars (km/s); v_0 = 0.5 km/s.

For establishing the formula, it is not necessary to know either anything about the complete planetary configuration of the extrasolar systems or about the terrestrial-like planets. But, for each open entity, selected values have to be given. N_d = 0.1 pc^{-3} and σ = 20 pc/Ma are the values known for our galaxy. For a possible temporary "sibling" cluster around our Sun, N_d and σ are of course unknown; therefore, the estimations are based on observations of relatively new-born clusters such as Hyades. t is determined by biology as discussed in 4.3.4 and in [26]. However, several unknowns render the calculations difficult, such as (i) the fraction of stars in our galaxy with planetary systems; (ii) the fraction of those with habitable zones (see Chap. 3, Franck et al.); (iii) the fraction of those with Earth-like planets; and (iv) the fraction of those that have developed life (e.g., microorganisms, based on DNA/RNA/proteins with high resistance to the hostile environment of space). To overcome this problem, Mileikowsky et al. [67] have undertaken the following steps: (i) they have chosen for all fractions mentioned above the most favorable and overoptimistic value for viable transfer, namely 1; (ii) they have chosen a heavy bombardment in the early stage of the extrasolar system being orders of magnitude more intense than that in the early stage of our solar system; (iii) they have chosen very long survival times (e.g., 1-100 Ma) of microbes in ejecta in space. Table 4.4. shows that the transfer of ejecta from an extrasolar system to the Earth calculated by use of Eq. (4.1), is strongly enhanced by high densities, i.e. short distances (e.g., less than 1 pc) and/or by a very low relative speed between the emitting planetary system and the receiving system (e.g., less than 0.5 km/s) [67, 68].

In case of the conditions of the general galactic star field, n, the number of ejecta from all planetary systems of the galaxy, reaching the Earth within the early 500 Ma is utterly small, namely about 10^{-9}. Even if the bombardment by comets and asteroids would be thousand times more intense (given by term R), no ejecta would reach the Earth either, with n being about 10^{-5}. In the case of clusters of "sibling" stars, assuming the conditions as observed in the Hyades cluster, which has at its present age of 625 Ma $N = 2$ and $\sigma = 0.25$ km/s, n would be about 10^{-4}. Even assuming a 20 times higher star density than observed in Hyades, the number of extrasolar ejecta hitting the Earth is still far below 1. Hence, assuming impact ejecta as the mode of transportation of putative life from an extrasolar planet to the Earth, the data show that from the general galactic star field no life-bearing ejecta have reached the Earth within the first 05-0.6 Ga. If the sun was part of a group of stars in a "sibling" cluster, viable transfer from one of the sister systems cannot be completely ruled out, however, the probability remains very low.

The longest survival time assumed is $t=100$ Ma (Table 4.4.). Recently, revival of bacterial spores from a 250 Ma old salt deposit has been claimed [20]. However, as discussed above, this finding is still controversial: it has been argued that the spores may be the result of recent contamination, which have penetrated the salt through small invisible cracks. This objection is mainly based on the high genetic similarity of the revived spores with contemporary species of *Bacillus maremortis*, which live in the very saline Dead Sea. However, it can be deduced from Eq. (4.1) that the number of viable transfers from planets of the general galactic field to the Earth would remain low even on the base of a survival time in space of 250 Ma.

From this example calculated for the case of extrasolar ejecta reaching the Earth, it can be concluded that the probabilities are too low to allow transport of viable microorganisms from one solar system to another by impact ejecta. The chances for such interstellar exchange of life may be increased for very close sister solar systems which are born from one parent molecular cloud, where they form a cluster of "sibling" stars all with planetary bodies, which all – in these calculations – were overoptimistically assumed to be populated by microbes.

Table 4.4. Number of impacts n on Earth during time interval T assuming maximum allowed microbial travelling time t from ejection to impact, expulsion rate R, density if expelling extrasolar systems N, and stellar velocity dispersion σ from the general galactic star field and from clusters of "sibling"stars

Case	T (Ma)	t (Ma)	R (Ma^{-1})	N (pc^{-3})	σ (km/s)	n
From all planetary systems in the general galactic star field	500	10	10^8	0.1	20	1.6×10^{-9}
	500	100	10^{11}	0.1	20	1.6×10^{-5}
Our Sun within a cluster of "sibling" stars	100	1	10^8	40	0.2	1.6×10^{-3}
	100	10	10^8	40	0.2	1.6×10^{-2}
Our Sun within cluster like Hyades	625	1	10^8	2	0.2	1.6×10^{-4}

4.7   References

1   J. Oro, Nature **190**, 389 (1961).
2   G. Horneck, A. Brack, in: S.L. Bonting (Ed.) *Advances in Space Biology and Medicine*, Vol. 2, JAI Press, Greenwich, CT, 1992, pp. 229.
3   C.F. Chyba, T.C. Owen, W.-H. Ip, in: T. Gehrels (Ed.) *Hazards due to Comets and Asteroids*, Univ. of Arizona Press, Tucson, 1994, pp. 9.
4   H.J. Melosh, A.M. Vickery, Nature **338**, 487 (1989).
5   V.R. Oberbeck, G. Fogleman, Origins Life Evol. Biosphere **20**, 181 (1990).
6   H.J. Melosh, Nature **332**, 688 (1988).
7   B. Gladman, J.A. Burns, M. Duncan, P. Lee, H. Levinson, Science **271**, 1387 (1996).
8   P. Warren, Icarus **111**, 338 (1994).
9   O. Eugster, A. Weigel, E. Polnau, Geochim. Cosmochim. Acta **61**, 2749 (1997).
10   H. Richter, Schmidts Jahrbuch Ges. Med. **126**, 243 (1865).
11   S. Arrhenius, Die Umschau **7**, 481 (1903).
12   G. Horneck, Planet. Space Sci. **43**, 189 (1995).
13   D.S. McKay, E.K. Gibson, K.L. Thomas-Keprta, H. Vali, C.S. Romanek, S.J. Clemett, X.D.F. Chillier, C.R. Maedling, R.N. Zare, Science **273**, 924 (1996).
14   M.A. Moreno, Nature **336**, 209 (1988).
15   J.D. O'Keefe, T.J. Ahrens, Science **234**, 346 (1986).
16   A.M. Vickery, H.J. Melosh, Science **237**, 738 (1987).
17   G. Horneck, D. Stöffler, U. Eschweiler, U. Hornemann, Icarus **149**, 285 (2001).
18   P. Weber, J.M. Greenberg, Nature **316**, 403 (1985).
19   R.J. Cano, M. Borucki, Science **268**, 1060 (1995).
20   R.H. Vreeland, W.D. Rosenzweig, D.W. Powers, Nature **407**, 897(2000).
21   G. Horneck, H. Bücker, G. Reitz, Adv. Space Res. **14**(10), 41 (1994).
22   M. Schidlowski, Nature **333**, 313 (1988).
23   C.R. Woese, O. Kandler, M.L. Wheelis, Proc. Natl. Acad. Sci. USA **87**, 4576 (1990).

24 D. Stöffler, 31st Lunar and Planetary Science Conference, Houston, Texas, CD ROM #1170, 2000.

25 H.J. Melosh, Icarus **59**, 234 (1984).

26 C. Mileikowsky,   F.A. Cucinotta,   J.W. Wilson,   B. Gladman,   G. Horneck, L. Lindegren, H.J. Melosh, H. Rickman, M. Valtonen, J.Q. Zheng, Icarus **145**, 391 (2000).

27 C. Mileikowsky, E. Larsson, B. Eiderfors, 25th General Assembly of the European Geophysical Society, The Hague, 19-23 April 1999.

28 R.M.F. Mastrapa, H. Glanzberg, J.N. Head, H.J. Melosh, W.L. Nicholson, 31st Lunar and Planetary Science Conference, Houston TX, CD ROM #2045, 2000.

29 G. Horneck, Origins Life Evolut. Biosphere **23**, 37 (1993).

30 G. Horneck, Adv. Space Res. **22**(3), 317 (1998).

31 G. Horneck, in: L.J. Rothschild, A. Lister (Eds.) *Evolution on Planet Earth: The Impact of the Physical Environment*, Academic Press, in press.

32 W.L. Nicholson, N. Munakata, G. Horneck, H.J. Melosh, P. Setlow, Microbiol. Molec. Biol. Rev. **64**, 548 (2000).

33 D. Labs, H. Neckel, P.C. Simon, G. Thuillier, Solar Phys. **107**, 203 (1987).

34 G. Horneck, D.D. Wynn-Williams, R. Mancinelli, J. Cadet, N. Munakata, G. Rontó, H.G.M. Edwards, B. Hock, H. Wänke, G. Reitz, T. Dachev, D.P. Häder, C. Brillouet, Proc. 2nd Europ. Symp. on Utilisation of the Internat. Space Station, ESTEC, Noordwijk, ESA SP-433, 1999, pp. 459.

35 P.J. Piggot, C.P. Jr. Moran, P. Youngman, (Eds.) *Regulation of Bacterial Differentiation*. American Society for Microbiology, Washington, D.C., 1994.

36 P. Setlow, Ann. Rev. Microbiol. **49**, 29 (1995).

37 E.I. Friedmann, Science **215**, 1045 (1982).

38 D.D. Wynn-Williams, Microbial Ecology **31**, 177 (1996).

39 L.J. Rothschild, L.J. Giver, M.R. White, R.L. Mancinelli, J. Phycol. **30**, 431 (1994).

40 R.L. Mancinelli, M.R. White, L.J. Rothschild, Adv. Space Res. **22**(3), 327 (1998).

41 B.E.B. Moseley, in: K.C. Smith (Ed.) *Photochemical and Photobiological Reviews* Vol. 7, Plenum, New York, 1983, pp. 223.

42 V. Mattimore, J.R. Battista, J. Bacteriol. **178**, 633 (1996).

43 M.D. Nussinov, S.V. Lysenko, Origins Life Evol. Biosphere **13**, 153 (1983).

44 G. Horneck, U. Eschweiler, G. Reitz, J. Wehner, R. Willimek, K. Strauch, Adv. Space Res. **16**(8), 105 (1995).

45 C.S. Cox, Origins Life Evol. Biosphere **23**, 29 (1993).

46 G. Horneck,   H. Bücker,   G. Reitz,   H. Requardt,   K. Dose,   K.D. Martens, H.D. Mennigmann, P. Weber, Science **225**, 226 (1984).

47 N. Munakata, M. Saitou, N. Takahashi, K. Hieda, F. Morohoshi, Mutat. Res. **390**, 189 (1997).

48 K. Dose, A. Bieger-Dose, M. Labusch, M. Gill, Adv. Space Res. **12**(4), 221 (1992).

49 K. Dose, A. Bieger-Dose, R. Dillmann, M. Gill, O. Kerz, A. Klein, H. Meinert, T. Nawroth, S. Risi, C. Stridde, Adv. Space Res. **16**(8), 119 (1995).

50 G. Horneck, Nucl. Tracks Radiat. Meas. **20**, 185 (1992).

51 J. Kiefer, M. Kost, K. Schenk-Meuser, in: D. Moore, P. Bie, H. Oser (Eds.) *Biological and Medical Research in Space,* Springer, Berlin, 1996, pp. 300.

52 H. Bücker, G. Horneck, in: O.F. Nygaard, H.I. Adler, W.K. Sinclair (Eds.) *Radiation Research*, Academic Press, New York, 1975, pp. 1138.

53 M. Schäfer, H. Bücker, R. Facius, D. Hildebrand, Rad. Effects **34**, 129 (1977).

54 R. Facius, G. Reitz, M. Schäfer, Adv. Space Res. **14**(10), 1027 (1994).

55 K. Baltschukat, G. Horneck, Radiat. Environm. Biophys. **30**, 87 (1991).
56 F.A. Cucinotta, J.W. Wilson, R. Katz, W. Atwell, G.D. Badhwar, M.R. Shavers, Adv. Space Res. **18**(12), 183 (1995).
57 J.W. Wilson, L.W. Townsend, J.L. Shinn, F.A. Cucinotta, R.C. Costen, F.F. Badavi, S.L. Lamkin, Adv. Space Res. **10**(10), 841 (1994).
58 G. Horneck, P. Rettberg, E. Rabbow, W. Strauch, G. Seckmeyer, R. Facius, G. Reitz, K. Strauch, J.U. Schott, J. Photochem. Photobiol. B: Biol. **32**, 189 (1996).
59 P. Rettberg, G. Horneck, W. Strauch, R. Facius, G. Seckmeyer, Adv. Space Res. **22**(3), 335 (1998).
60 C. Lindberg, G. Horneck, J. Photochem. Photobiol. B:Biol. **11**, 69 (1991).
61 B. Gladman, J. A. Burns, Science **274**, 161 (1996).
62 G. Horneck, Planet. Space Sci. **48**, 1053 (2000).
63 T. Lindahl, Nature **362**, 709 (1993).
64 A. Brack, ESA Bulletin **101**, 101 (2000).
65 G. Horneck, in: P. Ehrenfreund, C. Krafft, H. Kochan, V. Pirronello (Eds.) *Laboratory Astrophysics and Space Research*, Kluwer, Dordrecht, 1999, pp. 667.
66 G. Horneck, G. Reitz, P. Rettberg, M. Schuber, H. Kochan, D. Möhlmann, L. Richter, H. Seidlitz, Planet. Space Sci. **48**, 507 (2000).
67 C. Mileikowsky, F.A. Cucinotta, J.W. Wilson, B. Gladman, G. Horneck, L. Lindegren, H.J. Melosh, H. Rickman, M. Valtonen, J.Q. Zheng, in preparation.
68 M.J. Valtonen, K.A. Innanen, Astrophys. J. **255**, 367 (1982).

Part II

Water and Life

5 Water, the Spring of Life

André Brack

Primitive life probably appeared with the first chemical systems able to transfer their molecular information via self-reproduction and also to evolve. By analogy with contemporary life, it is generally believed that life on Earth emerged in liquid water from the processing of organic molecules. Schematically, the premises of primitive life can be compared to parts of "chemical robots". By chance, some parts self-assembled to generate robots capable of assembling other parts to form identical robots. Sometimes, a minor error in the building generated more efficient robots, which became the dominant species. The number of parts required for the first robots as well as the nature of these chemical robots are still unknown. The traces of the primitive chemical robots have been probably erased for ever on Earth by the combined action of plate tectonics, the permanent presence of liquid water, the unshielded solar ultraviolet radiation and atmospheric oxygen. Despite its harmful ability to erase fossils, liquid water exhibits interesting peculiarities, which contributed to the birth of life on Earth. It is therefore generally considered as a prerequisite for terrestrial-type life to appear on a planet. Some possible contributions of water to the origin of life are presented here.

5.1 Water as a Diffusion Milieu

Assembling parts of the primitive molecular robots must have occurred at a reasonable rate. Solid state life is generally discarded, the parts being unable to migrate and to be easily exchanged. A gaseous phase would allow fast diffusion of the parts, but the limited inventory of stable volatile organic molecules would constitute a severe restriction. A liquid phase, i.e. as liquid water, offers the best environment for the diffusion and the exchange of organic molecules. Other solvents can be considered such as liquid ammonia, hydrogen sulphide, sulphur oxide, as well as hydrocarbons, organic acids or alcohols. Compared to any of these possible solvents, water exhibits an excellent stability towards heat and UV radiation degradation.

Liquid water is a fleeting substance that can roughly persist above 0 °C and under a pressure higher than 6 mbars (Fig. 5.1). Salts dissolved in water (brines) depress the freezing point. For instance, the 5.5% (by weight) salinity of the Dead Sea depresses the freezing point of sea water by 2.97 °C. Large freezing point depressions are observed for 15% LiCl (23.4 °C) and for 22% NaCl (19.2 °C). Monovalent and divalent salts are essential for terrestrial life, because they are required as co-catalysts in many enzymatic activities. Usually, the tolerated salt concentrations are quite low (<0.5%),

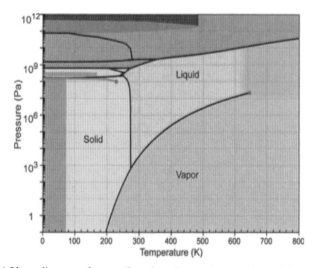

Fig. 5.1 Phase diagram of water (from http://www.sbu.ac.uk/water/phase.html).

because high salt concentrations disturb the networks of ionic interactions that shape biopolymers and hold them together. However, salt-loving microorganisms (both eucaryotes and procaryotes) known as extreme halophiles tolerate a wide range of salt concentrations (1-20%) and some procaryotes have managed to thrive in hypersaline biotopes (salines, salt-lakes) up to 25-30% sodium chloride.

The size of a planet and its distance from the star are two basic characteristics that will determine the presence of liquid water. If a planet is too small, like Mercury or the Moon, it will not be able to retain any atmosphere and, therefore, any liquid water. If the planet is too close to the star, the mean temperature rises due to starlight intensity. Any seawater present would evaporate delivering large amounts of water vapor to the atmosphere, thus contributing to the greenhouse effect. Such a positive feedback loop could lead to a runaway greenhouse: all of the surface water would be transferred to the upper atmosphere where photodissociation by ultraviolet light would break the molecules into hydrogen, which would escape into space, and oxygen, which would recombine into the crust. If a planet is far from the star, it may host liquid water providing that it can maintain a constant greenhouse atmosphere. However, water would provoke its own disappearance. The atmospheric greenhouse gas, CO_2 for instance, would be dissolved in the oceans and finally trapped as insoluble carbonates by rock-weathering. This negative feedback would lower the surface pressure and the temperature to an extent that water would be largely frozen. A volcanically active planer could recycle the carbon dioxide by breaking down subducted carbonates. The size of the Earth and its distance from the Sun are such that the planet never experienced either a runaway greenhouse or a divergent glaciation.

The origin of primitive Earth's water remains difficult to define. The source of the primitive terrestrial water can be found in a combination of H_2O trapped in the rocks that made the bulk of the planet's mass ("internal reservoir") and a late accreting veneer of extraterrestrial material ("external reservoir"). The formation of the internal reservoir is a natural consequence of the accretion of the planet. A volatile-rich veneer may have replaced the atmosphere produced by the planet's early accretion which was subsequently blown off by the giant impact between the Earth and a Mars-sized planetary embryo that generated the Moon and caused the early atmosphere to undergo massive hydrodynamic loss. The abundance and isotopic ratios of the noble gases neon, argon, krypton, and xenon suggest that meteorites alone or in combination with planetary rocks could not have produced the Earth's entire volatile inventory and that a significant contribution from icy planetesimals was required [1, 2].

The surface temperature of the early Earth was very much dependent upon the partial pressure of CO_2 in the atmosphere. The dominant view in recent years has been that the early atmosphere was a weakly reducing mixture of CO_2, N_2, and H_2O, combined with lesser amounts of CO and H_2 [3, 4]. An efficient greenhouse effect induced by a high CO_2 partial pressure was the most likely mechanism capable of compensating for the faint young Sun, the luminosity of which was about 30% lower than today. The dominant sink for atmospheric CO_2 was silicate weathering on land and subsequent formation and deposition of carbonate sediments in the ocean. However, a negative feedback system probably maintained the Earth's surface temperature above freezing: if the temperature were to drop below freezing, silicate weathering would slow down and volcanic CO_2 would accumulate in the atmosphere. Walker [5] suggested that the CO_2 partial pressure could have been as high as 10 bars prior to the emergence of the continents because of limited silicate weathering and storage in continental carbonate minerals. Under these conditions, the early Earth could have been as hot as 85 °C. Recent analysis of samarium/neodymium ratios in the ancient zircons indicate that the continents grew rapidly, so a dense CO_2 atmosphere may have been restricted to the first few hundred million years of the Earth's history.

5.2   Water as a Selective Solvent

According to its molecular weight, water should be a gas under standard terrestrial conditions by comparison with CO_2, SO_2, H_2S, etc. Its liquid state is due to its ability to form hydrogen bonds. This is not restricted to water molecules, since alcohols exhibit a similar behavior. However, the polymeric network of water molecules via H-bonds is so tight that the boiling point of water is raised from 40 °C (temperature inferred from the boiling point of the smallest alcohols) to 100 °C. Hydrogen bonds are formed between water molecules and organic molecules, providing that the latter contain -OH, -NH, -SH groups in addition to carbon and hydrogen. As a consequence of this affinity, many CHONS organic molecules are soluble in water. Hydrocarbons cannot form hydrogen bonds with water. As a consequence, they are much less soluble in water.

In addition to the H-bonding capability, water exhibits a large dipole moment (1.85 debye) as compared to alcohols (<1.70 debye). This large dipole moment favors the dissociation of ionizable groups such as $-NH_2$ and -COOH leading to ionic groups, which can form additional H-bonds with water molecules, thus improving their solubility. Water is also an outstanding dielectric (ε = 80). When oppositely charged organic groups are formed, their recombination is unfavored, because the attraction force of reassociation is proportional to $1/\varepsilon$. This is also true for mineral ions, which have probably been associated with organic molecules since the beginning of life's adventure.

5.3   Water as a Clay Producer

Clay minerals are formed by water alteration of silicate minerals. As soon as liquid water appeared in the surface of the primitive Earth, clay minerals probably accumulated and became suspended in the primitive ocean. The importance of clay mineral in the origins of life was first suggested by Bernal [6]. The advantageous features of clays for Bernal were (i) their ordered arrangement, (ii) their large adsorption capacity, (iii) their shielding against sunlight, (iv) their ability to concentrate organic chemicals, and (v) their ability to serve as polymerization templates. Since the seminal hypothesis of Bernal, many prebiotic scenarios involving clays have been written and many prebiotic experiments have used clays. So far, the most impressive results have been obtained by Ferris at al. [7]. Oligomers up to 55 long were obtained for both nucleotides and amino acids in the presence of montmorillonite for nucleotides and of illite for amino acids.

5.4   Water Structures the Biopolymers

Prebiotic organic molecules can be classified into two groups, CH containing molecules (hydrocarbons) and CHONS containing molecules. In the presence of liquid water, hydrocarbons, which cannot form hydrogen bonds with water molecules, escape water molecules as much as possible. They are hydrophobic. CHONS, especially those, which bear ionizable groups, form hydrogen bonds with water molecules and have therefore some affinity for water. They are hydrophilic. When these two species coexist within the same molecules, the duality generates interesting chain geometries. When they are separated by long distances such as in fatty acids or phospholipids, micelles, vesicles or liposomes are formed due to clustering of the hydrophobic groups. Over short distances, hydrophobic and hydrophilic groups generate chain conformations, which depend strongly on the sequence as illustrated by synthetic polypeptides.

Strictly alternating homochiral poly (L-leucyl-L-lysyl), hereafter referred to as poly(Leu-Lys), is soluble in water. At neutral pH, the lysyl side-chain amines are ionized as NH_3^+ groups. Due to charge repulsion, the chain cannot adopt a regular conformation. Addition of salt to the solution, for instance 0.1 M NaCl, produces a screening of the charges and allows the polypeptide to adopt a β-sheet structure [8]. Because of the alternating sequence, all hydrophobic residues are confined to one side of each strand. The chains aggregate into asymmetrical bilayers with a hydrophobic interior and a hydrophilic exterior because of hydrophobic side-chain clustering (Fig. 5.2). When the properties of water are modified by addition of increasing amounts of alcohol, a β- to α-structure transition is obtained, probably by relaxation of the water constraint on the hydrophobic groups [9].

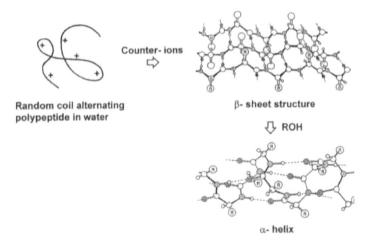

Random coil alternating   β- sheet structure
polypeptide in water

α- helix

Fig. 5.2 Conformations of alternating hydrophilic/hydrophobic polypeptides. The polypeptides adopt a random coil conformation in pure water because of charge repulsions. When the charges are screened by salts, they undergo a coil to β-sheet transition driven by the clustering of the hydrophobic groups. Addition of alcohol to the β-sheet releases the constraint of water and induces the formation of α-helices.

Due to bilayer formation, strictly alternating hydrophobic-hydrophilic sequences are thermostable. Non-alternating sequences form α-helices, which are thermolabile. Heating samples, in which α- and β-structures coexist increases the amount of β-structure with a loss of α-helix [9]. Alternating sequences are also more resistant to chemical degradation than α-helical sequences [10]. To get a β-sheet, and therefore a strong resistance to degradation, the hydrophobic amino acids must display their hydrophobicity to a marked degree. For instance, poly (αAbu-Lys) associating L-α amino butyric acid, NH_2-CH(CH_2-CH_3)-COOH, with L-lysine, does not form β-sheets and was found to be 15 times more sensitive to mild hydrolysis than poly (Leu-Lys) [11].

Aggregation of alternating sequences to form β-sheets is possible only with homochiral (all-L or all-D) polypeptides. For instance, racemic alternating poly (D,L-Leu-D,L-Lys) is largely unable to adopt the β-structure and remains mostly unstructured [12]. When increasing amounts of L-residues are introduced into the racemic alternating polypeptide, the proportion of β-sheets increases and there is a good relationship between the percentage of β-form and the amount of L-residues in the polymer. The *molecules* can be described as a mixture of β-sheets and disordered segments. Those segments containing six or more homochiral residues aggregate into stable nuclei of optically pure β-sheets surrounded by the more fragile heterochiral disordered segments [13, 14] (Fig. 5.3). The samples were subjected to mild hydrolysis. The kinetic measurements showed two rate constants in agreement with the existence of two conformational species. After partial hydrolysis, the remaining polymeric fraction was enriched in the dominant enantiomer [15]. The enantiomeric excess (absolute

Fig. 5.3 Example of enantioselective self-organization of peptide chains in water. Only all-L or all-D sequences self-organize into homochiral β-sheet nuclei. The heterochiral sequences are unable to adopt a regular conformation.

difference between the % of L and D) had increased from 54 to 68%. An identical effect has been observed with homochiral helices, since they are more stable than the disordered segments. Bonner et al. [16] raised the enantiomeric excess from 45 to 55% by partial hydrolysis of heterochiral polyleucine.

The β-sheet conformation of poly (Leu-Lys) exhibits a catalytic activity. It strongly accelerates the hydrolysis of oligoribonudeotides [17]. The β-sheet geometry plays a determinant role in the observed activity, since basic polypeptides, which do not adopt this conformation such as poly (D,L-Leu-D,L-Lys) are inefficient. Even a β-forming basic decapeptide is large enough to exhibit an efficient hydrolytic activity [18].

5.5   Water as a Driving Power for Chemistry

Water molecules are good nucleophiles, which compete efficiently with other nucleophiles. Water is feared by organic chemists because of its ability to spontaneously hydrolyse energy-rich chemical bonds. However, under certain conditions, hydrolysis favors unexpected pathways as exemplified with the activating agent N,N'-carbonyldiimidazole (Im-CO-Im, CDI). In organic solvents, CDI is known to activate carboxylic acid group R-COOH to give the corresponding acylimidazolide R-CO-Im. Added to free amino acids in an organic solvent, CDI led to aminoacylimidazolides, a carboxylic acid activation. In the presence of water, aminoacylimidazolides suffer partial hydrolysis. They also polymerise to afford oligopeptides including substantial amounts of diketopiperazine, a cyclc dipeptide, which is a dead end [19]. The formation of acylimidazolides directly in water seems unlikely and most organic chemists would discourage, *a priori*, the use CDI for the polymerisation of amino acids in water. Ehler and Orgel [20] did the experiment and obtained oligopeptides via the intermediary formation of N-imidazoyl carbonyl amino acid, an amino activation. By continuous extraction of the reaction mixture with chloroform, we isolated N-carboxyanhydrides, which are excellent candidates for the selective polymerization of proteinaceous amino acids in water [11, 21] (Fig. 5.4). L-leucine treated with CDI in water afforded optically pure oligo-L-leucines in 70% yield with an average molecular weight of 8. Oligomers up to the 11-mer were identified with glutamic acid. A mixture of amino acids containing both natural and non-natural amino acids and close to that found in the Murchison meteorite was treated with CDI in water. The condensate was found to be enriched in natural amino acids [11] (Fig. 5.5).

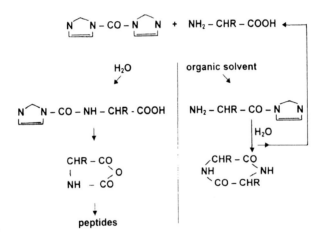

Fig. 5.4 Liquid water favors the formation of peptides (small proteins) from amino acids (left). When the reaction is started in an organic solvent, an non productive pathway is observed (right).

5.6   Water as a Heat Dissipator

Deep-sea hydrothermal systems may also represent likely environments for the synthesis of prebiotic organic molecules [22] and perhaps for primitive life [23]. Experiments have been carried out in order to test, whether amino acids can be formed under conditions simulating the hydrothermal vents. Yanagawa and Kobayashi [24] treated methane and nitrogen under simulated hydrothermal vent conditions (325 °C and 200 kg/cm^2) and obtained glycine and alanine in low yields (10^{-4}%). Hennet et al. [25] obtained glycine, alanine, serine, aspartic and glutamic acids, isoleucine when treating mixtures of hydrogen cyanide, formaldehyde and ammonia at 150 °C and 10 bar in the presence of pyrite-pyrrhotite-marquetite. The amino acid diversity is roughly similar to that reported for electric discharges, but the yields are significantly higher (6.9% for glycine). Hydrothermal vents are often disqualified as efficient reactors for the synthesis of bio-organic molecules because of the high temperature. However, the products that are synthesized in hot vents are rapidly quenched in the surrounding cold water thanks to the good heat conductivity of the solvent. When fluid containing glycine repeatedly circulated through the hot (225 °C) and cold (0 °C) regions in a laboratory reactor that simulated a hydrothermal system, glycine peptides up to hexaglycine were obtained [26].

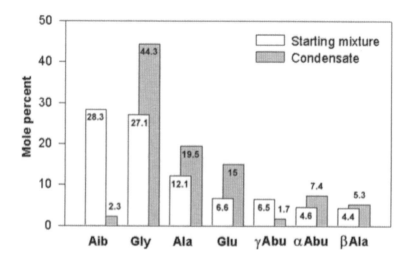

Fig. 5.5 Condensation in water of a mixture of amino acids similar to that found in the Murchison meteorite. The condensate is strongly enriched in biological amino acids i.e. glycine (Gly), alanine (Ala) and glutamic acid (Glu). The non-biological amino acids - α-amino isobutyric acid (Aib), γ-amino butyric acid (γAbu), α-amino butyric acid (αAbu) and β-alanine (βAla) - are not (or only weakly) selected.

5.7   Life Without Water: The Case for Titan

Water played a major role in the appearance and evolution of life. However, other possible biogenic liquids must not be ruled out. For instance, Saturn's moon Titan may host an ocean of methane and ethane. Titan is the only object in the solar system to bear a resemblance to our own planet in terms of atmospheric pressure (1.5 bar) and carbon/nitrogen chemistry [27]. Titan's atmosphere was revealed mainly by the Voyager 1 mission in 1980, which yielded the bulk composition (90% molecular nitrogen and about 1-8% methane). The Infrared Space Observatory (ISO) satellite has recently detected tiny amounts of water vapor in the higher atmosphere, but Titan's surface temperature (94 K) is much too low to allow the presence of liquid water. A great number of trace constituents were also observed in the atmosphere in the form of hydrocarbons (acetylene, ethane, ethylene, etc.), nitriles (HCN, C_2N_2, HC_3N) and oxygen compounds (CO and CO_2), attesting to a very active organic chemistry running in the atmosphere. This chemistry probably produces asymmetric hydrocarbons. More than 10 chiral hydrocarbons ranging from 3 to 7 carbon atoms can be foreseen with molecular weights lower then 100 Dalton. Racemic mixtures could conceivably spontaneously resolve in one-handed conglomerates or in open systems with input of soluble atmospheric organics and output of insoluble compounds. In a later stage, these homochiral molecules might initiate an exotic carbon-based life provided a hydrocarbon ocean is present on Titan's surface.

The Cassini-Huygens spacecraft launched in October 1997 will arrive in the vicinity of Saturn in 2004 and perform several flybys of Titan making spectroscopic, imaging, radar and other measurements. A descent probe will penetrate the atmosphere and will systematically study the organic chemistry in Titan's geofluid. During 150 minutes, *in situ* measurements will provide detailed analysis of the organics present in the air, in the aerosols and at the surface (see Chap. 24, Foing). Unfortunately, no instrument has been designed to measure the handedness of the organic molecules.

5.8   Conclusion

Water plays a crucial role in modern biology and the love story probably started with the origin of life. It is therefore legitimate to search for liquid water when searching for extraterrestrial life. Even if somewhat egocentric, and therefore restrictive, searching for liquid water is the easiest way to recognize any extraterrestrial life. Concerning the possible existence of another water-based life, the early histories of Mars and Earth clearly show some similarities. The existence of large valley networks and different channels strongly suggests that liquid water was once stable on the surface of Mars. If organic molecules were brought to Mars by cometary grains or meteorite impacts, then an aqueous organic chemistry might have developed on Mars until liquid water disappeared, which is assumed to have happened globally about 3.5 billion years ago (for a more detailed assessment of liquid water in the history of Mars see Chap. 6, Jaumann et al.) . The search for organic molecules and fossilized primitive life below the Martian surface is probably the most fascinating prospect for Mars exploration [29]. The Voyager images of Europa showed that Europa's ice shell might be separated from the silicate interior by a liquid water layer. The last images and magnetism measurements from the Galileo spacecraft strengthen the case for the presence of an ocean below the surface [30, 31]. If liquid water is present within Europa, it is quite possible that it includes organic matter derived from thermal vents. Terrestrial-like prebiotic organic chemistry and primitive life may therefore have developed in Europa's ocean. Other "biogenic" liquid must not be excluded. For instance, Saturn's moon Titan hosts perhaps an ocean of methane and ethane and an exotic form of life, which might be very difficult to identify.

Searching for a second genesis of a water-based life, either artificial in a test tube or natural on another celestial body, will be one of the most exciting challenge for the coming decades.

5.9   References

1   T.C. Owen, A. BarNun, Icarus **116**, 215 (1995).
2   T.C. Owen, in: A. Brack (Ed.) *The molecular origins of life: assembling pieces of the puzzle*, Cambridge University Press, Cambridge 1998, pp. 13.
3   J.F. Kasting, Science **259**, 920 (1993).
4   J.F. Kasting, L.L. Brown, in: A. Brack (Ed.) *The molecular origins of life: assembling pieces of the puzzle*, Cambridge University Press, Cambridge, 1998, pp. 35.
5   J.C.G. Walker, Origins Life Evol. Biosphere **16**, 117 (1985).
6   J.D. Bernal, in: Routledge and Kegan (Eds.) *The Physical Basis of Life*, Paul, London, 1951.
7   J.P. Ferris, A.R. Hill, R. Liu, L.E. Orgel, Nature **381**, 59 (1996).
8   A. Brack, L.E. Orgel, Nature **256**, 383 (1975).
9   A. Brack, G. Spach, J Amer. Chem. Soc. **103**, 6319 (1981).
10   A. Brack, G. Spach, in: Y. Wolman (Ed.) *Origin of Life*, Kluwer Acad. Publ., 1981, pp. 487.
11   A. Brack, Origins Life Evol. Biosphere **17**, 367 (1987).
12   A. Brack, G. Spach, J. Mol. Evol. **13**, 35 (1979).
13   G. Spach, A. Brack, J. Mol. Evol. **13**, 47 (1979).
14   A. Brack, G. Spach, Origins Life **11**, 135 (1981).
15   A. Brack, G. Spach, J. Mol. Evol. **15**, 231 (1979).
16   W.A. Bonner, N.E. Blair, F.M. Dirbas, Origins Life **11**, 119 (1981).
17   B. Barbier, A. Brack, J. Amer. Chem. Soc. **114**, 3511 (1992).
18   A. Brack, B. Barbier, Origins Life Evol. Biosphere **20**, 139 (1990).
19   A.L. Weber, J.C. Lacey Jr., Biochim. Biophys. Acta **349**, 226 (1974).
20   K.W. Ehler, L.E. Orgel, Biochim. Biophys. Acta **434**, 233 (1976).
21   A. Brack, BioSystems **15**, 201 (1982).
22   N.G. Holm, E.M. Andersson, in: A. Brack (Ed.) *The molecular origins of life: assembling pieces of the puzzle*, Cambridge University Press, Cambridge, 1998, pp. 86.
23   K.O. Stetter, in: A. Brack (Ed.) *The molecular origins of life: assembling pieces of the puzzle*, Cambridge University Press, Cambridge, 1998, pp. 315.
24   H. Yanagawa, K. Kobayashi, *Origins Life Evol. Biosphere* **22**, 147 (1992).
25   R.J.-C. Hennet, N.G. Holm, M.H. Engel, Naturwissenschaften **79**, 361 (1992).
26   E-i. Imai, H. Honda, K. Hatori, A. Brack, K. Matsuno, Science **283**, 831 (1999).
27   F. Raulin, in: A. Brack (Ed.) *The molecular origins of life: assembling pieces of the puzzle*, Cambridge University Press, Cambridge, 1998, pp. 365.
28   J.P. Lebreton, in: *Huygens: Science Payload and Mission*, ESA-SP 1177, ESA Pub. Noordwijk, The Netherlands 1997.
29   F. Westall, A. Brack, B. Hofmann, G. Horneck, G. Kurat, J. Maxwell, G.G. Ori, C. Pillinger, F. Raulin, N. Thomas, B. Fitton, P. Clancy, D. Prieur, D. Vassaux, Planet. Space Sci. **48**, 181 (2000).
30   M.H. Carr, J. Geophys. Res. **100**, 7479 (1995).
31   M.G. Kivelson, K.K. Khurana, C.T. Russel, M. Volwerk, R.J. Walker, C. Zimmer, Science **289**, 1340 (2000).

6 Geomorphological Record of Water-Related Erosion on Mars

Ralf Jaumann, Ernst Hauber, Julia Lanz, Harald Hoffmann and Gerhard Neukum

In 1972, for the first time the Mariner 9 photographs showed large erosional features of giant channels and branching networks of small valleys on Mars which were revealed in more detail by the Viking orbiter images obtained in 1975.

The topography of Mars exhibits a clear dichotomy [1-3] which divides the surface into a southern highland hemisphere, as heavily cratered as the lunar highlands raising several thousands of meters above the zero level, and relatively smooth northern lowlands well below this level. The southern highlands have survived the heavy bombardment prior to 3.8 billion years with only minor modifications and record the early history of the planet, whereas the younger northern lowlands represent the later evolution of the Martian surface. Networks of small valleys are very common in the southern highlands. The highland/lowland boundary region is characterized by large channeled outflow systems [4-10] and retreating erosion. The northern plains are textured and fractured in a way that has been attributed to the interaction with ground ice and sedimentation as a result of large flood events, finally modified by wind [7]. The ejecta of Martian impact craters 2-50 km in diameter is, unlike the craters on the Moon and Mercury, commonly arrayed in discrete lobes which are defined by low ridges. The most plausible explanation of these features is that impact craters of a certain size penetrated the cryosphere and ejected material mixed with water and/or ice, causing the flow-like deposition of ejecta materials fluidized by volatiles [11-19]. The floors of some craters are partly filled, mantled and modified by fluvial, periglacial and aeolian processes.

The south polar region is surrounded by cratered highlands, whereas the northern circumpolar plains are largely covered with younger aeolian deposits building large dune fields and mantles [20]. At each pole a thick sequence of layered deposits extends outward to about 80° latitude, containing alternating light and dark sheets capped by H_2O and CO_2 ice [20]. Incised valleys curl outward from the poles, clockwise in the south and counter-clockwise in the north. Mainly the channels and valley networks but also the ground textures in the northern plains exhibit features which seem to be formed by running water and/or moving ice [4, 6-9, 21-23]. However, the present-day conditions on Mars, with average temperatures well below freezing and atmospheric pressures at or below the 6.1 mbar triple point vapor pressure of water, are far from supporting liquid water on the surface.

Many attempts have been made to explain the Martian erosion without including water such as formation by lava flows, wind, debris flows and liquid hydrocarbons [24-29], but none of these theories is able to comprehensively describe the formation of all the associated features. Therefore, erosion on Mars is most likely to be driven by water, either in liquid form or as ice (see for example [7, 8, 10]). This led to the assumption that liquid water might have been stable on the surface in the past history of Mars (see for example [5, 6, 9, 10]). Thus, besides Earth, Mars is the only planet with a record of resurfacing processes and environmental changes that indicate the past operation of a hydrologic cycle. Although the large-scale morphology of the Martian channels and valley networks show remarkable similarities with fluid-eroded features on Earth, there are major differences in their size, small-scale morphology, inner channel structure and source regions, indicating that the erosion on Mars has its own characteristic genesis and evolution. The details of the channel forming processes are still unknown.

6.1 Outflow Channels

Outflow channels are several to tens of kilometers across, reaching lengths of a few hundred to thousands of kilometers with gradients of channel floors (Fig. 6.1) ranging from 0 to 2.5 m/km [10]. Tributaries are rare, but branching downstream is common resulting in an anastomosing pattern of channels [7]. The channels tend to be deeper at

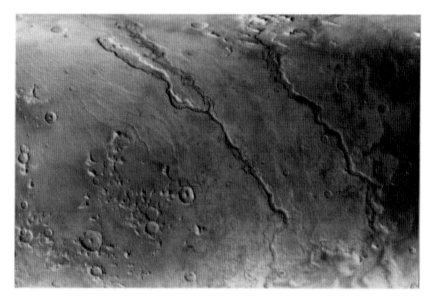

Fig. 6.1 Outflow channels near the highland volcano Hadriaca Patera. The close proximity of the collapse features and the volcano (caldera at upper left of the image) suggests an origin by volcano-ice interactions [e.g., 95]. From left to right: Dao Vallis, Niger Vallis, and Harmakhis Vallis, each one more than 1 km deep and 8-40 km wide. MOC wide angle image, width ~800 km, centered at 40 °S, 270 °W, north toward left, illumination from lower left (NASA/JPL/Malin Space Science Systems).

their source than downstream and in general have a high width to depth ratio and low sinuosities [7, 10]. Isolated upland remnants separating channels, mesa-like remnants at channel borders, hanging valleys, transected divides, and expansions and constrictions of the channel beds are common. A variety of bedforms are located on the floors of outflow channels [5, 9, 10, 30-35] such as residual or recessional headcuts, scour marks, longitudinal grooves, which parallel the presumed flow direction, teardrop-shaped islands, inner channels, terraces, cataracts and plucked zones, similar to scabland topography, which form by stripping away surface materials due to high-velocity turbulent fluid flow [5]. Outflow channels are common in the vicinity of the Chryse-Acidalia basin, west of the Elysium volcano complex in Elysium Planitia, in the eastern part of Hellas basin and along the western and southern border of Amazonis Planitia. All these channels are denoted as outflow channels, even though their occurrence, source regions and geological context differ from area to area. By far the largest channels occur around the Chryse Basin. Outflow channels in this area start in the cratered Xanthe Terra uplands (Fig. 6.2) or on high volcanic plains and converge in the Chryse Basin mainly from the south but also from the southeast and west. On the Chryse plains they broaden and finally disappear at about 40 °N latitude. Most of the outflow channels around Chryse emanate fully developed from circular to elliptical depressions termed chaotic terrain [36]. These areas are 2-5 km below the surrounding undisturbed terrain and mostly covered with large jumbled blocks of material from the former surface. This indicates that the chaotic terrain have formed by collapse rather than by removal of material from the above, indicating the involvement of groundwater in the channel formation process. Many of the chaotic terrain east of the Valles Marineris merge westward with the canyon system and northwards with outflow channels. However, particularly to the north of the Valles Marineris, channels tend to emanate from completely enclosed canyons containing chaotic terrain like Echus, Hebes, Juventas and Ganges Chasma. The context of the Chryse channels source regions with the Valles Marineris canyon system suggests a close correlation between canyon tectonics and channel formation. In the Elysium region large outflow channels originate at fractures oriented radial to the west and northwest flanks of the Elysium volcano complex. The channels mostly start at elongated depressions. Outflow channels also occur at the eastern flank of the Hellas basin adjacent to the old volcano Hadriaca Patera.

6.2   Valley Networks

Open branching valleys, in which tributaries merge downstream, are common in the southern highlands. The number of branches is low with large undissected areas between individual branches. The networks themselves are spaced apart leaving large areas of undissected highland between them [23], indicating that there has been no or only little competition between adjacent drainage basins [37]. The main characteristics of individual branches are their U-shaped form with flat floors and steep walls. They differ from channels by the absence of bed-forms, which are indicators for fluid flow [9]. Some valleys are several hundreds of kilometers in length and a few tens of kilometers wide with short accordant tributaries such as Nirgal Vallis. However, the majority of the valleys is typically no longer than 200 km and only a few kilometers wide. The valley networks are mainly located in the cratered uplands, the oldest Martian terrain, at elevations ranging from 2 km to over 5 km. Some valley networks drain into craters, but most of them start at local heights and drain as winding streams with relatively large junction angles of branches into low areas between large impact crater where they terminate against smooth plains [38]. At the upland/lowland boundary regions, the valley networks tend to drain northward (Fig. 6.3). Valleys in the Noachis

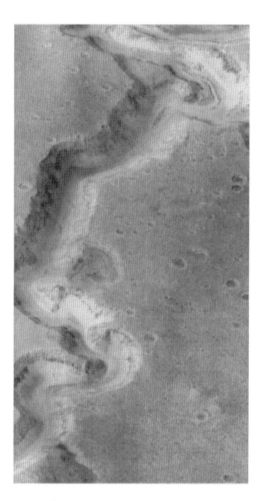

Fig. 6.2 Nanedi Vallis is a valley system in the ancient cratered plains of Xanthe Terra. Terraces and the small inner channel within the canyon (near top of image) suggest continuous flow and down-cutting. Other morphologic features, however, indicate a formation by collapse: the lack of small contributing valleys on the surrounding plateau, box- or amphitheater-headed short side valleys, and the form of the meanders. It seems possible that both sustained flow and collapse formed the valley. Image width is ~10 km, illumination is from the left (NASA/JPL/Malin Space Science Systems/RPIF/DLR).

region are mostly oriented northwest into the low-laying region of chaotic terrain east of Valles Marineris. Around Hellas the valley networks are focussed radial to the basin and valley networks in the south pole region drain north towards the Hellas and Agyre Basins. Another set of channel networks occurs on volcanoes. Although channels on volcanoes usually are caused by lava or volcanic density flows so called nuée ardentes [39], there is evidence that valleys on Alba Patera, Ceraunius Tholus and Hecates Tholus as well as on Hadriaca, Thyrrena and Apollinaris Patera are compatible with a fluvial origin [40-42]. These valleys are narrower and shallower as their counterparts in the uplands and show a more dense drainage pattern with small junction angles. At some places the walls of canyons are deeply incised by V-shaped valleys that lack deposits at their mouth within the canyon [7]. In particular at the upland/lowland boundary between about 290°-360 °W and 30°-50 °N a specific type of terrain is exposed. In this area the northern plains interfere in a complex way with the highlands. Numerous islands with 1-2 km high escarpments are isolated from the uplands [43]. The isolated mesa-like islands are surrounded by aprons with well-developed flow fronts [44]. Flat-floored, broad(up to 20 km wide) and steep-walled channels reach deep into the uplands. These fretted channels branch upstream as do valley networks. Some of these channels have longitudinal ridge-like features on their floors, indicating compressional features caused by merging aprons from the walls [44] and removal of material through the channel by mass wasting downstream [38]. Furthermore, processes of undermining and collapse seem to have been involved in forming the fretted channels [45].

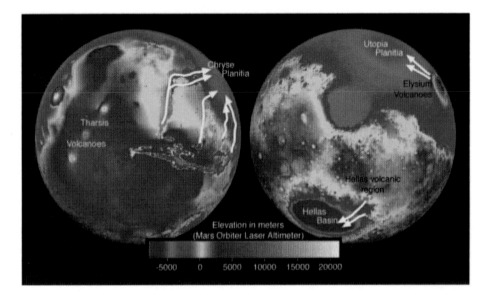

Fig. 6.3 Global elevation map based on MOLA data. White arrows show the main flow directions of water or ice within the outflow channels. Note the spatial proximity of channel source areas to volcanic centers in Tharsis, Elysium, and Hellas.

6.3   Periglacial and Permafrost Landforms

6.3.1   Debris Flow and Terrain Softening

Steep topographic scarps often dissect the terrain at the transition from the southern cratered uplands towards the northern lowlands. At their base, well-defined flow fronts with convex-upward surfaces extend into the low-lying plains [43]. These *lobate debris aprons* have been attributed to gelifluction and/or frost creep [46]. The water content in this mixture of rock and ice might be derived from the atmosphere [44], or from ground ice by sapping or scarp collapse [45]. Where flow fronts from opposing walls collide, e.g., in valleys, linear ridges and grooves cover the whole valley floor. This so-called *lineated valley fill* resembles terrestrial median moraines on glaciers (Fig. 6.4).

A certain type of surface modification by viscous creep of material is found in both hemispheres at latitudes between 30° and 60°. The effects are most obvious at craters: Fresh impact craters have clearly defined, crisp rims. In some places, however, crater rims appear to be subdued or rounded and small craters are only sparsely visible [47]. In other places, material has obviously moved down the inner crater wall by creep, not

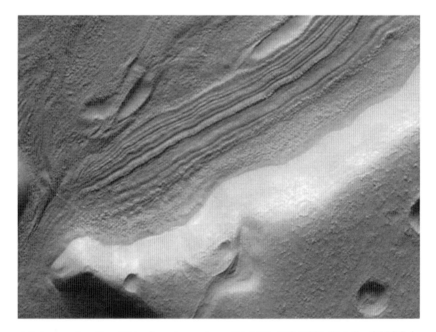

Fig. 6.4 *Lineated valley fill* in fretted terrain at 34.4 °N, 302.0 °W. Detail of MOC image 46704, image height 5.5 km (resolution 13.2 m/pixel), illumination is from above (NASA/JPL/Malin Space Science Systems).

by fall, and forms what is called *concentric crater fill*. These and other types of *terrain softening* are mostly seen in a latitudinal belt between 30° and 60° [7, 48, 49]. This might be due to the fact that ice is not in equilibrium with the present atmosphere at lower latitudes [49]. On the other hand, at very high latitudes > 60° the temperature are very low and, correspondingly, the high strain rate of ice prevents flow.

The viscous flow of material on Mars has been compared to terrestrial *rock glaciers* [e.g., 50, 51], which are accumulations of angular rock debris and move down due to deformation of internal ice or frozen sediments [e.g., 50, 52]. Squyres [50] calculated that near surface flow (in a viscous layer less than 5 km deep) is probably responsible for the formation of the flow features.

6.3.2   Ground Ice

The huge difference between the water volumes inferred by the morphological evidence of water release on the surface (chaotic terrain, outflow channels) on one hand and the present inventory of volatiles (mainly polar caps) on the other hand implies that large amounts of water infiltrated the Martian megaregolith. The old megaregolith where the ground ice is mainly found, is a global blanket of fractured and blocky materials up to 2 km thick as a consequence of impact cratering during the primordial bombardment. It can be inter-bedded with ice layers, lava flows or sedimentary deposits [31]. If the pores are filled with ice, and if a decrease of the porosity with increasing depth is taken into account, the Martian ground ice would correspond to a global layer of water at least 400 m deep [53].

Given a mean annual surface temperature of –60 °C, a 6 mb atmospheric pressure and a dry periglacial–type climate, the present ground ice on Mars should be stable at some depth over virtually the entire planet. Its total thickness depends on the mean annual surface temperature, on the geothermal heat flow, the thermal conductivity of the material, and the ground ice melting temperature. Permafrost thicknesses from 3 to 7 km near the poles and between 1 and 3 km near the equator seem possible [54]. Liquid water should exist under the ground ice, at least at mid-latitudes.

6.3.3   Rampart Craters

The particular morphology of some impact crater ejecta blankets provides very strong evidence for subsurface volatiles on Mars. On the Moon with its dry crust, the fragments ejected by an impact follow ballistic trajectories. Consequently, they form a continuous blanket with a rough surface near the crater, slowly grading outward into secondary craters. In a completely different and characteristic pattern, the ejecta of so-called *rampart craters* on Mars are marked by distinct lobes. Because they seem to have flown around pre-existing obstacles, it has early been suggested that the entrainment of subsurface volatiles - most probably water - has caused the ejecta to flow over the surface [7, 11, 55].

Not all impact craters show lobate ejecta blankets, and rampart craters are not distributed randomly over the Martian surface. This observation has led to several efforts to map the volatile content in the subsurface based on rampart crater morphologies.

Very small craters with diameters less than a few kilometers never have lobate ejecta. The impacting bodies obviously were not large enough to penetrate the uppermost dry surface layer of the planet. Kuzmin et al. [56] used the minimum diameter of craters with rampart ejecta to map the global distribution of the ground ice table, i.e. the depth, at which water is present. They found smaller onset diameters for rampart formation at higher latitudes, indicating water at shallower levels.

Other workers have investigated the ratio of the ejecta blanket versus the diameter of rampart craters. The basic idea is that a larger water content should result in a larger ejecta blanket (having flown farther outward) for a given crater diameter. A correlation of larger ratios with higher latitudes has been found [12, 56-59], again indicating more near-surface water in mid- and high latitudes, mainly in the northern plains [7]. Particularly conspicuous concentrations of rampart craters with high ratios of ejecta to diameter have been noted in large topographic depressions or basins more than 4000 m below the Martian datum near the mouth of outflow channels in Utopia and Acidalia Planitiae [60, 61].

6.3.4   Polygons

Giant polygons have been found in several locations on Mars: The best examples can be observed near the mouths of outflow channels in Acidalia, Elysium, and Utopia Planitiae [62-64], and less developed or smaller features were identified in the Valles Marineris [65] and in very high latitudes north of 70 °N [59]. The diameters of the giant polygons are in the range of 30 km [e.g., 62, 66-68], and the average width and depth of troughs bordering the interior of polygons in Utopia Planitia is 2 km and 30 m, respectively [69]. Their genesis was attributed to various models, all involving cracking of the surface in response to tension. Based on terrestrial analogy, McGill and Hills [70] proposed cooling and dessication of wet sediments deposited at the termination of outflow channels to account for polygon formation.

Many recent Mars Observer Camera (MOC) images show well-developed polygons with relatively small diameters of 10-100 m (Fig. 6.5). They were already identified in Viking images and are interpreted to be the result of thermal contraction in ice-rich soils based on the similar scales as compared to terrestrial ice-wedge polygons [68, 71-73]. Theoretical considerations by Mellon [74] support the formation of polygons by thermal contraction at mid and high latitudes.

6.3.5   Thermokarst

Thermokarst features form by the melting of ground ice and subsequent surface collapse. They are typically found in periglacial alluvial plains or in glaciofluvial outwash plains. On Mars, possible thermokarst was first noted by Sharp [43] and Anderson et al. [75] who noted the similarity of some Martian depressions to thermokarst features like *thaw lakes* or *alases* in Alaska and Siberia. In more recent studies based on high-resolution Viking Orbiter and MOC images, Costard and Kargel [61, 76] support the thermokarst interpretation and report particularly high concentrations in

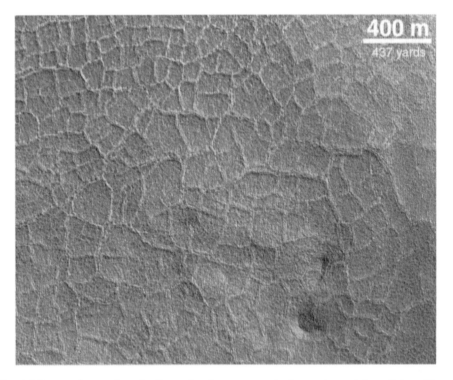

Fig. 6.5 Patterned ground on the floor of an impact crater in the northern plains of Mars. The polygons might have formed by repeated freeze-thaw cycles of subsurface ice. MOC image release MOC2-150 (NASA/JPL/Malin Space Science Systems).

Utopia Planitia, the same region where polygonally fractured ground and small rampart craters are frequently found. Another thermal erosion process has been proposed as a typical situation along most outflow channels. *Fluvial-thermal erosion*, occurring along most arctic rivers, might enlarge outflow channels by a combination of thermal and mechanical erosion along frozen riverbanks [77]. Lateral erosion produces thermokarstic subsidence that favors collapsed blocks.

6.4   Glacial Landforms

6.4.1   Eskers and Moraines

Glaciation was considered to be responsible for producing specific landforms on Mars for a long time [e.g., 53, 78-81]. However, it was not before the early 1990's when the first integrated sequence of glacial features in Argyre, Hellas, the south polar region, and the northern plains was reported. Kargel and Strom argued in a series of papers [e.g., 82-84] that only glaciation can account for these features in a simple and unifying way, and they point out that otherwise many different and unusual processes would have to work independently to explain them.

According to Kargel and Strom, long curvilinear ridges in mid and high latitudes provide striking morphological evidence for ancient glaciation. A formation as *eskers* seems plausible [e.g., 83, 85-88]: Eskers are elongated sinuous ridges of glacifluvial sand and gravel and form as infillings of sub-glacial, en-glacial, or supra-glacial rivers.

6.5   Volcano Ice Interactions

Early in the Viking mission, several workers noted landforms which might be related to volcano-ice or volcano-groundwater interactions [e.g., 78, 89-91]: Table mountains at high northern latitudes resemble Canadian tuyas and Icelandic flat-topped mountains with steep sides which developed from sub-glacial volcanic eruptions and are the sub-glacial equivalents to shield volcanoes. Elsewhere on Mars, ridges and mountains have been compared to terrestrial *móberg* ridges consisting of sub-glacially erupted material.

The interaction of volcanic heat and ground ice might be responsible for the close spatial proximity of braided channels with streamlined islands to the Elysium volcanoes [92, 93]. Chapman [94] found evidence for móberg ridges, table mountains, and jökulhlaup deposits and proposed the existence of a paleo-ice sheet northwest of the Elysium volcanoes in Utopia Planitia. Northeast of Hellas, the source regions of some outflow channels (Fig. 6.1) close to the ancient highland volcano Hadriaca Patera also suggest that channel formation has been triggered by volcanic activity [95].

6.6   Chronology of Martian Erosional Processes

The stratigraphic position of geologic units on Mars is estimated by the means of superposition and intersection relations and by the concentration of impact craters superposed on geologic units [20]. Three major time systems are defined, throughout which the surface has been formed: Noachian, Hesperian and Amazonian. Due to the lack of samples from Mars, the assignment of absolute ages to the epochs is based on crater densities depending on cratering rates [20, 96-98] and thus is model dependent. Different models define the Amazonian-Hesperian boundary between 1.8-3.5 Ga ago and the Hesperian-Noachian boundary between 3.5-3.8 Ga ago [96, 98, 99].

The relatively small size of Martian valleys, the modification by aeolian processes and the not sufficient coverage of high-resolution imagery makes it difficult to use crater counting for age determination. However, in some places, particularly on the floors of large outflow channels, it was possible to estimate crater retention ages [21, 96, 100, 101]. Neukum and Hiller [96] performed crater counts on units (surroundings and floors) of Chryse channels such as Kasei, Ares, Tiu and Maja Vallis, Elysium channels such as Hrad Vallis and Elysium Fossae, one channel NE of Hellas and Ladon Vallis north of the Agyre basin, taking into account data from other authors [21, 44, 102]. As expected, they estimated the channel surroundings to be significantly older than the measured channel ages. The number of craters with 1 km diameter N(1) per square kilometer are for the surroundings either >4×10^{-2} (km^{-2}) representing the early crust or 2×10^{-3} (km^{-2}) to 2×10^{-2} (km^{-2}), representing a younger resurfacing period. This translates to crater model ages as proposed by Neukum and Hiller [96] of 3.6 to 4.1 Ga and to more than 4.2 Ga, respectively. In the following, crater model ages will refer to the Neukum and Hiller [96] Model I. However, it is mentioned that these ages are model dependent and may be accurate only within a factor of two for younger units. The estimated ages of channel floors range from N(1) = 1×10^{-2} (km^{-2}) to 1×10^{-4} (km^{-2}) or 3.9 to 0.4 Ga, the bulk lying between 3.8 and 3.1 Ga. The very young ages measured most probably represent post-erosional fillings. There is evidence that channel resurfacing was going on during two separate phases with the cratering age of 3.1 Ga marking the end of the first phase [96]. The valleys that belong to the older group such as Maja, Ares, Ladon and Hard Vallis have crater retention ages ranging from 3.9 Ga to 3.1 Ga. The younger group exhibits valley units with crater model ages of 3.1 Ga and about 0.4 Ga. The data from Neukum and Hiller [96] also show a sequence of the circum-Chryse valles with most of the Kasei Vallis floors being oldest followed by the mouth of Maja Vallis, Ares Vallis, Tiu Vallis and the mouth of Kasei as the youngest unit in this area. Marchenko et al. [101] suggest four major stages of fluvial activities and resurfacing in the Chryse region, indicating episodic events of flooding. Robinson et al. [100] concluded that some areas of Ares Vallis were formed prior to 3.5 Ga ago and have been later reworked between 2.2 Ga ago and 1.6 Ga ago. However, channel floors are modified by post-fluvial mass wasting and aeolian processes, making it difficult to clearly address younger floor units to flood events. Neukum and Hiller [96] concluded that the major channel formation took place between 3.8 Ga and 3.1 Ga ago. Channels on volcanoes are too small for crater counting on their floors, but they dissect volcanic units for example at the flanks of Alba Patera as young as 0.5 Ga. The resolution and coverage of existing imagery restrict the available age data, which can be derived by crater counting. Thus, only a small amount of channels and almost no valley networks are dated by this method. In order to get a better statistics of the distribution of channels in time, we used the relative relationship of superposing units and intersections. A channel that is superposed on a geologic unit must be contemporaneous with or younger than that unit. Branches of a channel that dissect others must be younger. This method provides maximum ages and can be used as a first statistical estimate of the distribution of channels and valley networks throughout time. The analysis of intersections of different branches of the Granicus Vallis in the Elysium region indicates that at least four episodic events have been involved in the formation of the system. Nelson and Greeley [103] mapped the circum-Chryse channels in detail and found the youngest units of different channels sequenced, ranging from Mawrth and Ares Vallis being the oldest, followed by Tiu and Simud Vallis to Shalbatana Vallis and the youngest channel units exposed at the mouth of Kasei Vallis, indicating that the last flood events of each channel getting younger from east to west around Chryse.

Based on the geologic maps of Scott and Tanaka [104], Greeley and Guest [105], Tanaka and Scott [79] and Viking imaging data, we examined the geologic relationships between channels and valley networks and the surrounding units they dissect. For a total of 65 outflow channels and individual channel branches and 276 of the largest valley networks in the uplands the maximum relative ages have been estimated. Most of the investigated outflow channels (70%) dissect Hesperian units with about the half of the outflow channels having Upper Hesperian maximum ages. Amazonian units are eroded by about 25% of the outflow channels. The majority of valley networks in the uplands (about 63%) has an Upper Noachian maximum age. The formation of valley networks declined in the Hesperian (25%) and only about 12% of the valley networks, mostly valleys on volcanoes, dissect Amazonian units. Scott and Dohm [106] measured intersection relations for several hundred networks and found about 70% to be older than Hesperian and Carr [38] suggested that more than 90% of the valley networks have a maximum age older than Hesperian. Thus outflow channels and valley networks are not only separated spatially but are also delayed in time, indicating a major change in the erosion processes on Mars at the Noachian/Hesperian boundary. However, the evaluation of new high resolution MOC images gave evidence for very young small lava flows [107] and very young small-scale mass wasting processes on the walls of craters and valleys probably involving water [108].

6.7   Standing Water

A major point in the ongoing discussion about the "watery" history of Mars is whether there have ever been large bodies of standing water on the surface of the planet and if so when. This is especially important for the question of the possible existence of former life on Mars.

The existence of large outflow channels and extended valley networks as well as the widespread occurrence of permafrost features and features, resembling glacial and periglacial landforms makes the assumption that Mars had or still has a large inventory of water at least probable. Different climatic conditions with temperatures above freezing and a denser atmosphere, as they are proposed for the early history of Mars, would thus make large bodies of water on the Martian surface almost inevitable.

Several authors have proposed the existence of an ancient ocean in the northern lowlands of Mars (see e.g., [109-111]). Parker et al. [109, 110] mapped two discontinuous contacts near and generally parallel to the southern boundary of the northern lowlands, based on breaks in slope, albedo boundaries and textural contacts that they interpreted to be shorelines of a former polar ocean. Head et al. [111] tried to identify those contacts in MOLA data and found the outer contact 1, exhibiting a wide range of elevations (~11 km!), to be an unlikely candidate for a shoreline. A more plausible candidate seemed to be the inner contact 2 with an elevation range of ~4.7 km and the most substantial variations occurring in regions where post-contact 2 activity has occurred (i.e. Tharsis, Elysium and Arabia). However, Malin and Edgett [112] who studied MOC-images of the proposed shoreline found that none of the coastal landforms common on Earth are evident.

A very promising feature supporting the theory of former standing bodies of water on Mars are outcrops of layered sedimentary rocks (Fig.6.6) recently imaged by MOC (see [113]). The outcrops occur in topographic lows mainly between ± 30° latitude, but there are numerous examples at higher latitudes as well. The geomorphic settings in which they appear are different and can be grouped into four types: crater interior (e.g., Becquerel Crater, Holden Crater, Gale Crater), inter-crater terrain (e.g., Terra Meridiani, Arabia Terra), chaotic terrain (e.g., Iani Chaos, Aram Chaos) and chasm interiors (e.g., Capri Chasma, Ganges Chasma, Ophir Chasma etc.).

Paleolacustrine environments had already been widely recognized on the surface of Mars especially putative crater lakes, which often exhibit well developed sedimentary landforms (see e.g., [114-121]). Aqueous sedimentary deposits in Gusev Crater were described by Cabrol et al. [118] such as deltaic structures from the inflowing Ma'adim Vallis as well as streamlined terraces, layers and channels observed on the central sedimentary deposits in Gale Crater [119, 120]. Ori et al. [121] described Gilbert-type deltas at the mouth of channels debouching into craters in Ismenius Lacus and Memnonia Fossae.

Finally, the Argyre and Hellas basins, the two largest impact craters on Mars, have also often be discussed as depositional basins with areas ranging as high as 3×10^6 km^2 for the Hellas Basin (see e.g., [115, 122, 123]). Numerous channels and valley networks terminate in the areas and large sedimentary deposits as well as local debris flows within the deposits, attested to concentrations of groundwater in the basins, have been identified.

Many of the described features as well as crater counting of the deposits and surroundings indicates that some lakes might have been active for long time periods of at least thousands of years [e.g., 121] up to 1-2 Ma [118]. This lead to the assumption that the climatic settings were different from those prevailing today.

6.8 Origin of Martian Surface Erosion and Implications for the Palaeoclimate

The source regions of the outflow channels are directly related to tectonics and in the Elysium and Hellas region to volcanic features. This evidence together with the fluid-like erosion features of the channels indicates that there is a close correlation with fast groundwater release triggered by melting of ground ice, tectonical thinning of the cryosphere and thereby provision of easy access of groundwater to the surface. Compared to alternate hypothesis such as lava erosion, wind erosion or erosion by other fluids like liquid hydrocarbons, the hypothesis, which accounts for most of the observed channel features is the catastrophic release of groundwater or drainage of surface lakes [10]. The dimensions of the outflow channels indicate discharges in the order of 10^7-10^9 m^{3}s^{-1} [5, 31, 124, 125]. This is about two orders of magnitude larger as the largest known flood events on Earth, such as the peak discharges of the Channeled Scabland flood of eastern Washington or the Chuja Basin flood in Siberia [10, 126]. Several processes are discussed that may cause large floods on Mars. Groundwater under high artesian pressure confined below a permafrost zone may break out, triggered by events, which disrupt the permafrost seal, such as impacts or Marsquakes, either by breaking the surface or sending a large pressure pulse through the aquifer [7, 31, 38]. This hypothesis needs a thick permafrost layer in order to build up the artesian pressure and is consistent with climatic conditions similar to those of present day Mars. Another argument for this mechanism is the low elevation of the outflow channel source regions compared to the high groundwater level of the regional aquifer system [7], which provides hydrostatic pressure. After groundwater release, floods would decline as the artesian pressure drops and finally come to an end when the aquifer reseals by freezing and water can no longer reach the surface. The duration of massive flood events is short and has been estimated to last a few days to several weeks depending on the discharge rates [see 7]. If the aquifer was recharged by groundwater flow from distant sources, the flood could reoccur, resulting in episodic events of catastrophic groundwater release.

Fig. 6.6 Hundreds of *layers* of the same thickness characterize the interior of an old impact crater in Arabia Terra. The repeated changes between single layers indicate repeated changes in the depositional environments. A plausible process which may have formed these deposits is sedimentation in a standing body of water. Image width is about 3 km (NASA/Malin Space Science Systems/RPIF/DLR).

Another water release process may be the melting of ground ice by volcanic heat that provides local and regional accumulation and migration of groundwater [81]. The outflow channels in proximity to the Elysium volcanoes and Hadriaca Patera near the Hellas basin are possible examples for this mechanism. A higher thermal gradient will weaken the cryosphere, providing easy access of the surface. Groundwater will also move preferentially along faults as indicated by the emergence of channels from fractures in Elysium and in the Memnonia Fossae region where Mangala Vallis originates at about 16 °S, 149 °W from a graben and continues northwards for several hundred of kilometers into the Amazonis basin.

Melting by volcanic heat and fracture controlled groundwater movement might also have contributed to the groundwater accumulation around Tharsis, particularly in the Valles Marineris region. The canyon floors are 4 to 8 km below the surroundings. Even at a present thickness of the cryosphere of about 2.7 km at the equator [127] water could have leaked into the canyons. New data from MOC give evidence that this mechanism works also in smaller valleys [108]. Given a global aquifer system, it is expected that water will pool in the canyons. As already mentioned above, layered sediments in several canyons such as Echus Chasma and Ganges Chasma, give evidence for bodies of standing water within large depressions [113, 128, 129]. Malin and Edgett [113] proposed two climatic scenarios for the formation of the layered deposits. The first one is based on the theory of a warmer early Mars and an environment capable of sustaining liquid water on its surface. The second scenario proposes the modulation of atmospheric pressure by astronomical perturbations, combined with catastrophic modulation of sediment sources giving rise to conditions recorded by the layered sedimentary deposits.

Catastrophic release of the lakes will also form outflow channels [128]. Kasei Vallis is suggested to be formed by catastrophic drainage of a former lake in Echus Chasma [125]; a large channel in Amazonis Planitia might have drained a lake in southern Elysium Planitia [115] and Anderson [130] and Chapman [94] suggest that Granicus Vallis, west of Elysium Mons, developed under thick ice deposits as a result of volcanic activity. From Ganges Chasma a broad depression trends south to the chaotic source region of Shalbatana Vallis indicating subsurface drainage of a possible former lake [38].

The large outflow channels emptied into the northern lowlands and their floods must have left extended deposits and probably large lakes. The water released by the outflow channels to the northern plains is estimated to amount at least 6×10^6 km^3 but probably more [131]. Baker et al. [81] calculated that it needs 6×10^7 km^3 of water to fill up the northern plains to the 0 km-contour. The northern plains cover an older rougher underground that survived as hills and knobs commonly outlining old pre-plain impact craters. The plains are complex deposits probably formed by many processes such as sedimentation from outflow channels, volcanism and mass wasting from the adjacent highlands modified by impact craters. Many of the surface features seem to be formed by the action of ground ice.

Outflow channels, their source regions and termination areas in the northern lowlands provide the best evidence for surface water on Mars and a widespread groundwater system. The outflow channels could form by running water under current climatic conditions, because the large amount of released water will prevent freezing and sublimation at least during the relatively short periods of flooding [31, 38].

Valley networks in the uplands are much smaller than outflow channels indicating smaller corresponding discharges. Therefore, it must be assumed that smaller branches of the networks will freeze very rapidly and cut off downstream motion resulting in interrupting the further development of the valleys [10]. However, the pattern of dissection produced by valley networks, the drainage development and the general valley morphologies have been judged to be similar to those of terrestrial rivers and a similar origin has been implied. This led to the assumption that the valley networks mainly have formed by running water, and most of the discussion of their origin focused on the relative roles of surface runoff and sapping [4-6, 23, 132-134]. Consequently, these assumptions had implications on the evolution of the Martian climate [81, 135-137] (for summary see [7, 10]). The valley networks are almost entirely restricted to the old uplands and the simplest explanation is that the valleys are old themselves and the climatic requirements for valley formation were met early in the planet's history and rapidly declined during the subsequent evolution. A warmer, wet Mars with a dense atmosphere at the time after the heavy bombardment is supposed to provide the conditions for valley formation by running water. Carr [38] argued that due to the low drainage densities of valley networks, which are orders of magnitudes lower than on Earth, and the short length of the valleys combined with the probably long time they needed to form, the processes, which cause the valleys are extremely insufficient compared with terrestrial fluvial processes. In addition, the total absence of meter-scale flow-features and dissections in the valleys support a subsurface rather than an atmospheric source for the valley formation [138]. Based on the evaluation of high resolution MOC images, Malin and Carr [138] concluded that the valleys have been formed by fluid erosion, but that in most cases the source has been groundwater. Relatively young small-scale alcove-like gullies combined with small channels and aprons in the walls of impact craters (Fig. 6.7), south polar pits and two of the larger valleys indicate even recent groundwater seepage and probably short-term surface runoff under almost current climatic conditions [108]). These evidences do not necessarily contradict the hypothesis of an early warm Mars, but constrain a hydrologic cycle on Mars mostly to subsurface processes. On the other hand, erosion rates declined at the end of the Noachian [38, 139, 140) and climate change at this time is one plausible explanation. Furthermore, channels on volcanoes such as on Alba Patera are interpreted to be formed by groundwater sapping in poorly consolidated pyro-clastic deposits such as ash as the result of hydro-thermal activities [41]. Gulick and Baker [42] suggested that a combination of several genetic processes like water erosion, volcanic density flows and lava might have been important in the formation of those channels. Carr [38] pointed out that mass wasting in poorly consolidated ash with only little involvement of water could have formed the channels on volcanoes. Nevertheless, the channels on volcanoes are fairly young and demonstrate that the involvement of water in Martian erosion appears still lately in the planets' history.

Although it is agreed that water has been involved during the formation of the valley networks, the process itself is only poorly understood. Carr [38] stated that the flat-floored fretted channels contain abundant evidence for mass wasting based on subsurface erosion, seepage of groundwater and creep flow of water-lubricated and subsequently ice-lubricated debris movement downstream. Because of the resemblance of many of the valley networks to fretted channels, and the lack of evidence for fluvial activities other than plani-metric shape, Carr [38] suggested that the valleys formed like the fretted channels, mainly by water lubricated mass wasting.

The high resolution MOC images so far demonstrate the need of recognizing small-scale features in order to fully understand the processes, which caused fluid-like Martian erosion. However, until high-resolution stereo data with considerable areal coverage, as expected from future missions such as Mars Express, are available, the problem of valley and channel formation and the implications on the climate remains debatable.

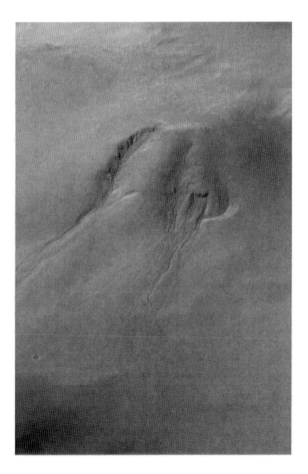

Fig. 6.7 *Gullies* in the inner wall of an impact crater. At the base of the gullies, debris aprons are clearly visible. These erosional landforms probably formed by a combination of groundwater seepage, surface runoff, and debris flow. They are geologically very young, as indicated by the lack of small impact craters on the channels and aprons. Image width is about 3 km, illumination is from upper left (NASA/Malin Space Science Systems/RPIF/ DLR).

6.9   References

1   D.U. Wise, M.P. Golombek, G.E. McGill, J. Geophys. Res. **84**, 7934 (1979).
2   G.F. Davies, R.E. Arvidson, Icarus **45**, 339 (1981).
3   D.E. Wilhelms, S.W. Squyres, Nature **309**, 138 (1984).
4   R.P. Sharp, M.C. Malin, Geol. Soc. Amer. Bull. **86**, 593 (1975).
5   V.R. Baker (Ed.) *The channels of Mars,* Univ. of Texas Press, Austin, 1982.
6   M.H. Carr (Ed.) *The surface of Mars,* Yale Univ. Press, New Haven, 1981.
7   M.H. Carr (Ed.) *Water on Mars,* Oxford University Press, New York, 1996.
8   B.K. Lucchitta, D.M. Anderson, in: *Reports of Planetary Geology Program – 1980,* NASA TM-81776, 1980, pp. 271.
9   Mars Channel Working Group, Geol. Soc. Amer. Bull. **94**, 1035 (1983).
10   V.R. Baker, M.H. Carr, V.C. Gulick, C.R. Williams, M.S. Marley, in: H.H. Kieffer, B.M. Jakosky, C.W. Snyder, M.S. Matthews, (Eds.) *Mars,* University of Arizona Press, Tucson, 1992, pp. 493.
11   M.H. Carr, L.S. Crumpler, J.A.Cutts, R. Greeley, J.E. Guest, H. Masursky, J. Geophys. Res. **82**, 28, 4055 (1977).
12   P.J. Mouginis-Mark, J. Geophys. Res. **84**, B14, 8011 (1979).
13   P.J. Mouginis-Mark, Icarus **45**, 60 (1981).
14   P.J. Mouginis-Mark, Icarus **71**, 268 (1987).
15   L.A. Johansen, in: *Reports of Planetary Geological Program, 1978–1979,* NASA TM-80339, 1979, pp. 123
16   R. Battistini, Rev. Geom. Dyn. **33**, 25 (1984).
17   K.H. Wohletz, M.F. Sheridan, Icarus **56**, 15 (1983).
18   J.S. Kargel, Proc. Lunar Planet. Sci. Conf. XVII, 1986, pp. 410.
19   F.M. Costard, Earth, Moon, and Planets **45**, 265 (1989).
20   K.L. Tanaka, D.H. Scott, R. Greeley, in: H.H. Kieffer, B.M. Jakosky, C.W. Snyder, M.S. Matthews (Eds.) *Mars,* University of Arizona Press, Tucson, 1992, pp. 354.
21   H. Masursky, J.V. Boyce, A.L. Dial Jr., G.G. Schaber, M.E. Strobell, J. Geophys. Res. **82**, 4016 (1977).
22   V.R. Baker, R.C. Kochel, Proc. Lunar Planet. Sci. Conf. IX, 1979, pp. 3181.
23   D.C. Pieri, Science **210**, 895 (1980).
24   M.H. Carr, Icarus **22**, 1 (1974).
25   E. Schonfeld, Eos Trans. AGU **57**, 948 (1976).
26   J.A. Cutts, K.R. Blasius, J. Geophys. Res. **86**, 5075 (1981).
27   D. Nummedal, in: *Reports of Planetary Geology and Geophysics Program, 1977–1988,* NASA TM–79729, 1978, pp. 257.
28   D. Nummedal, D.B. Prior, Icarus **45**, 77 (1981).
29   Y.L. Yung, J.P. Pinto, Nature **288**, 735 (1978).
30   V.R. Baker, J. Geophys. Res. **84**, 7985 (1979).
31   M.H. Carr, J. Geophys. Res. **84**, B6, 2995 (1979).
32   M.H. Carr, Icarus **56**, 187 (1986).
33   B.K Lucchitta, J. Geophys. Res. **87**, 9951 (1982).
34   P.D. Komar, Geology **11**, 651 (1983).
35   P.D. Komar, J. Geol. **92**, 133 (1984).
36   R.P. Sharp, J. Geophys. Res. **78**, 4063 (1973).
37   M.H. Carr, G.D. Clow, Icarus **48**, 91 (1981).
38   M.H. Carr, J. Geophys. Res. **100**, 7479 (1995).
39   C.E. Reimers, P.D. Komar, Icarus **39**, 88 (1979).

40 L. Wilson, P.J. Mouginis-Mark, Nature **330**, 354 (1987).
41 P.J. Mouginis-Mark, L. Wilson, R.J. Zimbelman, Bull. **50**, 361 (1988).
42 V.C. Gulick, V.R. Baker, J. Geophys. Res. **95**, 14325 (1990).
43 R.P. Sharp, J. Geophys. Res. **78**, 4073 (1973).
44 S.W. Squyres, Icarus **34**, 600 (1978).
45 B.K. Lucchitta, Proc. Lunar Sci. Conf. XIV, J. Geophys. Res. **89**, Suppl., B409 (1984).
46 M.H. Carr, G.G. Schaber, J. Geophys. Res. **82**, 28, 4039 (1977).
47 D.G. Jankowski, S.W. Squyres, Icarus **106**, 365 (1993).
48 S.W. Squyres, J. Geophys. Res. **84**, B14, 8087 (1979).
49 S.W. Squyres, M.H. Carr, Science **231**, 249 (1986).
50 S.W. Squyres, Icarus **79**, 229 (1989).
51 A. Colaprete, B.M. Jakosky, J. Geophys. Res. **103**, E3, 5897 (1998).
52 D.I. Benn, D.J.A. Evans (Eds.) *Glaciers and glaciation.* Arnold, London, 1998, 734 pp.
53 S.M. Clifford, in: *Reports of Planetary Geology and Geophysics Program, 1990,* NASA TM–82385, Washington, D.C., 1980, pp. 405.
54 F.P. Fanale, J.R. Salvail, A.P. Zent, S.E. Postawko, Icarus **67**, 1 (1986).
55 M.H. Carr, in: D.J. Roddy, R.O. Pepin, R.B. Merrill, R.B. (Eds.) *Impact and Explosion Cratering,* Pergamon Press, New York, 1977, pp. 593.
56 R.O. Kuzmin, N.N. Bobina, E.V. Zabalueva, V.P. Shashkina, Solar System Res. **22**, 3, 121 (1988).
57 K.R. Blasius, J.A. Cutts, in: *Reports of Planetary Geology and Geophysics Program, 1980,* NASA TM–82385, Washington D.C., 1980, pp. 147.
58 F.M. Costard, Proc. Lunar Planet. Sci. Conf. XIX, Lunar and Planetary Institute, Houston, 1988, pp. 211.
59 S.W. Squyres, S.M. Clifford, R.O. Kuzmin, J.R. Zimbelman, F.M. Costard, in: H.H. Kieffer, B.M. Jakosky, C.W. Snyder, M.S. Matthews, (Eds.) *Mars,* University of Arizona Press, Tucson, 1992, pp. 523.
60 F.M. Costard, Proc. Lunar Planet. Sci. Conf. XXV, Lunar and Planetary Institute, Houston, 1994, pp. 287.
61 F.M. Costard, J.S. Kargel, 5th International Conference on Mars, LPI Contribution No. 972, Lunar and Planetary Institute, Houston (CD-ROM), 1999, Abstract #6088.
62 J.C. Pechmann, Icarus **42**, 185 (1980).
63 B.K. Lucchitta, H.M. Ferguson, C. Summers, J. Geophys. Res. **91**, B13 (1986), Proc. Lunar Sci. Conf. XVII, E166–E174.
64 B.K. Lucchitta, H.M. Ferguson, C. Summers, Proc. Lunar Planet. Sci. Conf. XVII, Lunar and Planetary Institute, Houston, 1986, pp. 498.
65 K.R. Blasius, J.A. Cutts, J.E. Guest, H. Masursky, J. Geophys. Res. **82**, 28, 4067 (1977).
66 G.E. McGill, Proc. Lunar Planet. Sci. Conf. XVI, Lunar and Planetary Institute, Houston, 1985, pp. 534
67 G.E. McGill, Geophys. Res. Lett. **13**, 705 (1986).
68 B.K. Lucchitta, in: *Permafrost:* Proc. 4th Int. Conf., Nat. Acad. Press, 1983, pp. 744.
69 H. Hiesinger, J.W. Head, J. Geophys. Res. **105**, E5, 11999 (1999).
70 G. McGill, L.S. Hills, J. Geophys. Res. **97**, E2, 2633 (1992).
71 T.A. Mutch, R.E. Arvidson, E.A. Binder, E.C. Guinness, E.C. Morris, J. Geophys. Res. **82**, 28, 4452 (1977).

72   N. Evans, L.A. Rossbacher, in: *Reports of Planetary Geology Program, 1980*, NASA TM–82385, NASA, Washington, D.C., 1980, pp. 376.

73   G.A. Brook, in: *Reports of Planetary Geology Program, 1982*, NASA TM–85127, Washington, D.C., 1982, pp. 265.

74   M.T. Mellon, Proc. Lunar Planet. Sci. Conf. XXVIII, Lunar and Planetary Institute, Houston, 1997, pp. 933.

75   D.L. Anderson, L.W. Gatto, F. Ugolini, in: F.G. Sanger (Ed.), *Permafrost:* 2nd Int. Conf., National Academy of Sciences, Washington, D.C., 1973, pp. 449.

76   F.M. Costard, J.S. Kargel, Icarus **114**, 93 (1995).

77   F. Costard, J. Aguirre-Puente, R. Greeley, N. Makhloufi, J. Geophys. Res. **104**, E6, 14091 (1999).

78   C.C. Allen, J. Geophys. Res. **84**, B14, 8048 (1979).

79   K.L. Tanaka, D.H. Scott, *Geologic map of the polar region of Mars, scale 1:15,000,000.* U.S.G.S. Misc. Inv. Series Map I–1802–C, 1987.

80   D.H. Scott, J.R. Underwood, in: G. Ryder, V.L. Sharpton (Eds.) Proc. LunarPlanet. Sci. Conf. XXI, Lunar and Planetary Institute, Houston, 1991, pp. 627.

81   V.R. Baker, R.G. Strom, V.C. Gulick, J.S. Kargel, G. Komatsu, V.S. Kale, Nature **352**, 589 (1991).

82   J.S. Kargel, R.G. Strom, Proc. Lunar Planet. Sci. Conf. XXI, Lunar and Planetary Institute, Houston, 1990, pp. 597.

83   J.S. Kargel, R.G. Strom, Geology **20**, 3 (1992).

84   J.S. Kargel, V.R. Baker, J.E. Begét, J.F. Lockwood, T.L. Péwé, J.S. Shaw, R.G. Strom, J. Geophys. Res. **100**, E3, 5351 (1995).

85   M.H. Carr et al. (Eds.) *Viking orbiter views of Mars.* U.S. Government Printing Office, Washington, D.C., 1980, 136 pp.

86   A.D. Howard, in: *Reports of Planetary Geology Program, 1979–1980*, NASA TM–84211, 1981, pp. 286.

87   J.S. Kargel, R.G. Strom, Proc. Lunar Planet. Sci. Conf. XXIII, Lunar and Planetary Institute, Houston, 1991, pp. 683.

88   S.M. Metzger, Proc. Lunar Planet. Sci. Conf. XXIII, Lunar and Planetary Institute, Houston, 1992, pp. 901.

89   H. Frey, B.L. Lowry, S.A. Chase, J. Geophys. Res. **84**, B14, 8075 (1979).

90   C.A. Hodges, H.J. Moore, J. Geophys. Res. **84**, B14, 8061 (1979).

91   P.J. Mouginis-Mark, L. Wilson, J.W. Head, J. Geophys. Res. **87**, B12, 9890 (1982)

92   P.J. Mouginis-Mark, L. Wilson, J.W. Head, S.H. Brown, J.L. Hall, K.D. Sullivan, Earth, Moon, and Planets **30**, 149 (1984).

93   P.J. Mouginis-Mark, Icarus **64**, 265 (1985).

94   M.G. Chapman, Icarus **109**, 393 (1994).

95   S.W. Squyres, D.E. Wilhelms, A.C. Moosman, Icarus **70**, 385 (1987).

96   G. Neukum, K. Hiller, J. Geophys. Res. **86**, 3097 (1981).

97   W.K. Hartmann, Geophys. Res. Lett. **5**, 450 (1978).

98   G. Neukum, D.U. Wise, Science **194**, 1381 (1976).

99   W.K. Hartmann, R.G. Strom, S.J. Weidenschilling, K.R. Blasius, A. Woronow, M.R. Dence, R.A.F. Grieve, J. Diaz, C.R. Chapman, E.N. Shoemaker, K.L. Jones, in: *Basaltic Volcanism on the Terrestrial Planets,* Pergamon, New York, 1981, pp. 1049.

100 C.A. Robinson, G. Neukum, H. Hoffmann, A. Marchenko, A.T. Basilevsky, G.GL. Ori, Proc. Lunar Planet. Sci. Conf. XXVII, 1996, pp. 1083.

101 A.G. Marchenko, A.T. Basilevsky, H. Hoffmann, E. Hauber, A.C. Cook, G. Neukum, Solar System Res. **32**, 425 (1998).

102 M.C. Malin, J. Geophys. Res. **81**, 4825 (1976).

103 D.M. Nelson, R. Greeley, J. Geophys. Res. **104**, 8653 (1999).

104 D.H. Scott, K.L. Tanaka, *Geologic map of the western equatorial region of Mars, scale 1:15,000,000*. U.S.G.S. Misc. Inv. Series Map I–1802–A, 1986.

105 R. Greeley, J.E. Guest, *Geologic map of the eastern equatorial region of Mars, scale 1:15,000,000*. U.S.G.S. Misc. Inv. Series Map I–1802–B, 1987.

106 D.H. Scott, J.M. Dohm, Proc. Lunar Planet. Sci. Conf. XXIII, 1992, pp. 1251.

107 W.K. Hartmann, D.C. Berman, J. Geophys. Res. **82**, 4067 (2000).

108 M.C. Malin, K.S. Edgett, Science **288**, 2330 (2000).

109 T.J. Parker, R.S. Saunders, D.M. Schneeberger, Icarus **82**, 111 (1989).

110 T.J. Parker, D.S. Gorcine, R.S. Saunders, D.C. Pieri, D.M. Schneeberger, J. Geophys. Res. **98**, 11061 (1993).

111 J.W. Head, H. Hiesinger, M.A. Ivanov, M.A. Kreslavsky, S. Pratt, B.J. Thomson, Science **286**, 2134 (1999).

112 M.C. Malin, K.S. Edgett, Geophys. Res. Lett. **26**, 3049 (1999).

113 M.C. Malin, K.S. Edgett, Science **290,** 1927 (2000).

114 D.H. Scott, M.G. Chapman, Proc. Lunar Planet. Sci. Conf. XXII, 1991, pp. 669.

115 D.H. Scott, M.G. Chapman, J.W. Rice, J.M. Dohm, Proc. Lunar. Planet. Sci. Conf. XXII, 1992, pp. 53.

116 R.A. De Hon, Earth Moon Planets **56**, 95 (1992).

117 H.E. Newsom, G.E. Britelle, C.A. Hibbitts, L.J. Crossey, A.M. Kudo, Icarus **44**, 207 (1996).

118 N.A. Cabrol, E.A. Grin, R. Landheim, R.O. Kuzmin, R. Greeley, Icarus **133**, 98 (1998).

119 N.A. Cabrol, E.A. Grin, H.E. Newsom, R. Landheim, C.P. McKay, Icarus **139**, 235 (1999).

120 N.A. Cabrol, E.A. Grin, Icarus **142**, 160 (1999).

121 G.G. Ori, L. Marinangeli, A. Baliva, J. Geophys. Res. **105**, 17,629 (2000).

122 D.A. Crown, K.H. Price, R. Greeley, Icarus **100**, 1 (1992).

123 K.L. Tanaka, G.J. Leonard, J. Geophys. Res. **100**, 5407 (1995).

124 P.D. Komar, Icarus **37**, 156 (1979).

125 M.S. Robinson, K.L. Tanaka, Geology **18**, 902 (1990).

126 V.R. Baker, Geological Soc. of America, Boulder, SP–144, 1973.

127 S.M. Clifford, J. Geophys. Res. **98**, E6, 10973 (1993).

128 J.F. McCauley, *Geologic map of the coprates quadrangle of Mars, scale 1:5,000,000*. U.S. Geol. Surv. Misc. Inv. Series Map I–897, 1978.

129 S.S. Nedell, S.W. Squyres, D.W. Andersen, Icarus **70**, 409 (1987).

130 W.M. Anderson, LPI Tech. Rep. **92**–08. 1 (1992).

131 M.H. Carr, S.C. Wu, R. Jordan, F.J. Schafer, Proc. Lunar Planet. Sci. Conf. XVIII, 1987, pp. 155.

132 D.C. Pieri, Icarus **27**, 25 (1976).

133 J.E. Laity, M.C. Malin, Geol. Soc. Amer. Bull. **96**, 203 (1985).

134 V.R. Baker, Spec. Pap. Geol. Soc. Am. **252**, 235 (1990).

135 M.H. Carr, Icarus **56**, 476 (1983).

136 S.W. Squyres, Proc. Lunar Planet. Sci. Conf. XX, 1989, pp. 1044.

137 S.W. Squyres, J.F. Kasting, Science **265**, 744 (1994).

138 M.C. Malin, M.H. Carr, Nature **397**, 589 (1999).

139 M.H. Carr, Proc. Lunar Planet. Sci. Conf. XXIII, 1992, pp. 205.

140 R.A. Craddock, T.A. Maxwell, J. Geophys. Res. **98**, 3453 (1993).

7 Europa's Crust and Ocean: How Tides Create a Potentially Habitable Physical Setting

Richard Greenberg, B. Randall Tufts, Paul Geissler and Gregory V. Hoppa

As understanding of the physical setting on Jupiter's moon Europa has advanced, the likelihood that it may provide a potentially habitable environment for native or imported organisms has increased. The surface was discovered to be water ice in 1972 [1]. That tidal heating might maintain a sub-surface layer of water in the liquid state was recognized in 1979 [2], stimulating both popular and scientific speculation of the possibility that the satellite might support life. Most recently, interpretation of images and other data obtained by the Voyager and Galileo spacecraft has indicated that indeed there is an ocean underlying the crust of ice at the surface, and has suggested that the ocean is linked to the surface through a variety of processes that may provide habitable niches.

A full-disk view of Europa (Fig. 7.1) provides hints of some of the most important characteristics of the satellite. The visible surface is ice, but orange-brown markings (exaggerated here) due to traces of other substances reveal global-scale lineaments and splotches of a wide range of sizes and shapes. Even at this low image resolution, the lineaments and splotches, respectively, provide indications of the two major categories of surface terrains, tectonic (resulting from crustal cracking) and chaotic (likely the product of crustal melting), which provide the basis for inferring the habitability of Europa. Although one major impact feature is clearly visible in this global view, and recognizable by its rays of ejecta, the general paucity of impact craters indicates that this surface is very young. Estimates of impact rates and the paucity of craters show that the surface has been entirely renewed by tectonics and chaos within the past 100 million years [3]. Resurfacing has been so active that almost nothing now on the surface of Europa was there at the time dinosaurs roamed the Earth.

As discussed in more detail below, Europa's surface appears to have formed by two major classes of resurfacing processes: tectonics and chaos formation, each erasing what was there before. Where tectonics have been most recently dominant, fragments of earlier terrain (either chaotic or tectonic) are visible between the more recent lineaments. Where chaos formation occurs, older terrain (either chaotic or tectonic) can often be seen on the surfaces of rafts within the matrix of modified surface.

It is likely that both major types of resurfacing processes (tectonic and chaotic) involve interaction of a liquid water ocean with the surface. This chapter describes how the character of tectonic terrain indicates that the ice is thin and that oceanic water contributes to the surface morphology. This chapter also discusses how the appearance of chaotic terrain, combined with the indication from tectonics that the ice is thin, suggests that the crust of ice has been melted through from below. The few existing craters have been used to estimate the thickness of the ice crust to be only a few kilometers thick [4], although it is likely quite variable and non-uniform. The key types of features, and the processes that create them, all involve linkages between the ocean and the surface.

This emerging picture contrasts with the thick-ice model, in which ridges, bands, and chaotic terrain are attributed to solid state convection or diapirism within a ductile zone of the ice crust [5, 6]. In such a model, the ocean, if any, would be isolated from the surface by an ice crust >10 km thick. That thick-ice model came from initial geological interpretation of Galileo images, based on qualitative appearance of selected features and analogies with terrestrial geology. It also depended on thermal models developed shortly after the Voyager encounter, which had conservatively demonstrated the possibility of a liquid ocean. Parameter estimates that would have given very thin ice were avoided, as such choices might have seemed too radical at that time, when the idea of significant tidal heating of the Galilean satellites was still new and evidence for thin ice was far in the future.

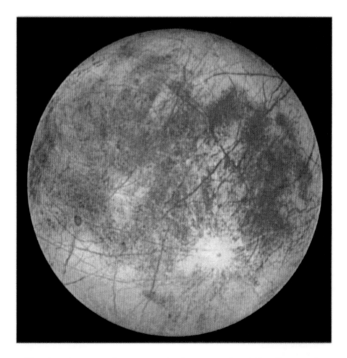

Fig. 7.1 The trailing hemisphere of Europa. The X just to the right of center is near longitude 270°, just north of the true center of the trailing hemisphere. (The direction of Jupiter defines 0° longitude.) Just south of the X is a dark splotch indicating Conamara Chaos (see Fig. 7.3). Over 1000 km further south is the impact crater Pwyll, a white splotch with an extensive ray system.

The thick-ice model led to a relatively pessimistic view of the possibility of life on Europa. Because the icy crust itself would have been inhospitable, attention focused on the ocean. However, with the ocean isolated from the surface, consideration of energy requirements turned to endogenic sources, and to the possibility that sub-oceanic vents might support life on Europa. This model was inspired by the analogy of the hydrothermal vents at mid-ocean ridges on Earth that harbor fairly isolated eco-systems, for which the metabolic energy comes from chemolithotrophic reactions at high-temperature black-smoker chimneys, as well as at relatively lower temperature vents. No direct observational evidence exists for such sub-oceanic volcanism on Europa, but a plausible case can be made for tidal heating rates great enough to allow such volcanism [7]. However, such heating rates may be so great that they would be inconsistent with the thick-ice model.

Even if there were localized ocean-floor volcanism below a thick cap of ice, habi-tability would face another difficulty. Because the ocean would be isolated from the surface and from oxidants there, life would be impossible to sustain [8]. Life could flourish only if the ocean were linked to the surface; otherwise viable biomass would be limited.

The emerging view of an ice crust that allows the ocean to be linked to the surface is more promising from a biological perspective [9]. Europa's crust now appears to provide a variety of potentially habitable environmental niches. Moreover, it may also play a key role in making the ocean habitable as well, by overcoming the isolation from sources of oxidants. Individual niches in the crust such as specific tidally active cracks, probably are stable environments for thousands of years, providing reasonably secure settings for organisms and ecosystems to prosper. Over longer time-scales, such individual niches come and go, so they are not too secure: Organisms would need to adapt and spread, and would have opportunity and reasons to evolve. Europa's crust may be one of the most likely and accessible places for us to find extra-terrestrial life.

This result is based on quantitative investigations of tidal-tectonic processes and quantitative analyses of observations, especially correcting for observational selection effects that may have skewed initial impressions of the surface character of the satel-lite. In what follows, the various lines of evidence are outlined, showing how they have led to the emerging picture of Europa's history and physical structure. While the evidence for this broad picture is compelling, the model is not proven and important issues remain unresolved. Further geophysical modeling and data analysis will be required before one can be confident that the processes and structure are truly as de-scribed here. With that caveat, we can discuss how such a physical setting, produced by tidal processes and inferred from observations, may include environmental niches in the crust that meet the requirements for survival, spread, and evolution of life.

7.1   Tidally Driven Geology and Geophysics

The Galilean satellites Io, Europa, and Ganymede have orbital periods in a ratio of 1:2:4, resulting in periodic alignments that enhance their mutual gravitational pertur-bations and force significant orbital eccentricities (reviewed by [10, 11]). The eccen-tricities, in turn, cause regular variations in the height and orientation of tides raised on the satellites by Jupiter. The time-scale for these variations is the orbital period, ~3.5 days for Europa. These tidal variations drive the heat, rotation, and stress that are key to habitable environments on Europa.

7.1.1   Tidal Heating

Tidal working due to orbital eccentricity generates heat due to frictional dissipation. For Europa, tides may readily generate enough heat to maintain a liquid water ocean within the layer of H_2O that constitutes the outer 100 km (or more) of Europa [e.g., 2, 12]. Geophysical modeling of the style of tidal dissipation remains inconclusive. Some models include substantial heating in the rocky mantle [e.g., 7], perhaps generating volcanism at the base of the water layer, similar to sub-oceanic volcanism on Earth. Dissipation may occur in the liquid ocean, depending on its thickness profile, as it flows in response to the tide. It has also been suggested that much of the heating may originate in the frozen crust due to viscous working as it stretches over the tidal bulge [e.g., 13] or friction along cracks in the ice [14].

7.1.2   Tidally Driven Rotation

The response of Europa to Jupiter's tide-raising gravitational potential also causes a rotational torque. For a satellite in a circular orbit and/or with a substantial permanent mass asymmetry (such as the Moon's), rotation would be expected to be locked into a synchronous state, with one hemisphere always facing the planet. The length of a day on the satellite would be the same as the orbital period. If the rotation was changed in some way, tidal effects would quickly return it to the synchronous state. However, Greenberg and Weidenschilling [15] pointed out that, because Europa is not in a circular orbit and does not necessarily have a permanent mass asymmetry (e.g., it may have been too warm to support one), the tidal torque can drive it to a rotation rate slightly faster than synchronous.

To test this possibility, Hoppa et al. [16] measured the orientation of Europa as observed by Voyager 2 in 1979 and by Galileo 17 years later. No deviation from synchroneity was detected, thus placing a lower limit on the period of rotation relative to the direction of Jupiter of 12 000 years. The non-synchronous component of rotation is so slow that a day on Europa is indistinguishable from the 3.5 day orbital period over short periods of time, but the hemisphere facing Jupiter could turn away from the planet in as little as 6000 years.

Evidence that there has been substantial rotation even during the most recent small fraction of the age of Europa's surface comes from interpretation of three types of tectonic features that have been formed by tidal stress: (a) global or regional lineaments display crosscutting sequences that show systematic changes in orientation, consistent with migration of this real estate from west to east during rotation relative to the direction of Jupiter [17]; (b) strike-slip fault distributions are consistent with a theory of their formation by diurnal tidal working, but only if many formed to the west of their current locations [18]; and (c) the character and distribution of distinctive cycloidal crack features, also created by diurnal tidal variations, also suggests substantial rotation. In fact, on that basis Hoppa et al. [19] infer a period of rotation less than 250 000 years relative to the direction of Jupiter.

7.1.3   Tidal Stress on the Ice Crust

The global lineament patterns on Europa prove to correlate with a ridge morphology, and in fact, much of Europa is covered with ridge systems, the common denominator of which is the ubiquitous double ridge system (see Fig. 7.2). The global and regional ridge systems, and the time sequence in which they formed, appear consistent with formation that initiated with cracking of the crust in response to tidal tension [20-22]. The stresses involved are the diurnal variation due to the eccentric orbit, superimposed on the stress due to non-synchronous rotation, which builds up as the direction of Jupiter moves around relative to the surface of Europa over thousands of years.

These results have several important implications: First, they support the widely held assumption that global lineaments represent crack patterns. Second, they imply that a liquid layer tens of kilometers thick must lie beneath a thinner crust. Otherwise, the amplitude of the tide, and hence the tidal stress, would be inadequate for cracking ice. Third, they provide the evidence that Europa rotates non-synchronously, a source of local, time-variable environmental change on time-scales of 10^3-10^5 years.

The ubiquitous double ridges can also be understood in terms of diurnal tidal working [22]: As the diurnal tides open cracks, water flows to the float line, where it boils in the vacuum and freezes owing to the cold. As the walls of the crack close a

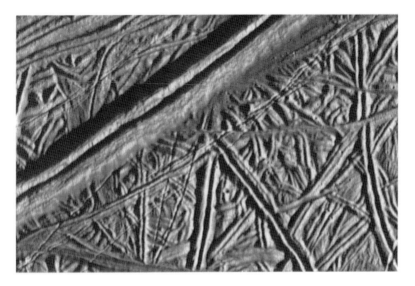

Fig. 7.2 Densely ridged terrain just north of Conamara at resolution 21 m/pixel, showing examples of densely packed and superposed double ridges. The broad double ridge is about 2 km across, and about 200 m high.

few hours later, a slurry of crushed ice, slush, and water is squeezed to the surface and deposited on both sides of the crack. Given the frequency of the process, ridges of typical size (100 m high and 1 km wide) can grow quickly, in as little as 20 000 years, consistent with the time that a crack is likely to remain active before rotating to a different stress regime.

Dilational lineaments where cracks appear to have opened, accompanied by corresponding displacement of adjacent terrain, are common on Europa. Many are the broad pull-apart features called "bands" [23-26]. Tufts et al. [27] have demonstrated a continuum of morphologies ranging from flat bands to dilational ridges, depending on the extent to which dilation was continuous or accompanied by varying degrees of diurnal tidal reclosing. Reconstructions of numerous dilational features demonstrate plate motion, requiring decoupling (e.g., by a fluid layer) from deeper structure, with continual infilling (to the surface) by fluid material from below.

Strike-slip (i.e. lateral shear) displacement provides important evidence regarding the structure of the crust. The most dramatic example is the 800 km long strike-slip fault along Astypalaea Linea [28], which displays shear offset ~40 km. Diurnal variation of tidal stress in the Astypalaea region drives the shear displacement by a mechanism analogous to walking [18]. The stress goes through a daily sequence starting with tensile stress across the fault, followed 21 hours later by right-lateral shear, followed 21 hours later by compression across the fault, followed 21 hours later by left-lateral shear. Because the left-lateral shear stress occurs after the fault is compressed, friction at the crack may resist displacement, while right-lateral stress occurs immediately after the crack is opened by tension. Thus, tides drive shear in the right-lateral sense in a ratcheting process. The mechanism is similar to walking, where one repeatedly separates a foot from the ground, moves it forward, compresses it to the ground, and pushes it backward, resulting in forward motion. On Europa this process moves crustal plates.

Hoppa et al. [18] showed widespread examples of strike-slip displacement to be consistent with this model. The success of the model in explaining the observed faulting argues strongly for a fluid decoupling layer under the ice: Such a layer is required for the daily steps in the "tidal walking" process, because it can deform on the time-scale of the diurnal tides. Moreover, this line of argument provides the best evidence that cracks must penetrate all the way down to that fluid layer.

Cracks in the form of cycloids (or chains of arcs) are ubiquitous on Europa and evidenced by various types of evolved lineaments. Hoppa and Tufts [29] and Hoppa et al. [30] have explained cycloidal cracking in terms of the tidal stress. Once a crack begins, it propagates across the surface perpendicular to the local tidal tension. As the day wears on, the orientation of the diurnal stress varies, so the crack propagates in an arc until the stress falls below a critical value necessary to continue cracking. A few hours later, the tension returns to a high enough value, now in a different direction, so crack propagation resumes at an angle to the direction at which it had stopped. Thus, a series of arcs is created (each corresponding to one day's worth of propagation) with cusps between them. By adjusting parameters (crack propagation rate, strength of ice, etc.), virtually all the characteristics of observed arcuate features can be reproduced.

In all cases a subsurface ocean is required in order to have adequate tidal amplitude to drive this process. Thus, the presence of cycloidal lineaments, and their correlation with theoretical characteristics, is perhaps the strongest currently available evidence for the existence of a liquid water ocean on Europa.

7.1.4   Chaotic Terrain

With evidence that the ice overlies liquid water, and that it is thin enough for cracks to penetrate (probably a few kilometers or less), next consider chaotic terrain which has the appearance of sites of melt-through: a textured lumpy matrix, often around rafts of displaced older surface (Fig. 7.3). Only modest concentration of tidal heat is needed to allow such melt-through [31]. Detailed characteristics of this broad class of features, as well as the general appearance are consistent with this interpretation of their formation [32].

Initial impressions of Galileo spacecraft images suggested that chaotic terrain is relatively recent and anomalous ([33] and references cited there). However, more complete mapping [34] has shown these impressions to be artifacts of the difficulty in recognizing older chaotic terrain that has been cut-up or covered by subsequent tectonic terrain. Just as chaos formation destroys pre-existing terrain (except for what remains on rafts), cracks and subsequent ridge formation can readily modify or overprint previously existing terrain, including pre-existing chaos areas. There are many examples of various degrees of such partial erasure of chaos areas [32, 34].

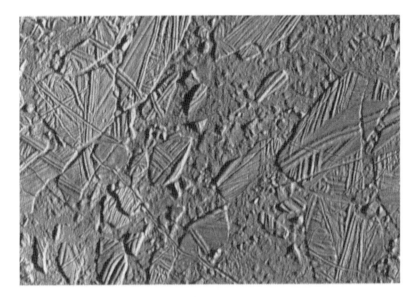

Fig. 7.3 A portion of Conamara Chaos (at 40 m/pixel). Here we see rafts of older tectonic terrain which have been slightly displaced in a lumpy frozen matrix, giving the appearance of a site of melting of the crust followed by refreezing. Two examples of cracks with double ridges have formed subsequently, providing examples of a widespread process by which tectonics have obliterated chaotic terrain to various degrees.

In fact, about 40% of Europa's surface is chaotic terrain. A competing interplay of comparable amounts of resurfacing by chaos formation and tectonics has been going on throughout the history of the observable surface. Each process, at different times and places, destroys the surface that was there before.

Another early impression, which has proven to be incorrect, was that the sizes of patches of chaotic terrain indicated a peak in the size distribution at about 10 km diameter. This appearance combined with terrestrial geological analogs was interpreted as representing the tops of cells of solid-state convection [6], supporting an argument that the ice is 10-20 km thick. However, the putative peak in the size distribution for these features is an artifact of observational selection. It has been shown to result from one's inability to recognize smaller patches of this terrain at the resolution of most images used in these studies [34, 35]. Therefore, the strongest evidence for thick ice (and for an isolated ocean) has been shown to be an artefact.

Thus, the two major terrain-forming processes on Europa are (1) melt-through, creating chaos, and (2) tectonic processes of cracking and subsequent ridge formation, dilation, and strike-slip. These two major terrain-forming processes have continually destroyed pre-existing terrain, depending on whether local or regional heat concentration was adequate for large-scale melt-through or small enough for refreezing and continuation of tectonism. The processes that create both chaotic and tectonic terrains all include transport or exposure of oceanic water through the crust to the surface.

7.2   A Habitable Setting

The emerging understanding of processes and conditions in Europa's crust supports earlier speculation that the satellite may be habitable by some forms of life. Whether organisms exist there or have existed there remains unknown, but the physical conditions and rates of change of those conditions seem consistent with a readily habitable environment. Let us summarize these conditions, and discuss their implications for life on the satellite. Of particular importance are the several time-scales for change that provide reasonable stability, while also providing need for adaptation and evolution.

The outer portion of Europa consists predominantly of water, at least tens of kilometers thick, and probably mostly liquid. The liquid is maintained by heat from tidal energy dissipation, with an outer crust of ice, most likely a few kilometers thick. At the base of the crust the temperature is 0 °C, while the surface, because it is exposed to space without a significant atmospheric blanket, is about 170 °C colder (i.e. about 100 K).

Cracks are formed in the crust due to tidal stress, resulting from the combination of non-synchronous rotation and diurnal variations. Many of the cracks penetrate from the surface down to the liquid water ocean. Even after a crack forms, the crust continues to be worked by the diurnal tides, so that the crack is continually opened and closed over the course of each day on Europa. Thus, on a time-scale of 10^{-2} years (a Europan day), water periodically flows up to the float line during the hours of opening, and is squeezed out during the hours of closing. Slush and crushed ice is forced to the surface, while most of the water flows back into the ocean. The daily tidal flow transports substances and heat between the surface and the ocean.

At the surface, oxidants are continually produced by disequilibrium processes such as photolysis by solar ultraviolet radiation, and especially radiolysis by charged particles. Significant reservoirs of oxygen have been spectrally detected on Europa's surface in the form of H_2O_2, H_2SO_4, and CO_2 [36-38]. In addition, molecular oxygen and ozone are inferred to be within the near-surface ice, on the basis of the existence of a tenuous (10^{-11} bar) oxygen atmosphere [39] and the detection of these substances on other icy satellites such as Ganymede [40, 41]. Moreover, impact of cometary material could also provide a source of organic materials and other fuels at the surface (see Chap. 1, Ehrenfreund and Menten) such as those detected on the other icy satellites [42]. In addition, significant quantities of sulfur and other materials may be continually ejected and transported from Io to Europa. Such substances are being given consideration as potentially important biogenic materials [e.g., 43].

In the ocean is a mix of endogenic and exogenic substances that may contain salt, sulfur compounds, organics [e.g., 44], etc. Evidence for such substances is found at the surface at sites where oceanic material most likely has been exposed, especially along major ridge systems and around chaotic terrain. In most color representations of Europa, this material appears as orange-brown. However, the filters used in constructing color images were shifted toward the infrared, so the colors represent trends with wavelength rather than true appearance. Images taken through these filters, at visible or near-IR wavelengths, are not diagnostic of composition. However, Galileo's Near-Infrared Mapping Spectrometer did provide potentially diagnostic spectra, which are variously (and with controversy) interpreted as hydrated, sulfur-based salts in frozen brines [45, 46], various granularities of water ice [47], or sulfuric acid and related compounds [37]. The orangish brown appearance at visible wavelengths may be consistent with organics or other unknowns. The ocean likely contains a wide range of biologically important substances.

Chemical disequilibrium among materials at various levels in a crack is maintained by surface production at the top and the oceanic reservoir at the bottom, while the tidal ebb and flow of water continually transports and mixes these substances vertically, all within the context of an ambient temperature gradient ~0.1 °/m.

How might this physical setting support and affect life (Fig. 7.4)? No organisms could survive very near the surface where bombardment by energetic charged particles in the Jovian magnetosphere would disrupt organic molecules [48] within ~1 cm of the surface, and organisms in a crack would probably need to stay much farther down to survive. Nevertheless, Reynolds et al. [49], Lunine and Lorenz [50], and Chyba [9] note that sunlight adequate for photosynthesis could penetrate a few meters, farther than necessary to protect organisms from radiation damage. Thus, as long as some part of the ecosystem of the crack occupies the appropriate depth, it may be able to exploit photosynthesis. Such organisms might benefit from anchoring themselves at an appropriate depth where they might photosynthesize. Other non-photosynthesizing organisms might anchor themselves at other depths, and exploit the passing daily flow. Their hold would be precarious, as the liquid water could melt their anchorage away. Alternatively, some might be plated over by newly frozen water, and frozen into the wall. The individuals that are not anchored, or that lose their anchorage, would go with the diurnal flow. Organisms adapted to holding onto the walls might try to reattach their anchors. Others might be adapted to exploiting movement along with the

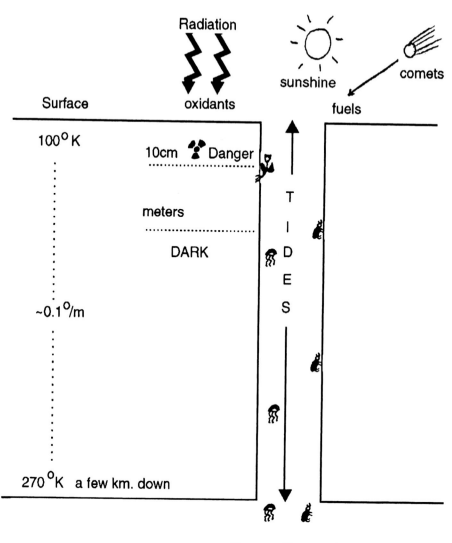

Fig. 7.4 Tidal flow through a working crack provides a potentially habitable setting, linking the surface (with its low temperature, radiation-produced oxidants, cometary organic fuels, and sunlight) with the ocean (with its brew of endo- and exogenic substances and relative warmth). Photosynthetic organisms (represented here by the tulip icon) might anchor themselves to exploit the zone between the surface radiation danger and the deeper darkness. Other organisms (the tick icon) might hold onto the side to exploit the flow of water and the disequilibrium chemistry. The hold would be difficult, with melting releasing some of these creatures into the flow, and with freezing plating others into the wall of the crack. Other organisms (jellyfish icon) might exploit the tides by riding with the flow. This setting would turn hostile after a few thousand years as Europa rotates relative to Jupiter, the local tidal stress changes, and the crack freezes shut. Organisms would need to have evolved strategies for survival by hibernating in the ice or moving elsewhere through the ocean.

mixing flow. A substantial fraction of that population would be squeezed into the ocean each day, and then flow up again with the next daily tide.

A given crack is likely active over thousands of years, because rotation is nearly synchronous and the crack remains in the same tidal-strain regime, allowing for a degree of stability for any ecosystem or organisms within it. However, our work has shown that over longer times Europa rotates non-synchronously, so that a given site moves to a substantially different tidal-strain regime in 10^3-10^5 years according to our evaluation of non-synchronous rotation. Thus, the tidal working of any particular crack is likely to cease by that age. The crack would seal closed, freezing any immobile organisms within it, while some portion of its organism population might be locked out of the crack in the ocean below.

For the population of a deactivated crack to survive, (a) it must have adequate mobility to find its way to a still active (generally more recently created) crack, or else or in addition, (b) the portion of the population that is frozen into the ice must be able to survive until subsequently released by a thaw. The first survival strategy, migration to an active crack, might be achieved in several ways, including flow with ocean currents, diffusion from crack to crack, and perhaps active propulsion.

The second strategy, survival in a frozen state until release, also has the potential for success. Our studies of "chaotic terrain" on Europa, which appears to have been created by near-surface melting, suggest that this phenomenon has been common and widespread throughout Europa's geological history. At any given location, a melt event probably has occurred every few million years, liberating frozen organisms to float free and perhaps find their way into a habitable niche. Also, in the time-scale for non-synchronous rotation (less than a million years), fresh cracks through the region would cross the paths of the older refrozen cracks, liberating organisms into a niche similar to where they had lived before. Survival in a frozen state for the requisite few million years seems plausible, given evidence for similar survival in Antarctic ice [51].

As described above, this model of the crust includes environmental settings that be suitable for life. Moreover, it provides a way for life to exist and prosper in the ocean as well, by providing access to oxidants and linkage between oceanic and crustal ecosystems, requirements emphasized by Gaidos et al. [9]. Any oceanic life would likely, in effect, be part of the same ecosystem as organisms in the crust. Components of the ecosystem might adapt to exploit sub-oceanic conditions such as possible sites of volcanism. However, if there is an inhabited biosphere on Europa, it most likely extends from within the ocean up to the surface. While one can only speculate on conditions within the ocean, we have observational evidence for conditions in the crust, and the evidence points toward a readily habitable setting.

This model is based on the surface manifestations of tectonic processing and chaotic terrain formation. The entire surface is very young, <2% of the age of the solar system according to the paucity of impact craters. Because the tectonic and melting processes described above were recent, they may well continue today. Moreover, extended periods of stability and even comfort within a niche are probably interrupted (as described above) by longer term changes in environmental conditions in the crust such as deactivation of individual cracks after thousands of years and later releasing of trapped organisms. Such change may provide drivers for adaptation and mobility, as well as opportunity for evolution.

Whether any organisms exist on Europa to exploit this setting is, of course, unknown and unlikely to be resolved for many years. However, a more immediate issue is whether data already in hand, or likely to become available soon, can test or refine this picture of a physical setting with the potential to support life. Several lines of future research are needed. We already have in hand extensive imagery from the Galileo spacecraft, which is amenable to geological analysis, particularly in further extraction of information regarding the time sequence of surface events. Geophysical modeling of the types reviewed here needs to be refined and extended. Studies of the interaction of geophysics and orbital evolution are also needed, because the history of Europa has been controlled by the linkage of these processes through tidal effects.

Continuing spacecraft exploration of Europa is likely in the future. The picture of Europa developed here has important implications for the exploration strategy. The current plan was conceived at a time when Europa's ocean was considered to be isolated from the surface. The Europa Orbiter mission would complete the 200-m resolution survey of the surface begun by Galileo. Subsequent missions would land on the surface, and eventually visit and sample the ocean. In that scenario, the main problem to be faced would be finding a way to penetrate through the ice; choice of the landing site would be less critical. However, that plan was predicated on a view that, if an ocean existed at all, it was isolated under thick ice.

With the increasing likelihood that the ice is thin and that the liquid ocean is linked to the surface in multiple ways, a very different strategy would be appropriate. If high-resolution reconnaissance by an orbiter could identify sites of active cracks or melt-through, a subsequent targeted lander could land where the action is (or was most recently), and sample fresh oceanic material. Even ridges, bands or chaotic terrain could contain frozen organisms. Thus, the priority and challenge becomes the pre-landing survey; the need to engineer a penetrator would be largely obviated. The nature of Europa may well be such that interior samples, possibly including organisms, may be readily accessible at or near the surface. The danger is that with a habitable biosphere linked to the surface, Europa's ecosystem may be vulnerable to contamination by our landers. Unless exploration is planned very carefully, we may discover life on Europa that we had inadvertently planted there ourselves.

Even if Europa proves to be sterile, the complex suite of geophysical processes and their unique relationships with geological and dynamical phenomena, make Europa one of the most active and exciting bodies in the solar system. On-going research and exploration will surely continue to surprise us.

Acknowledgements. Preparation of this chapter was supported by a grant from the U.S. National Science Foundation's Life in Extreme Environments program.

7.3   References

1   C.B. Pilcher, S.T. Ridgeway, T.B. McCord, Science **178**, 1087 (1972).
2   P. Cassen, R.T. Reynolds, S.J. Peale, Geophys. Res. Lett. **6**, 731 (1979).
3   K.L. Zahnle, L. Dones, H.F. Levison, Icarus **136**, 202 (1998).

4   E.P. Turtle , C.B. Phillips, G.C. Collins, A.S. McEwen, J.M. Moore, R.T. Pappalardo, P.M. Schenk, and the Galileo SSI Team, Lunar and Planet. Sci. Conf. XXX, 1999, CD-ROM (abstract).

5   J.W. Head, R.T. Pappalardo, R.J. Sullivan, J. Geophys. Res. **104**, 24, 223-24, 236 (1999).

6   R.T. Pappalardo, J.W. Head, R. Greeley, R.J. Sullivan, C. Pilcher, G. Schubert, W.B. Moore, M.H. Carr, J.M. Moore, M.J.S. Belton, D.L. Goldsby, Nature **391**, 365 (1998).

7   W.B. McKinnon, Eos, Transactions AGU (supplement) **80**(46), F94 (1999).

8   E.J. Gaidos, K.H. Nealson, J.L. Kirschvink, Science **284**, 1631 (1999).

9   C.F. Chyba, Nature **403**, 381 (2000).

10 R. Greenberg, in: T. Gehrels (Ed.) *Jupiter*, University of Arizona Press, Tucson, 1976, pp. 122.

11 R. Greenberg, in: D. Morrison (Ed.) *The Satellites of Jupiter*, University of Arizona Press, Tucson, 1982, pp. 65.

12 S.W. Squyres, R.T. Reynolds, P. Cassen, S.J. Nature **301**, 225 (1983).

13 G.W. Ojakangas, D.J. Stevenson, Icarus **66**, 341 (1986).

14 D.J. Stevenson, in: *Europa Ocean Conf.*, San Juan Capistrano Institute, San Juan Capistrano, Calif., 1996, pp. 69.

15 R. Greenberg, S.J. Weidenschilling, Icarus **58**, 186 (1984).

16 G.V. Hoppa, R. Greenberg, P. Geissler, B.R. Tufts, J. Plassmann, D.D. Durda, Icarus **137**, 341 (1999).

17 P.E. Geissler, R. Greenberg, G.V. Hoppa, P. Helfenstein, A. McEwen, R. Pappalardo, R. Tufts, M. Ockert-Bell, R. Sullivan, R. Greeley, Nature **391**, 368 (1998).

18 G.V. Hoppa, B.R. Tufts, R. Greenberg, P. Geissler, Icarus **141**, 287 (1999).

19 G.V. Hoppa, B.R. Tufts, R. Greenberg, T.A. Hurford, D.P. O'Brien, P.E. Geissler, Icarus, submitted.

20 P. Helfenstein, E.M., Icarus **61**, 175 (1985).

21 A.S. McEwen, Nature **321**, 49 (1986).

22 R. Greenberg, P. Geissler, G.V. Hoppa, B.R. Tufts, D.D. Durda, R. Pappalardo, J.W. Head, R. Greeley, R. Sullivan, M.H. Carr, Icarus **135**, 64 (1998).

23 B.K. Lucchitta, L.A. Soderblom, in: D. Morrison (Ed.) *The Satellites of Jupiter*, University of Arizona Press, Tucson, 1982, pp. 521.

24 P. Schenk, W.B. McKinnon, Icarus **79**, 75 (1989).

25 R.T. Pappalardo, R.J. Sullivan, Icarus **123**, 557 (1996).

26 B.R. Tufts, *Lithospheric displacement features on Europa and their interpretation*, PhD Thesis, University of Arizona, Tucson, 1998, 288 pp.

27 B.R. Tufts, R. Greenberg, G.V. Hoppa, P. Geissler, Icarus **146**, 75 (2000).

28 B.R. Tufts, R. Greenberg, G.V. Hoppa, P. Geissler, Icarus **141**, 53 (1999).

29 G.V. Hoppa, B.R. Tufts, *Formation of cycloidal features on Europa*, Lunar and Planet. Sci. Conf. XXX, 1999.

30 G.V. Hoppa, B.R. Tufts, R. Greenberg, P. Geissler, Science **285**, 1899 (1999).

31 D.P. O'Brien, P. Geissler, R. Greenberg, Bull. Amer. Astron. Soc. **32**, 1066 (2000).

32 R. Greenberg, G.V. Hoppa, B.R. Tufts, P. Geissler, J. Riley, S. Kadel, Icarus **141**, 263 (1999).

33 L.M. Prockter, A.M. Antman, R.T. Pappalardo, J.W. Head, G.C. Collins, JGR Planets **104**, 16,531 (1999).

34 J. Riley, G.V. Hoppa, R. Greenberg, B.R. Tufts, P. Geissler, J. Geophys. Res.-Planets **105**, E9, 22599 (2000).

35 G.V. Hoppa, R. Greenberg, J. Riley, B.R. Tufts, Icarus (2000), in press.
36 R.W. Carlson, M.S. Anderson, R.E. Johnson, W.D. Smythe, A.R. Hendrix, C.A. Barth, L.A. Soderblom, G.B. Hansen, T.B. McCord, J.B. Dalton, R.N. Clark, J.H. Shirley, A.C. Ocampo, D.L. Matson, Science **283**, 2062 (1998).
37 R.W. Carlson, R.E. Johnson, M.S. Anderson, Science **286**, 97 (1999).
38 W.D. Smythe, R.W. Carlson, A. Ocampo, D. Matson, T.V. Johnson, T.B. McCord, G.E. Hansen, L.A. Soderblom, R.N. Clark, Lunar Planet. Sci. Conf. XXIX, 1998, 1532.
39 D.T. Hall, D.F. Strobel, P.D. Feldman, M.A. McGrath, H.A. Weaver, Nature **373**, 677 (1995).
40 W.M. Calvin, R.E. Johnson, J.R. Spencer, Geophys. Res. Lett. **23**, 673 (1996).
41 K.S. Noll, R.E. Johnson, A.L. Lane, D.L. Dominque, H.A. Weaver, Science **273**, 341 (1996).
42 T.B. McCord, G.B. Hansen, R.N. Clark, P.D. Martin, C.A. Hibbitts, F.P. Fanale, J.C. Granahan, M. Segura, D.L. Matson, T.V. Johnson, R.W. Carlson, W.D. Smythe, G.E. Danielson, Science **278**, 271 (1998).
43 G.A. Vogel, Science **281**, 627 (1998).
44 J. Oró, S.W. Squyres, R.T. Reynolds, T.M. Mills, in: *Exobiology in Solar System Exploration*, NASA Spec. Publ., SP-512, 1992, pp. 103.
45 T.B. McCord, G.B. Hansen, F.P. Fanale, R.W. Carlson, D.L. Matson, T.V. Johnson, W.D. Smythe, J.K. Crowley, P.D. Martin, A. Ocampo, C.A. Hibbitts, J.C. Granahan, Science **280**, 1242 (1998).
46 T. McCord, G. Teeter, G. Hansen, M. Soeger, T. Orlando, Bull. Amer. Astron. Soc. **32**, 1068 (2000).
47 J.B. Dalton, R.N. Clark, Eos, Trans. AGU, **79**(45), Fall Meet. Suppl., F541 (1998).
48 E.S. Varnes, B.M. Jakosky, Lunar Planet. Sci. Conf. XXX, 1999, 1082 (abstract).
49 R.T. Reynolds, S.W. Squyres, D.S. Colburn, C.P. McKay, Icarus **56**, 246 (1983.
50 J.I. Lunine, R.D. Lorenz, Lunar Planet. Sci. Conf. XXVIII, 1997, 855.
51 J.C. Priscu, E.E. Adams, W.B. Lyons, M.A. Voytek, D.W. Mogk, R.L. Brown, C.P. McKay, C.D. Takacs, K.A. Welch, C.F. Wolf, J.D. Kirshtein, R. Avci, Science **286**, 2141 (1999).

8   Permafrost Model of Extraterrestrial Habitat

David A. Gilichinsky

Seven of nine planets of our Solar System (Fig. 8.1), as well as their satellites, comets, and asteroids are of a cryogenic nature, i.e. the permafrost is a common phenomenon in the cosmos. From an astrobiological point of view, the terrestrial permafrost, inhabited by viable cold adapted microorganisms may be considered as a model of extraterrestrial life. The microorganisms found in the Earth's permafrost provide a range of potential inhabitants on extinct space cryogenic bodies. Thus, in contrast to other disciplines, microbiology is looking not only for traces of extinct life on other planets, but, most intriguing for signs of existing life at the cellular level, especially within frozen materials.

8.1   The Importance of Permafrost

The ability of microorganisms, the most ancient life forms on the Earth, to live in a variety of natural environments continually forces us to redefine the limits of life in biosphere. Microorganisms not only have adapted to the cold and populate the main ecological niches, but also survive under conditions that seem absolutely unsuitable for life in large populations and a high diversity. More than 80% of the Earth's biosphere is permanently cold and cold adaptation of microorganisms would appear to be an important trait. Because of their ability to cope with low temperatures, cold adapted microbes, first isolated by Forster [1], can be regarded as Earth's most successful colonizers [2].

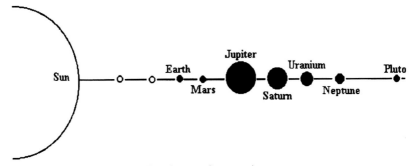

Fig. 8.1 The planets of cryogenic type.

A significant volume of the Earth's cold environments consists of permafrost (perennially frozen ground) - a naturally occurring material with a temperature below 0 °C, underling 20% of the land surface [3]: on the northern reaches of North America and Eurasia (Alaska, Siberia and Canada). In these regions, the permafrost reaches a thickness of more than 700-1000 m in the north, thinning toward the south. It also occurs in the ice-free regions of Antarctica and Greenland and surrounding Arctic and Antarctica as offshore permafrost. Alpine permafrost can be found in the high mountains of Europe, Asia, and America.

This considerable mass of frozen sediments, up to several hundreds of meters deep, harbors significant numbers of viable microorganisms. The first data on the existence of bacteria in permafrost appeared at the end of the 19th century, in relation to the discovery of mammoths and soil studies in Siberia [4, 5]. Later microbes were discovered in Holocene to late Pleistocene permafrost deposits in many Arctic regions, as well as viable cells in more ancient sediments were found in associated with Antarctic Dry Valley Drilling Project [6, 7]. In these studies the procedures and the application of drilling solutions did not guarantee the sterility of cores. Nevertheless, these authors have priority in raising the question of the possible preservation of living matter in permafrost.

In recent investigations, new aseptic methods of drilling, sampling, storage, transportation, and strict control, as well as new specialized tests, have shown that the microorganisms seen in permafrost samples could not have penetrated from the outside, but are present *in situ* [8-10]. They have been isolated (Fig. 8.2) from the cores up to 400-m deep in the Canadian Arctic and at the lowest ground temperatures (–27 °C) in Antarctica. The age of the microorganisms corresponds to the longevity of the permanently frozen state of the soils. The oldest viable cells date back to 2-3 million years (Ma) in northeastern Siberia, and probably older cells may be found in Antarctica. They are the only known forms of life that have retained viability over geological time periods and upon thawing renew physiological activity. Thus, the permafrost can be characterized as a unique physical-chemical complex, which can maintain life incomparably longer than any other known habitats. If we take into account the thickness of the deep permafrost layers, it is easy to conclude that they contain a total microbial biomass many times higher than that of the modern soil cover. This great mass of living matter, 10^2-10^7 cells per 1 g of soil (Table 8.1A.), is peculiar to permafrost strata only.

Therefore, permafrost is of great significance for research in cryobiology, biotechnology, ecology, molecular paleontology. Specially, the occurrence of a viable Cenozoic generation of microorganisms within the permafrost is intriguing for the newly emerging field of Astrobiology. If life should be found to be distributed beyond the planet Earth, the recovered organisms may possess unique mechanisms that probably might work for billions of years to allow them to maintain viability. Cameron and Morelli [11] first advanced the idea to look of Martian life by using terrestrial permafrost models.

8.2 Parameters of Permafrost Microbial Habitat

The continual subzero temperatures, as well as the main physical-chemical parameters, make permafrost one of the more stable and balanced of the natural environments. This determines the necessity to consider the traditional physical-chemical characteristics of permafrost as abiotic parameters of the habitat, what provides life support and ensures the formation of microbial communities that realize unknown possibilities of physiological and biochemical adaptation to prolonged cold and remain virtually invariable for millions of years. It may be suggested that the mechanisms of such adaptation are universal and operate within the broad limits of modification for natural ecosystems on and beyond the planet Earth.

8.2.1 Temperature

The temperature regime, the most fundamental aspect of any environment, acts as a regulator of all physicochemical reactions and biological processes [12]. In addition, the subzero permafrost temperatures are favorable to the preservation of biological systems, and is the main factor contributing to the long-term survival of cells. Evidently, the long-term impact of subzero temperatures should be regarded not as the extreme and limiting but rather as a stabilizing factor supporting the viability of microorganisms adapted to these conditions.

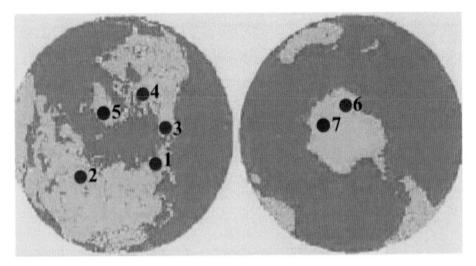

Fig. 8.2 Locations of microorganisms discovery in permafrost (red) and ice cores (blue). Arctic: 1 - Kolyma lowland; 2 - West Siberia; 3 - Alaska; 4 - Mackenzie Delta; 5 – Greenland Ice Sheet; Antarctica: 6 - McMurdo Dry Valleys; 7 - Vostok station on Ice Sheet [21].

Table 8.1. Viable microorganisms in permafrost

A. The number of viable aerobes in permafrost at different temperatures

Kolyma lowland, T= −10 to−12 °C		Mackenzie Delta, T= −5 to −6 °C		West Siberia, T= −1 to −2 °C		Dry Valleys, T= −18 to −25 °C	
depth, m	cells/g	depth, m	cells/g	depth, m	cells/g	depth, m	cells/g
21.0-62.4	2.2×10^5 $- 1.4 \times 10^7$	9.5-326.4	1.4×10^3 $- 1.6 \times 10^5$	2.0-18.5	1.4×10^2 $- 2.7 \times 10^3$	1.0-19.3	5.2×10^2 $- 6.2 \times 10^4$

B. The number of viable aerobic bacteria (cells/g) on the culturing temperatures

depth, m	temperature, °C			
	40	30	20	4
Antarctica, Tailor and Meyr Valleys, eolian, glacial and lake bottom sands; T= −20 °C				
0.5-16.3	$0-<10^1$	$0-1.1 \times 10^3$	$<10^2-5.9 \times 10^4$	$0-9.1 \times 10^3$
Antarctica, Mt. Feather (formation Sirius) and Beacon Valley (sands); T= −23 to −27 °C				
1.0-3.0	0	$0-1.5 \times 10^2$	$7.6 \times 10^1-6.3 \times 10^4$	0
Canadian Arctic, Mackenzie Delta, sands and loam				
9.5-41.2	0	$6.4 \times 10^1-1.9 \times 10^3$	$1.4 \times 10^3-1.6 \times 10^5$	$0-6.5 \times 10^3$
Kolyma lowland, marine and alluvial sands and loam				
5.7-55.2	0	$1.7 \times 10^1-5.3 \times 10^4$	$3.2 \times 10^1-1.1 \times 10^5$	$1.2 \times 10^2-2.3 \times 10^4$

C. The number of viable permafrost anaerobes (cells/g) growing at 15 °C

period (age, years)	depth, m	metanogens (CO_2+H_2)	denitrifing (NO_3+citrate)	sulfate reducers (SO_4+ lactate)
QIV ($5-10 \times 10^3$)	1.0	2.0×10^7	2.0×10^7	2.0×10^2
QIII ($1-4 \times 10^4$)	4.4-17.0	2.5×10^7	2.5×10^5 -2.5×10^6	0
QII ($1-6 \times 10^5$)	30.0-35.7	$(2.0-2.3) \times 10^7$	2.3×10^3 -2.0×10^6	$0-2.3 \times 10^2$
N_2-QI ($0.6-1.8 \times 10^6$)	37.2-64.3	$(2.0-2.5) \times 10^7$	2.5×10^3 -2.5×10^7	$0-2.0 \times 10^2$

In Arctic the mean annual permafrost temperature, as registered at high altitudes, is −10 to −12 °C, rising toward the southern border of the permafrost to −1 to −2 °C. In Antarctic polar desert, with its background of below freezing air temperatures even in summer, the maximal (−18 °C) mean annual permafrost temperature in free ice areas, such as the Dry Valleys, is registered at low hypsometric levels near the coast. This temperature decreases when moving inland and toward higher altitudes, reaching the lowest value on the Earth, –24 to –27 °C, however, higher that of Mars. But at some depths below the surface, where the amplitude of temperature oscillations goes down, the mean temperatures of Martian permafrost could be comparable with those of the Earth. And what is more, icy caps on Martian poles isolate the underlying permafrost from the ultra low temperatures prevailing in the atmosphere as ice sheets or glaciers on the Earth. The Martian dust in equatorial or moderate zones apparently plays the same role as finely dispersed sediments in terrestrial conditions - reduce sharply the temperature oscillations and the depth of their penetration. As a result, at depths below the annual temperature oscillations, the temperature of Martian permafrost (both, on the poles and at the equator) should be approximately –30 °C, not much different from the extreme values found on Earth. At these temperatures permafrost could present an inhabited environment identical to that of the Earth.

8.2.2   Iciness and Unfrozen Water

The permafrost is characterized by a multiphase state of water, which plays a dual role from the biological point of view [13]. The solid phase (ice) makes up 92-97% of total water volume in permafrost and serve as a cryoconservant for biological objects. The iciness of sandy-loam-based frozen ground in Arctic varies from minimal values of 10-20% to a wide range of maximal values, 40-50%, or more in the presence of ice wedges. At the higher ice contents the pores are completely filled with ice, obviously excluding any migration inside the stratum. The unexpectedly high (25-40% and more) content of ice in the subsurface coarse-grained Dry Valleys sands firmly cemented into a massive cryogenic structure, has no explanation. These data disprove the concepts that ground ice is unstable because of the active processes of sublimation at ultra low air temperature in Antarctica: –60 to –70 °C, as well as the thesis of the "dry Antarctic permafrost" and makes to have a new look at the regularities of the formation of frozen strata. In contrast, McKay with co-workers [14] think that some mechanism is recharging the ground ice from atmosphere. This is why one cannot exclude the high ice presence in Martian permafrost, too.

In permafrost 3-8% of the water is in an unfrozen state (W_{unf}), most commonly as thin films and, possibly, as brine pockets in saline soils. Among other parameters the unfrozen water play the leading role in the preservation of microorganisms [13]. These films, by coating the organo-mineral particles, protect the viable cells sorbed onto their surface from mechanical destruction by growing crystals of intrusive ice and make possible the mass transfer of microbial metabolic end-products within the permafrost, preventing the cell's biochemical death. Probably, they also may serve as a nutrient medium, because the unfrozen water contains high concentrations of various ions and molecules, as well as represent firmly bound, liquid, water with binding molecules; this is why permafrost may be considered as biologically dry environment.

The dependence of unfrozen water on the physical-chemical parameters of frozen strata has been studied by the many scientific schools [for review see 15]. The quantity of the unfrozen water and thickness of its films is independent of the total ground iciness, decreases with temperature [16, 17] and associates with texture composition: the more dispersed the sediments are, the larger is the W_{unf} in them and the thicker are its films, 5-75 Å (this is why the finely-dispersed Martian dust appears favorable for the presence of unfrozen water). At Arctic permafrost temperatures (−9 to −12 °C) the magnitude of $W_{unf.}$ in loam is 3-5%. The maximal values of (10%) are related to higher temperatures or saline marine facies. In sands, $W_{unf.}$ is minimal, tending to zero, although there a silt fraction in them that retains a measurable $W_{unf.}$. Only in Antarctic sands, because of the low temperatures, the $W_{unf.}$ values there are so small that the instrumental methods fail to record them.

8.2.3   Ice and Permafrost as a Habitat

The validity of unfrozen water as a main ecological niche, where the microorganisms might survive, can be demonstrated by a comparison between the numbers of viable cells recovered from permafrost and ice cores. In contrast to frozen soils with an abundance of microorganisms [18-20], the viable cells recovered from cores of the pure ice of the Antarctic Ice Sheet taken at the Vostok station [21] or Greenland Ice Sheets, is on the order of a few dozens per 1 ml of thawed water and increase [22] with increasing concentrations of dust particles in the core. It should be noted that reliable data are available only for young, not older than Holocene, ice layers. To date, no viable cells have been found in the fossil ices of Arctic - neither in intrusive ice nor in polygonal ice wedges [7, 23]. Also, it should be remembered that permafrost not only contains more information because of the high microbial numbers and diversity, it is also much older than ice sheets and glaciers. The oldest glacial ice, 400-700 thousand years old, is found at Vostok station in Antarctica and in Gyliya ice cap in the Tibetan mountains [24, 25], while the permafrost ages can reach millions of years.

The viable organisms in the Earth's Cryosphere - permafrost, ice sheets, and glaciers (Fig. 8.3) - represent a significant part of Biosphere thus, it might be called Cryobiosphere. Both pure ice and permafrost are at the same subzero temperatures. The only difference between the two is the unfrozen water associated with suspended solids. In nature, pure ice, lacking suspended solids and their associated $W_{unf.}$, has insufficient protection and transport abilities and cells are destroyed mechanically by ice crystals [26]. Permanently frozen fine-grained soils provides more favorable as microbial niche and represent the most inhabited part of the Earth Cryobiosphere [27, 20].

8.2.4   Gases and Supercooled Water

The composition of the gaseous phase of permafrost is different from that of atmospheric air. The pore space of frozen strata is occupied by oxygen, nitrogen, methane, carbon dioxide, etc. The concentration of O_2 and N_2 does not differ significantly from their values in the air, while the content of biogenic gases may be appreciably higher: CH_4 in Arctic permafrost varies 2 to 40 and CO_2 1-2 to 20 ml/kg. The facts suggest that these gases be held within the frozen layers in a clathrate form [28]. From a planetary science perspective it is important to note that CO_2 and CH_4 do not prohibit long-term cryopreservation of viable cells: CO_2 is expected to be a major constituent of the gas phase within Martian permafrost, and CH_4 in the permafrost of the satellites of the gas giants.

Any data about free water on Mars, because of the ultra low temperatures on the planet, means salt water. We have such model on the Earth. In Arctic, at some depths, super cooled (up to -10 °C) saline water lenses were found within permafrost. These lenses are formed after the retreat of the sea and the freezing of marine sediments. Because of the freezing out of salts from the deposits, the concentration of salts is up to 300 g NaCl per 1 liter. And even in these environments viable cells were found, both aerobic and anaerobic, and non-halophylic. It remains to be determined whether the salt tolerance of these cells may also be associated with cold tolerance.

8.2.5   Age and Radiation

The age of Mars is approximately the same as the age of the Earth. Therefore, the Earth and Mars may have experienced similar stages of development, including the emergence of early life forms. Indeed, the frozen subsurface environments of Earth and Mars seem to provide quite similar conditions for microbial habitation [27] (Fig. 8.4). From an exobiological point of view, the main difference between the terrestrial and outer planets permafrost, inhabited by microbes, is their age. The oldest continuously late Cenozoic permafrost, where the age of microorganisms corresponds to the longevity of sediments frozen state, found in the north hemisphere of the Earth dates back to about 3 Ma in northeastern Eurasia ([29], Fig. 8.5a, b), while on Mars, the age of permafrost is estimated at approximately 3-4 Ga. This difference in time scale could result in different patterns and have a significant impact on the possibility for life preservation. The Antarctic permafrost may be somewhat closer to Mars. The analysis of numerous publications, best reviewed by Wilson [30], indicates that permafrost has existed under Antarctic climatic conditions for last dozens of millions of years, greater than the duration of Arctic permafrost by a factor of ten [27]. To the moment, viable bacteria were found in bore holes in one prospective site, Mt. Feather [31]. The age of glacial deposits of the Sirius Group dates back at least 2 or, probably, 15 Ma [32, 33]. $^{40}Ar/^{39}Ar$ dates on ash layers suggest that the most ancient ground ice below these layers may be more than 4.5 Ma old in Arena Valley [34] and 8.1 Ma old in Beacon Valley [35]. Viable bacteria have been isolated in the frozen sediments and ice below this ash layer and if this age is correct, this is the oldest viable life on Earth (Fig. 8.5c, d).

Estimation of ground radiation has been made by McKay and Forman, using both elemental analysis of the radioactive elements in sandy and loam samples and direct *in situ* measurements in the bore holes on the Eurasian northeast. The dose received by the permafrost bacteria is about 2 mGy per year. Taking into account the age of bacteria, late Pliocene to late Pleistocene, the total dose received by cells would therefore range from 1 kGy in soils dozens of thousands years old to 6 kGy at over the 3 Ma age of microorganisms. Experiments of Vorobyova have shown that up to 30 kGy are required to sterilize soil at above-freezing temperatures and the results obtained by Japanese researchers, indicate that bacteria in soil have a much greater resistance to

Fig. 8.3 Permafrost (A, B) and glaciers (C) exposures. A, B: frozen layers with ice veins in Arctic tundra (photo by M.Grigor'ev) and forest-tundra; C: Taylor Glacier in Antarctica

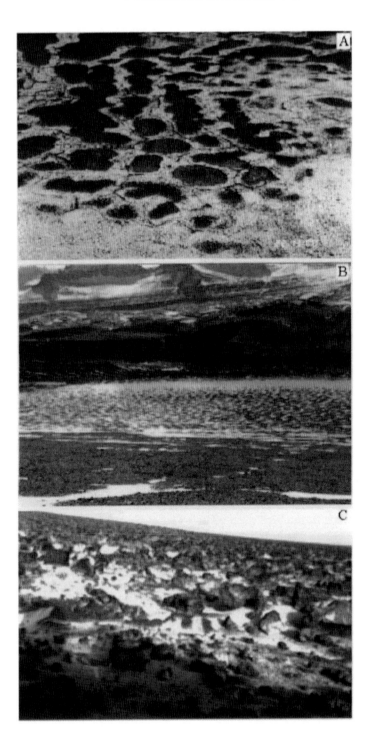

Fig. 8.4 Permafrost landscapes. A: Arctic in October; B: Antarctica in January; C: Mars in spring (NASA):

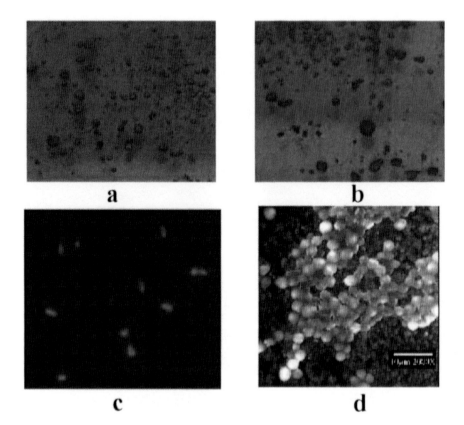

Fig. 8.5 The oldest viable microorganisms. The morphology of bacteria (a, b) from the late Pliocene Arctic permafrost; bacteria (c) and green algae Mychonastes (d) isolated from frozen sediments in Antarctica (DTAF, photo by E. Spirina and A. Shatilovich).

irradiation when frozen than when thawed. Our own results demonstrate that, at similar levels of ionizing radiation, a known quantity of viable cells and a total radiation dose of 5 to 50 kGy, most of the cells in frozen samples survived, while most of the cells in unfrozen samples died. This facts shows that freezing increased the cell's resistance to radiation and uniqueness of permafrost as an environment, where microorganisms display a high resistance to radiation. From these data the dose from radionuclides diffused through the permafrost is far from being sufficient for complete sterilization, i.e. is not fatal for viable cells, but should be large enough to cause some selection effect and to destroy the DNA of ancient cells. Their viability and growth on media implies the capacity for DNA repair. Probably DNA repair is occurring in the frozen environment, i.e. at the stable rate of damage accumulation, a comparable rate of repair also exists. From the biological point of view it is important that the permafrost preserves the cells from diffuse ground irradiation for thousands and millions of years. And what is more, the oldest, Pliocene cells, which had received an *in situ* maximal dose during the time of cryopreservation, were found to be more resistant to radiation than microorganisms from modern tundra soils.

8.3   Biodiversity in Permafrost

The permafrost microbial community has been described as "a community of survivors" [36], which have overcome the combined action of extremely cold temperature, desiccation and starvation. Under these conditions, the starvation-survival lifestyle is the normal physiological state [37]. According to their growth temperatures [38], the microbial community, even after its long-term existence within the permafrost, is not composed of psychrophilic bacteria (Table 8.1B.) [39]. However, it is resistant to sharp temperature transits through the freezing point and to freezing/thawing stress. Most of the isolated cells did not grow at temperatures higher than +30 °C, but were often capable of growth on Petri dishes at subzero temperatures as low −10 °C in the presence of cryoprotectants such as glycerol [7]. According to Russell [2], this correlates with the lower growth temperature limit of psychrophilic microorganisms. Because this term is not clearly delimited, one can define the isolated microbes as psychrotolerant organisms.

It is interesting to note that Arctic and Antarctic sediments contain about the same number of microorganisms (estimated from direct counts using epifluorescence microscopy [20]), although the Arctic sediments are rich in organic carbon, whereas in the Antarctic ones the content of carbon and unfrozen water is close to zero. This total number of microbial cells is relative constant from modern up to the oldest frozen layers. In other words, there is the minimal constant level of microbial colonization that forms in natural systems in the presence of only mineral soil particles and short term existence of free water. Possibly, this unexpected result shows that these minimal conditions are enough even for origin of life.

Overall, a number of different morphological and physiological groups of microorganisms (spore forming and spore-less, aerobic and anaerobic, prokaryotic and eukaryotic) have been found (Fig. 8.6). Morphologically, they are coccoid, coryneforming, nocardia- and rod-like Gram-positive or Gram-negative bacteria. The ancient permafrost microbial community is predominantly bacterial, as is the community in the depths of Antarctic Ice Sheet [21]. This is in contrast with modern soils, where the fungal mass is much greater than the bacterial mass. Prokaryotes with thick well preserved cell walls and capsules, surrounded by additional surface layers of low electron density are observed rather frequently [40]. Evidently, good preservation of cell structures is determined by the equilibrium-state of the unfrozen water inside and outside the cells. Violation of the equilibrium results in cell death. As a consequence, ice formation in the experiments on Petri dishes results in cell death [26]. Eukaryotic cells, while present, seem less able to survive long-term cryopreservation [40]. In Arctic permafrost, non-spore-formers predominate while in Antarctic permafrost spore-formers dominate. Recently, phototrophs, organisms that had preserved their photosynthetic apparatus in the full permafrost darkness were isolated [41]. These included cyanobacteria, perspective the most ancient organisms (the fossil cyanobacteria often found in meteorites and basalt deposits by ages 3.5 billion years) that use light of almost all wave lengths to synthesize organic material. Green algae and mosses represent the lower plants, and now from the buried Arctic soils were isolated even Protozoa. Viable seeds of high plants were found in the late Pleistocene thickness, and these seeds from Canadian Arctic still are able to grow [42].

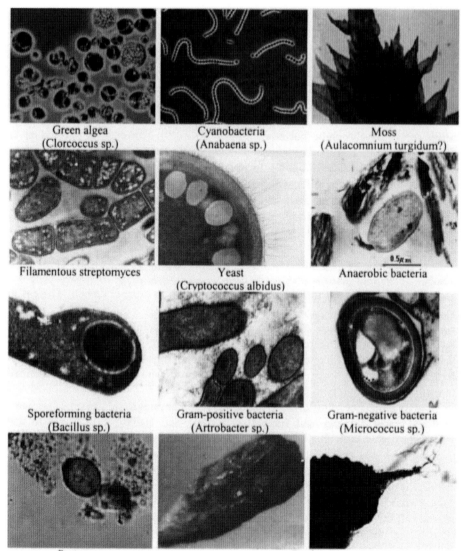

Green algae
(Clorcoccus sp.)

Cyanobacteria
(Anabaena sp.)

Moss
(Aulacomnium turgidum?)

Filamentous streptomyces

Yeast
(Cryptococcus albidus)

Anaerobic bacteria

Sporeforming bacteria
(Bacillus sp.)

Gram-positive bacteria
(Artrobacter sp.)

Gram-negative bacteria
(Micrococcus sp.)

Protozoa
(Ciliata?)

Seeds

Fig. 8.6 Permafrost biodiversity (photo by V. Soina, T. Vishnivetskaya, N. Suzina, E. Spirina and S. Gubin).

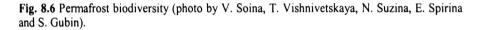

The presence of methane, as well as the value of the redox potential (+40 to –250), indicate that conditions in Arctic permafrost are mostly anaerobic; these permafrost thicknesses are primarily lacustrine-alluvial and marine deposits. The *Eh* of studied Antarctic sands (mostly, eolian and glacial origin) vary from +260 to +480, indicating that the conditions here are not as anaerobic as in Arctic; this is confirmed by the absence of biogenic methane. In any case, it is the specific character of permafrost soils, that, until the moment of transition to the permanently frozen state, they contain both aerobic and anaerobic microzones. This is why the ancient microbial communities consist of comparable numbers of anaerobic and aerobic species. The ratio of anaerobes to aerobes is different in different geological facies, i.e. the part of the community that dominates depends on the origin of sediments. In one study, based on alluvial deposits, aerobic groups dominated [18], while in another study, the number of anaerobes (methanogens, denitrifyiers, sulfate-reducers and iron-reducers) in swamp and lagoon soils was several orders of magnitude higher than that of aerobes [28]. The reduced state and viable anaerobic bacteria provides a model for possible anaerobic life forms on cryogenic planets without free oxygen and even in the absence of organic matter. These chemolithotrophic, psychrotolerant bacteria, with their ability to assimilate CO_2 in an environment similar to that of Mars, have been suggested as prospective analogues to the living forms that might be found in subsurface Martian layers [43]. This thesis supported by distribution of anaerobic bacteria in permafrost layers - it is determined by the anaerobic origin of the frozen strata, and if the community exists, their number is independent of permafrost age [28] (Table 8.1-C). For aerobes, after the prolonged exposure of the microorganisms to the permafrost environment, the ratio of readily culturable (hypometabolic) bacteria to those that may represent viable but nonculturable forms (deep resting cells) is determined by the extent and duration of exposure to subzero temperatures [20]. The greater the permafrost (not sediments) age, the lower the number of viable cells, the lower their qualitative and morphological diversity.

Some cells may have died, when subjected to the stresses of thawing and exposure to oxygen as the samples were suddenly melted in the laboratory for microbial study at relatively high temperatures. While these stresses are known to inhibit the recovery of a fraction of the community, strategies and techniques for the low-temperature recovery of bacteria from permafrost environments are only just beginning to be developed [44]. In isolations aerobic bacteria from Arctic permafrost at +20 °C, a significant fraction was found to be spore-formers [10]. This contrasts sharply with isolations carried out at +4 °C, where spore-formers were only rarely obtained. Perhaps this reflects an intriguing paucity of spore-forming bacteria in frozen environments [44]. The permafrost microbes have not received as much study and the need development of improved protocols for their recovery from ancient frozen environment, as preparation for the analysis of life in extraterrestrial materials.

For this reason we need to keep in mind that all above mentioned discussions are based on the culturable cells (analysis of Arctic sediments showed that only 0.1-5% of the cells counted by fluorescence microscopy grew on nutrient media and in Antarctica, the percentage is reduced to 0.001-0.01%) and, as is in other environments, a large portion of the permafrost microbial community remains uncultured. The phylogenetic diversity of the permafrost community has only recently begun to be addressed and the data based on DNA extracted from modern tundra soils suggest that the clones are phylogenetically diverse, and that many probably reflect new genera or families. Hence, most of this bacterial community has never been isolated and the physiology and function of its dominant members is unknown [45]. The first molecular data have recently been obtained from ancient permafrost bacterial communities, using

16S rDNA sequencing and the phylogenetic trees derived from this data indicate that the Arctic isolates fall into four major groups, partly determined by the age of the permafrost. Most are high-GC Gram-positive bacteria and β-proteobacteria. All γ-proteobacteria came from the samples, Ma old. Most of the low-GC Gram-positive bacteria came from the age, 3000-8000 years [10]. The communities also include members of Archaea, which have recently been detected in permafrost [46]. The results demonstrated good preservation of ancient genomic DNA in both, permafrost, and Greenland ice cores [47].

8.4   How Long the Life Might Be Preserved

To answer the question how long life might be preserved, we have to establish the principal mechanism of cell's behavior in permafrost conditions. Probably this mechanism, for example for spore-formers, might work for billions years. A key question regarding viable paleobacteria in permafrost is whether they are active at permafrost environment or whether are in a suspended "dormant" state. This is why the mechanisms for microbial survival during the both long-term anabiosis or low metabolic state must be studied. It is necessary to consider 2 ways of surviving viable cells. First, in Arctic, at temperatures around −10 °C, where cells are only in a cooled state, and the second, in Antarctic, at temperatures below −20 °C, where cells are preserved in a frozen state.

There are only a few data about metabolic activity of microorganisms below the freezing point, and only for recent microbial communities. Measurements show that Antarctic lichens may be active at temperature −17 °C [48]. Recently it was also obtained evidence of low rates of bacterial DNA and protein synthesis in Antarctic snow, which indicates that the organisms were metabolizing at ambient temperatures −12 to −17 °C [49]. The following facts provide indirect evidences that paleomicroorganisms can be active in the permafrost nutritional-temperature regime and allow to speculate on the possibility of biogeochemical processes *in situ*: the ability of immobilized enzymes in permafrost to become instantly activated [50]; the presence of usually metastable nitrites and ferrous sulfides [51, 52]; the simultaneous presence of methanogenic archaebacteria and methane [28, 46]. Experiments using ^{14}C acetate show that microbial ancient communities from 3 Ma old permafrost at temperatures down to −20 °C are able to form bacterial lipids [53]. However, the question about the metabolic state of microorganisms, microbiological and biogeochemical processes and life forms within permafrost still remains open.

8.5   The Perspectives for Future Studies

Permafrost is a depository of ancient biomarkers (Fig. 8.7), including biogenic gases, biominerals, biological pigments, lipids, enzymes, proteins, nucleotides, RNA and DNA fragments and molecules, microfossils, died and viable cells. They provide a range of analogues that could be used in the search for possible ecosystems and potential inhabitants on other planets. They might have been preserved and their biosignatures could be found at depths within permafrost on other planets, if life existed during the early stages of their development.

The existing data indicate the possibility to transport ingmicroorganisms embedded within the sryogenic Space bodies (meteorites) to the Earth. The longevity of life forms and their preservation obtained in the Arctic permafrost can be used as base to assess the survival of microorganisms during a hundreds of thousand of years transfer from Mars to Earth (see Chap. 4, Horneck et al.). These very microorganisms can be the Space Frontiers of Life confirming the possibility of panspermia. This is, why it is necessary to search for the presence of viable cells in the oldest Antarctic permafrost layers. The recent data from the Cape Rogers drilling project, show that early Oligocene sediments (38 Ma) old contain pollen spectra, indicating very cold conditions, and similar observations were made in studies in Australia. Such expedition to

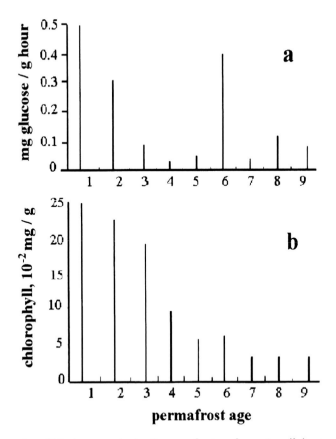

Fig. 8.7 Examples of biosignatures in Arctic permafrost. a: free extracellular enzymes: invertase (54); b: pigments: chlorophyll (41). Permafrost age: 1- modern tundra soil; 2- Holocene; 3, 4- late Pleistocene; 5, 6- middle Pleistocene; 7- early Pleistocene; 8- late Pliocene-early Pleistocene; 9- late Pliocene.

Antarctica is expensive, but thousand times cheaper than a mission to Mars. The limiting age - if one exists - within the Antarctic cores, where the viable microorganisms were no longer present, could be established as the age limit for life preservation within permafrost. Any positive results obtained from the Antarctic microbial data will extend the geological scale and increase the known duration of life preservation.

Acknowledgements. This work was supported by NSF grant LExEn-9978271 and by Russian Fund for Basic Research: grants 01-05-65043 and 01-04-49084.

8.6   References

1   J. Forster, Zentr. Bakteriol. Parasitenk. Infekt. Hyg. **2**, 337 (1887).
2   N. Russell, Philosophical Transactions of the Royal Society of London, Series B, **326**, 596 (1990).
3   T. Pewe, Encyclopaedia Britannica **20**, 172 (1995).
4   V. Omelyansky, Arkhiv biologicheskikh nauk **16**, N 4, 335 (1911) (in Russian).
5   B. Isachenko, Izvestiya Sankt-Peterburgskogo Botanicheskogo Sada **12**, N 5-6, 140 (1912) (in Russian).
6   T. Pewe, Geological Survey Professional paper, 835, United States Government Printing Office, Washington, 1975, pp. 139.
7   D. Gilichinsky, S. Wagener, Permafrost and Periglacial Processes **5**, 243 (1995).
8   D. Zvyagintsev, D. Gilichinsky, S. Blagodatsky, E. Vorobyova, G. Khlebnikova, A. Arkhangelov, N. Kudryavtseva, Microbiologiya **54**, 153 (1985) (in Russian).
9   D. Gilichinsky, G. Khlebnikova, D. Zvyagintsev, D. Fyodorov-Davydov, N. Kudryavtseva, Izvestiya Academii Nauk SSSR, series geol. **6**, 114 (1989) (in Russian).
10   T. Shi, R. Reevs, D. Gilichinsky, E.I. Friedmann, Microbial Ecology **33**, 169 (1997).
11   R. Cameron, F. Morelli, Antarctic Journal of the USA **9**, 113 (1974).
12   R. Herbert, in: R. Herbert, G. Codd (Eds.) *Microbes in extreme environments*, SGM Publisher Academy Press, 1986, pp. 1.
13   D. Gilichinsky, E. Vorobyova, L. Erokhina, D. Fedorov-Davydov, Adv. Space Res. **12**, N 4, 225 (1992).
14   C. McKay, M. Mellon, E.I. Friedmann, Antarctic Sci. **10**(1), 31 (1998).
15   D. Anderson, N. Morgenstern, *Permafrost:* 2nd International Conf. National Academy of Sciences, Washington, D.C, 1973, pp. 257.
16   Z. Nersesova, N. Tsytovich, *Permafrost:* Proc. International Conf. National Academy of Sciences of the USA, Washington, D.C., 1966, pp. 230.
17   D. Anderson, Nature **216**, 563 (1967).
18   G. Khlebnikova, D. Gilichinsky, D. Fedorov-Davydov, E. Vorobyova, Microbiologiya **59**, 148 (1990) (in Russian).
19   D. Gilichinsky, S. Wagener, T. Vishnivetskaya, Permafrost and Periglacial Processes **6**, 281 (1995).
20   E. Vorobyova, V. Soina, M. Gorlenko, N. Minkovskaya, N. Zalinova, A. Mamukelashvili, D. Gilichinsky, E. Rivkina, T. Vishnivetskaya, FEMS Microbiology Reviews **20**, 277 (1997).
21   S. Abyzov, in: E.I. Friedmann (Ed.) *Antarctic Microbiology*, Wiley-Liss, 1993, pp. 265
22   S. Abyzov, I. Mitskevich, M. Poglazova, Microbiologiya **67**, N 4, 547 (1998) (in Russian).

23   A. Kriss, N. Grave, Microbiologiya **13**, N 5, 251 (1944) (in Russian).
24   J. R. Petit, J. Jouzel, D. Raynaud, N. I. Barkov, J.-M. Barnola, I. Basile, M. Bender, J. Chappellaz, M. Davis, G. Delaygue. M. Delmotte, V.M. Kotlyakov, M. Legrand, V.Y. Lipenkov, C. Lorius, L. Pépin, E. Saltzman, M. Stievenard, Nature **399**, 429 (1999).
25   L. Tompson, T. Yao, E. Davis, K. Henderson, E. Mosley-Tompson, P.-N. Lin, J. Beer, H.-A. Synal, J. Cole-Dai, J. Boizan, Science **276**, 1821 (1997).
26   D. Gilichinsky, V. Soina, M. Petrova, Origins of Life and Evolution of the Biosphere **23**, 65 (1993).
27   D. Gilichinsky, Proc. SPIE 3111, 1997, pp. 472.
28   E. Rivkina, D. Gilichinsky, S. Wagener, J. Tiedje, J. McGrath, Biogeochemical Activity of Anaerobic Microorganisms from Buried Permafrost Sediments. Geomicrobiology **15**, 187 (1998).
29   A. Sher, Falls Church, American Geological Institute, 282 (Book Section), International Geology Review **16**, 7 (1974).
30   G. Wilson, Quaternary Science Reviews14, 101 (1995).
31   G. Wilson, P. Braddok, S. Foreman, E.I. Friedmann, E. Rivkina, D. Gilichinsky, D. Fyodorov-Davydov, V. Ostroumov, V. Sorokovikov, M. Wizevich, Antarctic Journal of USA **3**, 74 (1996).
32   P.-N. Webb, D. Harwood, Quaternary Science Reviews **10**, 215 (1991).
33   D. Marchant, C. Swisher, D. Lux, D. West, G. Denton, Science **260**, 667 (1993).
34   G. Denton, D. Sudgen, D. Marchant, B. Hall, T. Wilch, Geofrafiska Annaler **75A** (4), 155 (1993).
35   D. Sugden, D. Marchant, N. Potter, R. Souchez, G. Denton, C. Swisher III, J-L. Tison, Nature **376**, 412 (1995).
36   E.I. Friedmann, in: D. Gilichinsky, (Ed.) *Viable microorganisms in permafrost*, Russian Academy of Sciences, Pushchino, 1994, pp. 21.
37   R. Morita, *Bacteria in Oligotrophic Environments*, Chapman & Hall Microbiology Series, New York, 1997, 529 pp.
38   R. Morita, Bacteriological Review **39**, 144 (1975).
39   A.M. Gounot, in: Experimentia 42, Birkhauser Verlag, CH-4010 Basel, 1986, pp. 1192.
40   V. Soina, E. Vorobiova, D. Zvyagintsev, D. Gilichinsky, Adv. Space Res. **14** (6), 138 (1995).
41   T. Vishnivetskaya, L. Erokhina, D. Gilichinsky, E. Vorobyova, The Earth Cryosphere **2**, 71 (1997) (in Russian).
42   A. Porsild, C. Harington, G. Milligan, Science **158**, 113 (1967).
43   M. Ivanov, A. Lein, Doklady AN SSSR **321** (6), 1272 (1992) (in Russian).
44   T. Vishnivetskaya, S. Kathariou, J. McGrath, D. Gilichinsky, J. Tiedje, Extremophiles **4**, N 3, 165 (2000).
45   J. Zhou, M. Davey, J. Figueras, E. Rivkina, D. Gilichinsky, J. Tiedje, Microbiology (UK) **143** (12), 3913 (1997).
46   J. Tiedje, M. Petrova, C. Moyer, in: Abstracts, 8th International Symposium in Microbial Ecology, Halifax, Nova Scotia, 1998, p. 323.
47   E. Willerslev, A. Hansen, B. Christensen, J. Steffensen, P. Arstander, Proc. Natural Academy of Sciences USA **96**, 8017 (1999).
48   L. Kappen, B. Schroeter, C. Scheidegger, M. Sommerkorn, G. Hestmark, Adv. Space Res. **18**, 119 (1996).

49 E. Carpenter, S. Lin, D. Capone, Applied & Environmental Microbiology **66**, 4514 (2000).
50 E. Vorobyova, V. Soina, A. Mulukin, Adv. Space Res. **18**, 103 (1996).
51 H. Jansen, E. Bock, in: D. Gilichinsky, (Ed.) *Viable microorganisms in permafrost*, Russian Academy of Sciences, Pushchino, 1994, pp. 27.
52 C. Zigert, Mineralogical Journal **9**, 75 (1987) (in Russian).
53 E. Rivkina, E.I. Friedmann, C. McKay, D. Gilichinsky, Applied & Environmental Microbiology **66**, 3230 (2000).
54 N. Minkovskaya, *Enzyme activity in permafrost soil*, Abstract of PhD dissertation. Moscow State University, 1995 (in Russian).

9 Microbial Life in Terrestrial Permafrost: Methanogenesis and Nitrification in Gelisols as Potentials for Exobiological Processes

Dirk Wagner, Eva Spieck, Eberhard Bock and Eva-Maria Pfeiffer

The comparability of environmental and climatic conditions of the early Mars and Earth is of special interest for the actual research in astrobiology. Martian surface and terrestrial permafrost areas show similar morphological structures, which suggests that their development is based on comparable processes. Soil microbial investigations of adaptation strategies of microorganisms from terrestrial permafrost in combination with environmental, geochemical and physical analyses give insights into early stages of life on Earth. The extreme conditions in terrestrial permafrost soils can help to understand the evolution of life on early Mars and help searching for possible niches of life on present Mars or in other extraterrestrial permafrost habitats [1, 2].

9.1 Permafrost Soils and Active Layer

In polar regions huge layers of frozen ground are formed - termed permafrost - which are defined as the thermal condition, in which soils and sediments remain at or below 0 °C for two or more years in succession. Terrestrial permafrost, which underlay more than 20% of the world's land area, is above all controlled by climatic factors and characterized by extreme terrain condition and landforms. On Earth the permafrost thickness can reach several hundreds of meters, e.g., in East Siberia (Central Yakutia) about 600-800 m. During the relatively short period of arctic/antarctic summer only the surface zone of permafrost sediments thaws. This uppermost part of the permafrost (active layer) includes the so called Gelisols [3], which contains permafrost in the upper 100 cm soil depth. Gelisols are characterized by gelic material that have the evidence of cryoturbation and ice segregation. Permafrost soils may be cemented by ice which is typical for the Arctic regions, or, in the case of insufficient interstitial water, may be dry like the Antarctic polar deserts.

Permafrost can be divided into three temperature regimes (Fig. 9.1), which characterize the extreme living conditions: (i) The surface near upper active layer (0.2-2.0 m thickness) is subjected to seasonal freezing and thawing with an extreme temperature regime from about +15 °C to −35 °C, (ii) the correlated upper, perennially frozen permafrost sediments (10-20 m thickness) with smaller seasonal temperature variation of about 0 °C to −15 °C above the zero annual amplitude and (iii) the deeper permafrost sediments which are characterized by a stable temperature regime of about −5 °C to −10 °C [4].

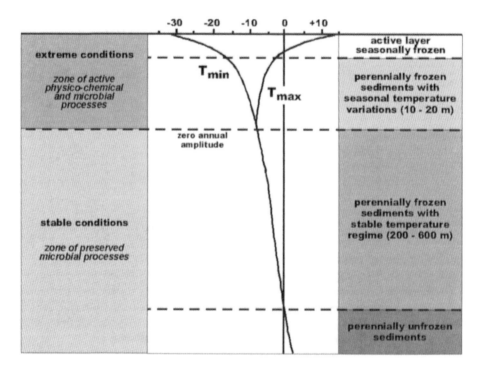

Fig. 9.1 Scheme of temperature amplitude in permafrost sediments (according to French [4], modified).

The main Gelisol-forming processes in permafrost landscapes are cryopedogenesis, which include freezing and thawing, frost stirring, mounding, fissuring and solifluction. The repeating cycles of freezing and thawing leads to cryoturbation features (frost churning) that includes irregular, broken or involuted horizons and an enrichment of organic matter and other inorganic compounds, especially along the top of the permafrost table. As a result of cryopedogenesis many Gelisols are influenced by a strong micro-relief (patterned ground, Fig. 9.2). The type of patterned ground has effects on soil formation and soil properties.

Ice wedge polygons for example of the Siberian lowlands (Fig. 9.3-A) which are typical for high arctic, are characterized by two different soil conditions: The Gelisols of the polygon center (*Historthels*) are water saturated and have a large amount of organic matter due to the accumulation under anaerobic conditions (Fig. 9.3-B). The Gelisols of the polygon border (*Aquiturbels*) show evidence of cryoturbation in more or less all horizons of the active layer (Fig. 9.3-C). These soils drain into the polygon center, which leads to dryer conditions in the upper layer of the border.

These examples demonstrate that the Gelisols of the active layer and upper permafrost sediments are the zone with active physico-chemical processes under extreme conditions. Therefore, microbial life in permafrost soils and sediments is influenced by extreme gradients of temperature, moisture and chemical properties. However, deeper permafrost layers characterize living conditions, which have been stable for long periods of time and microbial life is preserved (see Chap. 8, Gilichinsky).

9.2   Microbial Life under Extreme Conditions

Over 80% of the Earth's biosphere - including the polar regions - is permanently cold. Most natural environments have a temperature regime colder than 5 °C. Temperature is one of the most important parameters regulating the activity of microorganisms because it controls all metabolic activity of living cells [5]. The temperature in the upper zone of the cryolithosphere (active layer and upper permafrost sediments) ranged between –50 °C to +30 °C [6]. Especially permafrost soils are characterized by extreme variation in temperature. Previously the potential of growth as well as the molecular, physiological and ecological aspects of microbial life at low temperatures were investigated [7, 8]. Many microorganisms are able to survive in cold permafrost sediments, but this adaptation can be a tolerance or a preference. According to Morita [9] bacteria can be described by their temperature range of growth: psychrophiles (T_{min} <0 °C, T_{opt} ≤15 °C, T_{max} ≤20 °C), psychrotrophs (T_{min} ≤0 °C, T_{opt} ≥15 °C, T_{max} ≤35 °C) and mesophiles (T_{opt} 25-40 °C). The minimum temperature for growth of bacteria was recently reported with –20 °C [10], whereas the minimum temperature for enzyme activity was –25 °C [11].

The seasonal variation of soil temperature influences also the availability of pore water. The presence of unfrozen water is an essential biophysical requirement for the survival and activity of microorganisms in permafrost. Temperature below zero stands for an increasing loss of water. At the same time freezing of water leads to an increase of salt content in the remaining pore solution. However, in clayey permafrost soils liquid water was analyzed at temperatures up to –60 °C [12]. The most important feature of this water is the possible transfer of ions and nutrients [13]. Furthermore, McGrath et al. [14] showed that the intercellular water in fossil bacteria from permafrost soils was not crystallized as ice even at an extreme temperature of –150 °C.

For studying microbial life under extreme conditions it is also necessary to consider whether and where these conditions are changing or stable in a permafrost profile. The seasonal variation in soil temperature, particularly freeze-thaw cycles in the active layer, results in drastic changes of other environmental conditions like salinity, soil pressure, changing oxygen conditions (anoxic, microaerophilic, oxic) and nutrient availability. Therefore, besides the physico-chemical conditions of permafrost, the physiological properties of microorganisms are relevant for the adaptation to extreme conditions. On account of this potential, they developed strategies to resist salt stress, physical damage by ice crystals and background radiation [15]. Survival could be also possible by anabiosis (dormant stage of life) or by reduced metabolic activity in unfrozen waterfilms (see Chap. 8, Gilichinsky).

9.3   Microbial Key Processes

Terrestrial permafrost is colonized by high numbers of chemoorganotrophic bacteria as well as microbes like methanogenic archaea and nitrifying bacteria [16-18], which are highly specialized organisms. They are characterized by litho-autotrophic growth

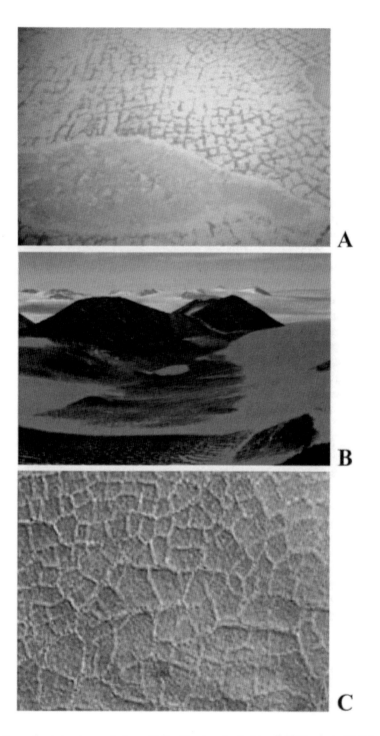

Fig. 9.2 Permafrost structures. **A:** Lena Delta, Russian Arctic (April 1999, photo W.Schneider, AWI); **B:** Haskard Highlands, Antarctica (December 1994, photo W.-D. Hermichen, AWI); **C:** Mars, Northern Hemisphere (May 1999, photo NASA).

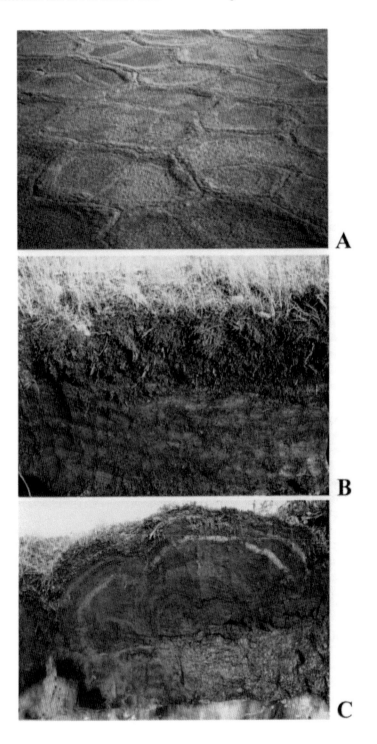

Fig. 9.3 Landscape and soils of polygon tundra, Lena Delta/Siberia. **A:** Low-centered polygons; **B:** *Typic Historthel* of the polygon center; **C:** *Glacic Aquiturbel* of the polygon border (photos L. Kutzbach, AWI).

gaining energy by the oxidation of inorganic substances. Carbon dioxide can be used as the only carbon source. Lithoautotrophic growth is an important presumption for long-term survival [19] of microbes in extreme environments like terrestrial permafrost or maybe on other planets of our Solar System.

9.3.1   Methanogenesis

Responsible for the biogenic methane production (methanogenesis) is a small group of microorganisms called methanogenic archaea [20]. Methanogenesis represents the terminal step in carbon flow in many anaerobic habitats, including permafrost soils, marshes and swamps, marine and freshwater sediments, flooded rice paddies and geothermal habitats. Although methanogens are widely spread in nature they show an extremely specialized metabolism. They are able to converse only a limited number of substrates (e.g., hydrogen, acetate, formate, methanol, methyl- amines) to methane. In permafrost soils two main pathways of energy-metabolism dominate: (i) the reduction of CO_2 to CH_4 using H_2 as a reductant and (ii) the fermentation of acetate to CH_4 and CO_2. In the case of CO_2-reduction organic carbon is not necessary for growth of methanogenic archaea [21].

At present, 68 species of methanogenic archaea are known including common genera like *Methanosarcina*, *Methanobacterium* and *Methanococcus*. Phylogenetically, they are classified as ARCHAEA [22], a group of microbes that are distinguished from BACTERIA by some specific characteristics (e.g., cell wall composition, coenzymes). They show a high adaptability at extreme environmental conditions like temperature, salinity and oxygen. Besides the mesophilic species, also thermophilic methanogens are known (see Chap. 11, Stetter). In newer times, more attention has been paid on the search for psychrophilic strains since many of methanogenic habitats belong to cold climates [23]. A lot of methanogens (e.g., *Methanogenium cariaci*, *Methanosarcina thermophila*) are able to adapt to high salinity by the accumulation of compatible solutes to equalize the external and internal osmolarity [24]. Although, they are regarded as strictly anaerobic organisms without the ability to form spores or other resting stages, they are found in millions of years old permafrost sediments [25] as well as in other extreme habitats like aerobic desert soils [26] and hot springs [27].

Because of the specific adaptations of methanogenic archaea to conditions like on early Earth (e.g., no oxygen, no or less organic compounds), they are considered to be one of the initial organisms from the beginning of life on Earth.

9.3.2   Nitrification

Nitrifying bacteria play a main role in the global nitrogen cycle by the transformation of reduced nitrogen compounds. Two groups of distinct organisms - the ammonia and nitrite oxidizers - are responsible for the oxidation of ammonia to nitrite and further to nitrate [28]. The genera of ammonia oxidizers have the prefix nitroso- whereas the nitrite oxidizers start with nitro-. The best known nitrifiers are *Nitrosomonas* and *Nitrobacter*. Up to now, 5 genera of ammonia oxidizers (with 16 species) and 4 genera of nitrite oxidizers (with 8 species) have been described [29, 30]. Phylogenetically, ammonia and nitrite oxidizers are affiliated to different subclasses of the *Proteobacteria* with the exception of *Nitrospira* (and maybe *Nitrospina*), which belong to a separate phylum [31, 32]. Although *Nitrosomonas* and *Nitrobacter* are usually the most isolated nitrifiers, they are not obligatelly the most abundant ones in a given habitat. For example, organisms of the genus *Nitrospira* seem to have a higher ecological importance than previously assumed since they were recognized as dominant nitrite oxidizers in several aquatic habitats (reviewed by Spieck and Bock [33]). These bacteria are the phylogenetic most ancient nitrifiers since they belong to a deep branching phylum. Here, a high diversity of new species was detected recently.

Nitrifiers exist in most aerobic environments where organic matter is mineralized (soils, compost, fresh- and seawater, waste water). In general, cell growth is slow with regard to the poor energy sources but can be adapted to changing environmental conditions. Especially for *Nitrobacter*, mixotrophic and heterotrophic growth with organic compounds is an alternative to the oxidation of nitrite. Nitrifiers are also active in low oxygen and anaerobic environments like sewage disposal systems and marine sediments where they are able to act as denitrifiers [34]. Although they form no endospores, they can survive long periods of starvation and dryness. Therefore, nitrifying bacteria were also detected in e.g., antarctic soils [35], natural stones [36], heating systems [32] as well as in subsurface sediments in a depth of 260 m [37]. Especially ammonia oxidizers form dense cell clusters, where cells are embedded in a dense layer of EPS (extracellular polymeric substances). These microcolonies may protect the cells against stress factors like dryness. Another protecting mechanism is the production or accumulation of compatible solutes (e.g., trehalose, glycine betaine or sucrose, see Chap. 12, Kunte et al.). Due to salt stress and dryness an increasing amount of compatible solutes was found in cells of *Nitrobacter* [38].

9.4 Methods for Analogue Studies of Microbial Processes in Terrestrial and Extraterrestrial Habitats

Increased attention has been paid on the processes of methanogenesis and nitrification in the last decades, because the involved bacteria [39] influence the global climate by the generation and transfer of climate-relevant trace gases like methane and gaseous nitrogenous oxides (NO, N_2O, NO_2). Recently, these microorganisms are also important in the area of astrobiology research because of their adaptation ability to extreme location-conditions and consequently for the search of extraterrestrial lives of particular interest.

In order to understand the described microbial key-processes in permafrost soils, it is necessary to know the microbial community which is involved in methanogenesis and nitrification. The most important members of this community are those archaea/bacteria, which are metabolic active under such extreme conditions in the soils. To learn about these microbes and their adaptation strategies, they must be isolated and characterized. For quantitative aspects, bacterial cell numbers and microbial biomass have to be determined using classical microbiological and modern molecular biological techniques.

9.4.1   Methanogenic and Nitrifying Populations

A polyphasic approach is needed to reveal the diversity, population dynamics and ecological significance of bacteria in permafrost soils and sediments. Enrichment and isolation of microorganisms is necessary for taxonomical and ecophysiological characterization of microbial populations in order to understand their adaptation strategies and potential to extreme environmental conditions.

Traditionally, cell numbers of methanogenic archaea and nitrifying bacteria were quantified by the most probable number (MPN) technique in selective chemolithoautotrophic media [28, 40]. The highest dilution serves as initial inoculum for cultivation studies like identification and characterization of the relevant bacteria. Viable methanogens and nitrifiers were detected in the Kolyma-Indigirka Lowland in northeast Siberia by Russian and German scientists [41]. The bacteria occurred in high cell numbers in the upper layers and in decreasing numbers in more ancient deposits. MPN counts of methanogenic archaea varied between 2.0×10^2 and 2.5×10^7 cells g^{-1}. Soina et al. [42] detected mesophilic nitrifying bacteria with 2.5×10^2 cells g^{-1} soil in a depth of 28 m. Lebedeva and Soina [17] found nitrifying bacteria in geological horizons up to 3 millions of years in a depth of 60 m. With increasing age of the sediments, psychrotrophic nitrifiers were found to be replaced by psychrophilic ones, although the permafrost communities are dominated by psychrotrophs [43]. Nevertheless, psychrophilic bacteria have a significant part in the microbial community in cold environments like permafrost soils [44]. The investigations of the methanogenic community on Taimyr Peninsula [45] and in the Lena Delta [46] gave hints for the adaptation to the low *in situ* temperatures. However, isolation of psychrophilic methanogens and nitrifiers from permafrost soils seems to be more complicated than in other physiological defined groups like acetogenic [47] and methane-oxidizing bacteria [48] as well as clostridia [49]. Cultivation at 5 °C of the slow growing microbes is hindered by prolonged lag-periods, which amounted to 2 - 14 months in the case of nitrite oxidizers [50]. Therefore, the organisms had to be incubated at higher temperatures of e.g., 17 °C. So far, there was only one methanogenic archaea isolated from Ace Lake/Antarctica, which showed psychrophilic growth characteristics [51].

In order to obtain pure cultures for physiological characterization (e.g., determination of the temperature optima) isolation of typical bacteria is required. This can be done by serial dilution in liquid growth media or deep-agar tubes (Agar-shakes), plating on agar plates under aerobic or anaerobic conditions and is also possible by percoll density gradient centrifugation [52]. However, separation of aggregated cells is problematically and requires further treatment. Identification of isolates and enriched organisms was performed by traditional light and electron microscopy with genus-specific morphology and ultrastructure as criteria. Classified by their spiral cell shape, the ammonia oxidizers isolated from soil samples taken during the expedition "Beringia" in 1991 and 1992 were identified as members of the genus *Nitrosospira* (or *Nitrosovibrio*). Among nitrite oxidizers, *Nitrobacter* was identified by its pleomorphic morphology and a polar cap of intra-cytoplasmic membranes [50]. In surface samples which were taken during the expedition "Lena 1999" [46], the coexistence of *Nitrobacter* and *Nitrospira* in enrichment cultures was demonstrated by their typical morphology (Fig 9.4) of pleomorphic short rods (with a diameter of 0.8 μm) respectively spiral rods (with a diameter of 0.2 μm).

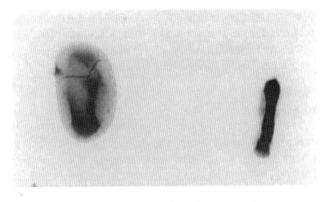

Fig. 9.4 Enriched bacteria in nitrite oxidizing medium from the active layer of a permafrost soil (Samoylov/Lena Delta) with a morphology similar to *Nitrospira* respectively *Nitrobacter*. Negative staining was performed with uranyl acetate. Magnification 20 300×.

Since the isolated organisms may not be the ecologically relevant ones, the development of new detection strategies was necessary to monitor the enrichment procedure. Modern microscopic techniques like CLSM (confocal laser scanning microscopy) in combination with fluorescent dyes enable specific or unspecific labeling of viable cells. In Fig. 9.5 an unspecific labeling of bacteria probably belonging to the genus *Nitrospira* is presented. Here, the organisms were affiliated by the formation of characteristic cell clusters. Like many ammonia oxidizers the *Nitrospira*-like bacteria were aggregated to micro-colonies.

New molecular techniques were developed for the detection of ecological relevant bacteria without cultivation [53, 54]. Fluorescence *in situ* hybridization (FISH) using population-specific gene probes targeting 16S rRNA enables direct microscopic enumeration of single cells (Fig. 9.6). Demanding low amounts of cell material, such methods are well suited for methanogenic archaea and nitrifying bacteria. Molecular 16S rDNA sequence analysis is required for phylogentic affiliation of new isolates.

An immunological approach for the identification of nitrite as well as ammonia oxidizers was developed by Bartosch et al. [55] and Pinck et al. [56]. They used monoclonal and polyclonal antibodies, respectively, recognizing the key-enzymes of these functional groups of bacteria as phylogenetic marker. Nitrite oxidizers enriched from permafrost sediments were identified immunologically as members of the genus *Nitrobacter* [55]. These nitrifiers originated from sediments with an age of 40 000 years. Further on, in the active layer of a permafrost soil from Samoylov Island/Lena Delta nitrite oxidizers of the genus *Nitrospira* were detected by Hartwig [57]. Depending on the substrate concentration, *Nitrospira* together with *Nitrobacte* (0.2 g $NaNO_2 l^{-1}$) or *Nitrobacter* alone (2 g $NaNO_2 l^{-1}$) could be enriched. Both genera of nitrite oxidizers could be distinguished in Western blot analysis by different molecular masses of the β-subunit of their nitrite oxidizing systems (Fig. 9.7). This protein of *Nitrobacter* has a molecular mass of 65 kDa, whereas in *Nitrospira* 46 kDa were determined [55].

Phospholipid analysis in microbial ecology is a further method to study the biomass, population structure, metabolic status and activity of natural communities [59]. Specific groups of microorganisms (like the nitrite oxidizers) contain characteristic phospholipid ester-linked fatty acids (PLFA), whereas methanogenic archaea are characterized by ether-linked glycerolipids [60]. Lipid biomarkers are important for the detection of single taxons. Such a characteristic new fatty acid (11-methyl-palmitate) was recently found in *Nitrospira moscoviensis* [61].

9.4.2   *In situ* Activity

The activity of microorganisms depends not only on their own physiological capability but is influenced also by habitat-qualities like the grain size or the availability of nutrients. That is why, besides the characterization of the microflora, their activity in

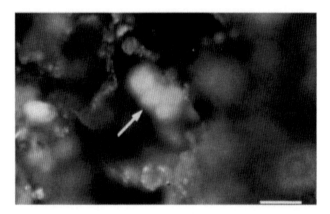

Fig. 9.5 With DAPI (4,6-diamidino-2-phenylindol) stained micro-colony of *Nitrospira*-like bacteria (arrow), enriched from the active layer of a permafrost soil (Samoylov/Lena Delta). Bar = 25 μm. (photo C. Hartwig, University of Hamburg).

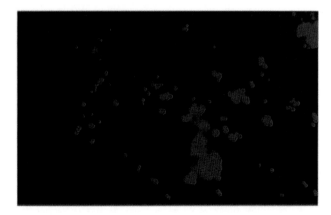

Fig. 9.6 Confocal microscopy of archaea from the family *Methanomicrobiales* enriched from permafrost soils (Lena Delta/Siberia). The culture was grown with H_2/CO_2 (80:20, v:v) at 10 °C. The hybridization was carried out with the oligonucleotide probe MG1200 [54] (photo S. Kobabe, AWI).

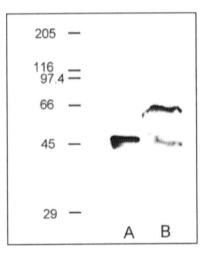

Fig. 9.7 Immunoblot of an enrichment culture derived from permafrost soil using monoclonal antibodies recognizing the β-subunit of the nitrite-oxidizing system. The values on the left are molecular masses in kDa. **A:** pure culture of *Nitrospira moscoviensis*, **B:** enrichment culture using 0.2 g NaNO$_2$ l^{-1} (modified from Bartosch et al. [58]).

the natural habitat is of importance for the understanding of life under extreme conditions. There are different methods for analyzing *in situ* activities, i.e. determination of concentration gradients [62], flux measurements [63] and assay of activity in soil samples [64]. A new technique for the determination of nitrification rates *in situ* is the introduction of microelectrodes (e.g., for ammonia and nitrate). These sensors make it possible to monitor metabolic reactions in the nitrogen cycle [65].

The activity of methanogenic archaea can be followed by the measurement of the metabolic end product CH$_4$ over a period of time. Methane generation takes place only under anaerobic conditions in the permafrost soils and sediments, for example in the water saturated soils of the polygon center. *In situ* rates of methanogenesis can only be obtained if the anaerobic food chain is not affected by the experimental procedure because methanogenesis depends on the substrates produced by other microorganisms. The *in situ* methane production can be investigated by incubation of soil samples from permafrost sites. Figure 9.8 shows the *in situ* methane production in dependence from the natural temperature gradient of a permafrost soil. For this investigation, fresh soil material was used. The prepared soil samples were re-installed in the same layers of the soil profile from which the samples had been taken [46].

The influence of soil texture on the activity of microorganisms can be examined by incubation experiments with model soils of a different grain size [66]. Changing temperature and pressure conditions as well as the impact of different substrates on microbial activity can be studied in special simulation experiments with undisturbed soil samples (see 9.4.4).

To estimate the nitrifying activity in permafrost sediments, the potential activity of soil bacteria was determined under optimal laboratory conditions. For that purpose, soil samples were taken from drill cores and transferred to the laboratory under sterile

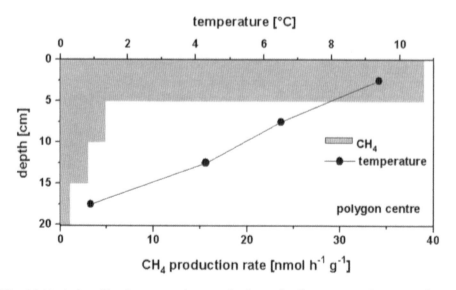

Fig. 9.8 Vertical profile of *in situ* methane production and soil temperature for a permafrost soil of the polygon center.

conditions. In the active layer and in sediments with an age of 40 000 years the nitrifying activity was higher at 28 °C in comparison to 17 °C, whereas in more ancient deposits (0.6-1.8 million of years and 2.5-3 million of years) the bacteria preferred lower temperatures of 17 °C (Lebedeva, pers. comm.).

Further investigations about nitrifying bacteria in permafrost sediments included measurements of ammonia, nitrite and nitrate as substrates respectively products of nitrification. Janssen [50] determined the concentrations of these nitrogen compounds in the soil samples by high-performance-liquid chromatography. The profiles showed that nitrite and nitrate were always found in the ppm range in sediments up to 150 000 years and occasionally in deeper layers. Ammonia concentrations amounted up to 100 ppm with increasing amounts in sediments with an age of 1-5 millions years. Nitrite and nitrate correlated with the presence of nitrifying bacteria although nitrifiers were also detected in samples without these nitrogen salts. The detection of the chemical unstable metabolic intermediate nitrite in correlation with the presence of viable ammonia oxidizers gave first evidence of modern microbial activity in permanently frozen sediments.

9.4.3   Isotopic Analysis: Carbon Fractionation via Microbial Processes in Permafrost

It is well known that microbial processes tend to fractionate the C-isotopes of organic matter in soils and sediments by favoring the lighter ^{12}C-carbon over the heavier ^{13}C-compounds. Methanogenesis for example leads to the strongest C-discrimination in nature with the result that soil organic matter will be enriched with ^{13}C-carbon (e.g., δ^{13}C-values of about -16‰ to -22‰) while the product of anaerobic decay - methane - will be depleted with ^{13}C (values of about -60‰ [67]). In anaerobic zones of permafrost soils with methane production the soil organic matter showed an absolute enrichment of ^{13}C-carbon of about 3.7‰ to 8.3‰ [45, 68]. Therefore, isotope-related analysis in combination with the microbial studies may be a powerful tool to search for traces of microbial life in extraterrestrial habitats, even if the applicability for extraterrestrial environments could not be examined until now sufficiently [69].

9.4.4   Simulation Experiments

The influence of environmental conditions on the activity and survival of microorganisms could be investigated by simulation experiments with bacterial cultures and with undisturbed soil material.

The thawing and freezing processes influence not only the soil temperature regime but also the availability of liquid water, the pressure conditions and the salinity of pore water. They produce also granular, platy and vesicular soil structure in the surface near horizons and a massive structure in the subsurface zones. Undisturbed soil samples (soil cores of different size) save the structure, pore system and stratification of the natural soil, which influence the interaction between microbes and soil matrix. Simulation of freeze-thaw cycles can help to understand: how will the microbial population be influenced by the natural permafrost system and by the interaction of the combined parameters?

The viability of the permafrost microflora under the environmental conditions of Martian atmosphere can be investigated by simulation experiments in special ice laboratories (Alfred Wegener Institute for Polar and Marine Research, AWI) and in a special Martian simulator (German Aerospace Center, DLR). Natural soil material and pure cultures of bacteria isolated from terrestrial permafrost habitats can be exposed to extreme cold temperatures (-60 °C), lower pressure (560 Pa), higher background radiation (UV 200 nm), drier soil moisture conditions and varying ice contents in comparison with well known terrestrial permafrost.

9.5   Conclusion and Future Perspectives

Microbial life in permafrost soils depends on available water (see Chap. 5, Brack). If inorganic compounds like hydrogen as well as ammonia or nitrite are present as substrates, conditions are favorable for the growth of methanogenic archaea and nitrifying bacteria. Since cell synthesis is carried out by the assimilation of carbon dioxide there is no further need for organic material. This mineralic basis resembles the situation on mars (e.g., C, H, O, N, P, K, Ca, Mg and S, reviewed by Horneck [70]). Lithoautotrophic bacteria are well investigated and ubiquitous distributed organisms on Earth. They survived even in terrestrial permafrost for several millions of years [41]. Here, they demonstrate the residue of the autochthon population within the paleosoils which was enclosed during deposition of fresh sediments. The frozen microorganisms in the deeper permafrost sediments are thought to have not evolved significantly during the past several million years because it was not necessary to adapt to their environment [16]. In contrast, the microbes living in the active layer and the transition permafrost sediments are influenced by extreme changes of live-decisive environmental conditions. Preserving their viability in such an extreme environment they had to develop different strategies to resist desiccation, freezing processes and nutrient-lack. The isolation and characterization of methanogenic archaea and nitrfying bacteria from permafrost soils should clarify the possible growth characteristic (psychrophile and psychrotroph) and ecological significance of these microbes.

The data obtained from future research on living conditions and adaptation strategies of microorganisms in terrestrial permafrost soils should be compared with the postulated environmental conditions on early Mars [1, 71]. They were characterized by liquid water, a moderate climate and a postulated biosphere which had been dominated by anaerobic processes and diversification of anaerobic organisms. Furthermore, the comparative system studies will serve for understanding the modern Mars cryosphere and other extraterrestrial permafrost habitats. This knowledge represents an essential basis for the understanding of the origin of life and the environmental development on extreme habitats.

Acknowledgements. The authors thank Dr V. Rachold and W. Schneider (Alfred Wegener Institute for Polar and Marine Research, Potsdam) for organization and logistic support of the expedition "Lena 1999" as well as Dr. D.A. Gilichinsky (Institute of Soil Science and Photosynthesis, Pushchino) for the leadership of the expedition "Beringia". The research was partly founded by the German Ministry of Science and Technology (System Laptev-See 2000, 03G0534G).

9.6   References

1   R.A. Wharton Jr., C.P. McKay, R.L. Mancinelli, G.M. Simmons Jr., Adv. Space Res. **9**, 147 (1989).

2   G. Horneck, Planet. Space Sci., in press.

3   Soil Survey Staff, *Keys to Soil Taxonomy.* Lincoln, Nebraska. U.S. Department of Agriculture, Soil Conservation Service. Pocahontas Press, Blacksburg, 1998.

4   H.M. French, *The periglacial environment.* Longman, Westminster, 1996.

5   R.A. Herbert, in: R.A. Herbert, G.A. Cod (Eds.) *The ecology and physiology of psychrophilic microorganisms, Microbes in extreme environments*, Academic Press, London 1996, pp. 1.

6   E.D. Yershov, *General Geochryology*, Cambridge University Press, Cambridge, 1998.

7   A.M. Gounot, J. Appl. Bacteriol. **71**, 386 (1991).

8   N.J. Russel, T. Hamamoto, in: K. Horokoshi, W.D. Grant, (Eds.) *Extremophiles: microbial life in extreme environments*, Wiley, New York, 1998, pp. 25.

9   R.Y. Morita, Bacteriol. Rev. **39**, 144 (1975).

10   E. Rivkina, E.I. Friedmann, C.P. McKay, D. Gilichinsky, Appl. Environ. Microbiol. **66**, 3230 (2000).

11   R.Y. Morita, in: J. Lederberg, (Ed.) *Encyclopedia of microbiology*, Academic Press, San Diego, 1992, pp. 625.

12 A.A. Ananyan, Merzlotnye Issledovaniya **10**, 267 (1970) (in Russian).
13 V.E. Ostroumov, C. Siegert, Adv. Space Res. **18**, 79 (1996).
14 J. McGrath, S. Wagener, D.A. Gilichinsky, in: D.A. Gilichinsky (Ed.) *Viable micro-organsisms in permafrost*, Pushchino Research Centre, Pushchino, 1994, pp. 48.
15 N.J. Russel, Phil. Trans. Roy. Soc. Lond. **326**, 595 (1990).
16 T. Shi, R.H. Reeves, D.A. Gilichinsky, E.I. Friedmann, (1997). Microb. Ecol. **33**, 169 (1997).
17 E. Lebedeva, V. Soina, in: D.A. Gilichinsky (Ed.) *Viable microorganisms in permafrost*, 1st Int. Conference on Cryopedology and Global Change, November 1992, Pushchino Research Centre, Pushchino, 1994, pp. 74.
18 D.A. Gilichinsky, E.A. Vorobyova, L.G. Erokhina, D.G. Fyordorov-Dayvdov, N.R. Chaikovskaya, Adv. Space Res. **12**(4), 255 (1992).
19 R.Y. Morita, Microbial Ecology **38**, 307 (2000).
20 J.L. Garcia, FEMS Microbiol. Rev. **87**, 297 (1990).
21 U. Deppenmeier, V. Müller, G. Gottschalk, Arch. Microbiol. **165**, 149 (1996).
22 W.B. Whitman, T.C. Bowen, D.R. Boone, in: A. Balows, H.G. Trüper, M. Dworkin, W. Harder, K.H Schleifer, (Eds.) *The Prokaryotes*, Springer Verlag, New York, 1992, pp. 719.
23 A.M. Gounot, in: R. Margesin, F. Schinner (Eds.) *Cold-adapted organisms*, Springer, Berlin, 1999, pp. 3.
24 D.E. Robertson, D. Noll, M.F. Roberts, J.A. Menaia, D.R. Boone, Appl. Environ. Microbiol. **56**, 563 (1990).
25 E. Rivkina, E.I. Friedmann, C.P. McKay, D.A. Gilichinsky, Geomicrobiology **15**, 187 (1998).
26 V. Peters, R. Conrad, Appl. Environ. Microbiol. **61**, 1673 (1995).
27 K.O. Stetter, G. Fiala, G. Huber, R. Huber, A. Segerer, FEMS Microbiol. Rev. **75**, 117 (1990).
28 S.W. Watson, E. Bock, H. Harms, H.P. Koops, A.B. Hooper, in: J.T. Staley, M.P. Bryant, N. Pfennig, J.G. Holt (Eds.) *Bergey's Manual of Systematic Bacteriology*, 1st Ed., Vol 3, Williams & Wilkins Co., Baltimore, 1989, pp. 1808.
29 E. Bock, H.-P. Koops, in: A. Balows, H.G. Trüper, M. Dworkin, W. Harder, K.H. Schleifer, (Eds.) *The Prokaryotes*, Springer Verlag, New York, 1992, pp. 2302.
30 H.-P. Koops, U.C. Möller, in: A. Balows, H.G. Trüper, M. Dworkin, W. Harder, K.H. Schleifer (Eds.) *The Prokaryotes*, Springer Verlag, New York, 1992, pp: 2625.
31 A. Teske, E. Alm, J.M. Regan, S. Toze, B.E. Rittmann, D.A. Stahl, J. Bacteriol. **176**, 6623 (1994).
32 S. Ehrich, D. Behrens, E. Lebedeva, W. Ludwig, E. Bock, Arch. Microbiol. **164**, 16 (1995).
33 E. Spieck, E. Bock, in: G.M. Garrity, S.T. Williams, J.T. Staley, D.J. Brenner, J.G. Holt, D.R. Boone, R.W. Castenholz, N.R. Krieg, K.H. Schleifer (Eds.) *Bergey's Manual of Systematic Bacteriology*, 2nd Ed., Vol 1, Williams & Wilkins Co., Baltimore, in press.
34 R. Stüven, E. Bock, Wat. Res. **35**(8), 1905 (2001).
35 K. Wilson, J.I. Sprent, D.W. Hopkins, Nature **385**, 404 (1997).
36 R. Mansch, E. Bock, Biodegradation **9**, 47 (1998).
37 J.K. Fredrickson, T.R. Garland, R.J. Hicks, J.M. Thomas, S.W. Li, K.M. McFadden, Geomicrobiol. J. **7**, 53 (1989).
38 L. Lin, *Kompatible Solute in nitrifizierenden Bakterien*, Ph.D. Thesis, University of Hamburg, Hamburg (1994).

39   R. Conrad, Microbiol. Rev. **60**, 609 (1996).

40   D. Wagner, E.M. Pfeiffer, FEMS Microbiol. Ecol. **22**, 145 (1997).

41   E. Rivkina, D. Gilichinsky, S. Wagener, T. Tiedje, J. McGrath, Geomicrobiol. **15**, 187 (1998).

42   V.S. Soina, E.V. Lebedeva, O.V. Golyshina, D.G. Fedorov-Davydov, D.A. Gilichinsky. Microbiologia **60**, 187 (1991) (in Russian).

43   I. Friedmann, in: D.A. Gilichinsky (Ed.) *Viable microorganisms in permafrost*, 1st Int. Conference on Cryopedology and Global Change, November 1992, Pushchino Research Centre, Pushchino, 1994, pp. 21.

44   N.S. Panikov, in: W.C. Oechel, (Ed.) *A kinetic approach to microbial ecology in arctic and boreal ecosystems in relation to global change. Global change and arctic terrestrial ecosystems*, Springer, Berlin, 1997, pp. 171.

45   V.A. Samarkin, A. Gundelwein, E.M. Pfeiffer, in: H. Kassens, H.A. Bauch, I.A. Dmitrenko, H. Eicken, H.-W. Hubberten, M. Melles, J. Thiede, L.A. Timokhov (Eds.) *Land-ocean systems in the Siberian arctic*, Springer, Berlin, 1999, pp. 329.

46   E.-M. Pfeiffer, D. Wagner, H. Becker, A. Vlasenko, L. Kutzbach, J. Boike, W. Quass, W. Kloss, B. Schulz, A. Kurchatova, V.I. Pozdnyakov, I. Akhmadeeva, in: V. Rachold, M.N. Gregoriev (Eds.) *Russian-German cooperation System Laptev Sea 2000: The Expedition Lean 1999*, Reports Polar Res. **354**, 22 (2000).

47   O.R. Kotsyurbenko, M.V. Simankova, A.N. Nozhevnikova, T.N. Zhilina, N.P. Bolotina, A.M. Lysenko, G.A. Osipov, Arch. Microbiol. **163**, 29 (1995).

48   M.V. Omelchenko, L.V. Vasilyeva, G.A. Zavarzin, Curr. Microbiol. **27**, 255 (1993).

49   G. Finne, J.R. Matches, Can. J. Microbiol. **20**, 1639 (1974).

50   H. Janssen, *Anreicherung, Isolierung und Charakterisierung nitrifizierender und methylotropher Bakterien aus Permafrostboden*, Diploma Thesis, University of Hamburg, Hamburg, 1994.

51   P.D. Franzmann, Y. Liu, D.L. Balkwill, H.C. Aldrich, E. Conway de Macario, D.R. Boone, Antarctica. Int. J. Syst. Bacteriol. **47**, 1068 (1997).

52   K.P. Putzer, L.A. Buchholz, M.E. Lidstrom, C.C. Remsen, Appl. Environ. Microbiol. **57**, 3656 (1991).

53   R.I. Amann, W. Ludwig, K.-H. Schleifer, Microbiol. Rev. **59**, 143 (1995).

54   L. Raskin, J.M. Stromley, B.E. Rittmann, D.A. Stahl, Appl. Environ. Microbiol. **60**, 1232 (1994).

55   S. Bartosch, I. Wolgast, E. Spieck, E. Bock, Appl. Environ. Microbiol **65**,4126 (1999).

56   C. Pinck, C. Coeur, P. Potier, E. Bock, Appl. Environ. Microbiol. **67**, 118 (2001).

57   C. Hartwig, *Anreicherung von Nitrospira aus Naturproben*, Diploma Thesis, University of Hamburg, Hamburg, 1999.

58   S. Bartosch, C. Hartwig, E. Spieck, E. Bock, *Detection of Nitrospira-like bacteria in various soils*, submitted.

59   J.R. Vestal, D.C. White, Bioscience **39**, 535 (1989).

60   S. Ohtsubo, M. Kanno, G. Miyahara, S. Kohno, Y. Koga, I. Miura, FEMS Microbiol. Ecol. **12**, 39 (1993).

61   A. Lipski, E. Spieck, A. Makolla, K.H. Altendorf, *Fatty acid profiles of nitrite-oxidizing bacteria reflect their phylogenetic heterogeneity*, submitted.

62   F. Rothfuss, R. Conrad, Biogeochem. **18**, 137 (1993).

63   T. Christensen, Biogeochem. **21**, 117 (1993).

64   O.R. Kotsyurbenko, A.N. Nozhevnikova, T.I. Soloviova, G.A. Zavarzin, Antonie van Leeuwenhoek **69**, 75 (1996).

65 A. Schramm, L.H. Larsen, N.P. Revsbech, N.B. Ramsing, R. Amann, K.-H. Schleifer, Appl. Environ. Microbiol. **62**, 4641 (1996).

66 D. Wagner, E.-M. Pfeiffer, E. Bock, Soil Biol. Biochem. **31**, 999 (1999).

67 H.W. Scharpenseel, E.M. Pfeiffer, in: R. Lal (Ed.) *Soil Processes and the Carbon Cycle*, CRC Press, Boca Raton, 1998, pp. 577.

68 E.-M. Pfeiffer, H. Janssen, in: J. Kimble, R.J. Ahrens (Eds.) *Proceedings of the meeting on the classification, correlation, and management of permafrost-affected soils*, Alaska, USA. USDA, Lincoln, 1994, pp. 90.

69 L.J. Rothschild, D. DesMarais, Adv. Space Sci. **9**, 159 (1989).

70 G. Horneck, Planet. Space Sci., **48**, 1053 (2000).

71 G. Horneck, Planet. Space Sci. **43**, 189 (1995).

10   Life in Cold Lakes

Erko Stackebrandt and Evelyne Brambilla

High radiation, low gravity, and low temperatures are among the physical parameters, in which non-terrestrial planets and their satellites are considered to differ most significantly from those permitting life on the planet Earth. Questions about whether or not extra-terrestrial life forms may indeed have been generated during the past 4 billion year old evolution of the solar system, and, if answered positively, whether these forms still exist or are available as fossil records, poses some of the greatest scientific challenges in modern biology. The spectrum of scientific problems to be responded to range from the search for and identification of alternatives to solutions of generating and maintaining life developed during the evolution of life on Earth, to hypotheses how evolution of (terrestrial-type) life may have been changed under the specific planetarian conditions, to the technologically demanding efforts to monitor and identify the degradation products of previously existing life, even under conditions that may not have been simulated on our planet.

Due to the necessity of working with terrestrial life forms, most exobiology studies have been performed on those organisms that represent descendants of ancestral organisms that evolved 3.8 to 3.0 billion years ago. Bacterial spores, radiation-resistant bacteria, and simple yeast and algae were analyzed most thoroughly with respect to the influence of radiation, dessication, and micro-gravity [1]. These physical-chemical studies are today complemented by studies searching for novel terrestrial microorganisms in environments, considered most inhabitable. These explorations led to the discovery of extreme halophilic, alkaliphilic, acidophilic, thermophilic and barophilic Bacteria and Archaea (see Chap. 11, Stetter), thus broadening our perception of the term "extremophilic". Moreover, novel molecular tools were introduced into biology and ecology which enable microbiologists to determine the natural relatedness and molecular diversity even of organisms that cannot yet be cultured in the laboratory.

One of the prime regions for unraveling the extent of microbial diversity and for determining the lower limits of life-supporting environmental parameters is the Antarctic continent. While initial surveys concentrated on the discovery of biotechnologically exploitable species, this environment is now the testing field for the development and application of methods to be used in future missions to other planets. Slopes of vulcanos, dry valleys, endolithic rocks, the coastal regions and perennial freshwater, saline and hypersaline lakes are among the sites targeted most frequently in the Antarctic. Indeed, many novel psychrotolerant and psychrophilic microbial species have been described [2-5]. The discovery of the Antarctic subglacial environment such as Lake Vostok [6] has fascinated scientists as it may represent an environment of the Neoproterozoic, a time period from 750-543 Ma ago. Microorganisms that may be found in the water body located beneath the 4 km of glacial ice may help scientists to understand the effect of high environmental stresses and to unravel evolutionary processes that occurred within and outside this highly sheltered ecological niche. Sensing methods to be developed for the exploration of Lake Vostok without its contamination may lead to the design of devices to be used in the search for life on Mars, the ice moon of Jupiter, Europa (see Chap. 7, Greenberg), and other elements of the solar system (see Chap. 24, Foing).

The introduction of culture-independent molecular screening techniques has allowed microbiologist to examine a spectrum of microbial diversity that is not necessarily reflected by the results of culturing studies. Together with culture studies, this method allows the assessment of a significantly broader range of diversity than obtained in the past. In the absence of information about conditions to optimally cultivate microorganisms, the culture-independent approach provides us with information about the phylogeny (meaning the determination of the course of natural relatedness during the evolution of living matter), of novel types of organisms from which their physiological role may be deduced.

This communication summarizes the results of three studies on the composition of microbial communities from Antarctic lakes [2, 3, 5]. The results point towards a rich, mainly cold-adapted prokaryotic diversity which may be used as baseline information for the assessment of microbial communities of other terrestrial subglacial environments. The tools described represent state-of–the-art technologies, which may be among those applied to monitoring life forms in subglacial lakes, possibly encountered on Mars and Europa.

10.1 Source of Samples and DNA Analyses

Samples were retrieved from a microbial mat of the shallow, moated area of the freshwater Lake Fryxell, McMurdo Dry Valleys [5], of the anoxic sediments of marine salinity meromictic (sample salinity in g kg^{-1}) Clear Lake (14), Pendant Lake (23), Scale Lake (28), Ace Lake (35), and Burton Lake (43) [2], as well as from sediments of the hypersaline Ekho Lake (150), Organic Lake (210) and Deep Lake (320) [3], Vestfolds Hills. Isolation of DNA, PCR amplification of 16S rRNA genes using Bacteria and Archaea-specific primer pairs, cloning, sequence analyses of PCR-amplified stretches of the rDNA and phylogenetic analyses has been described in detail by Bowman et al. [2, 3] and Brambilla et al. [5]. The basic principle of this approach is based upon the comparative analysis of the nucleic acid sequence of a gene which is present in all organisms known, from the evolutionary very ancient bacteria, to plants, animals and humans. The gene product has been part of the translation apparatus of the first living organism that evolved on Earth (or possibly somewhere else, e.g., Mars [7]) and its function has not been changed for about 4 billion years. The sequence is composed of almost invariable parts, conservative enough to detect the main lines of descent, while, at the same time, it contains variable regions that allow determination of closely related organisms. Any organism or nucleic acid, no matter where it will be found, that has maintained the basic conservative structure of this gene, will be identified as a member of terrestrial life forms; its analysis will then decide about whether the sequence belongs to a known or to a novel evolutionary line of descent (see also Chap. 11, Stetter).

Between about 45 and 320 clones were analyzed from the mat and sediment samples and the partial 16S rDNA sequences compared to those of cultured and as-yet uncultured organisms, deposited in public data bases. Due to methodological differences the sequences of non-overlapping stretches can only be compared with each other indirectly by phylogenetic analysis with the almost complete sequences of reference organisms. As most partial sequences do not indicate the presence of hitherto unrecognized phyla, inter-laboratory comparison at least at the level of orders, often families, sometimes even species, is possible. Clones sharing higher 98% sequence similarity formed taxonomic units, named phylotypes [2, 3] or potential species [5].

10.2 The Rich Diversity of Bacteria and Archaea

The lakes compared in this study differ significantly from each other in physicochemical parameters. The meromictic lakes from the Vestfold Hills system have different salinity, and water temperature (Deep Lakes annual sediment temperature range from –14 °C to –18 °C, while Ekho Lakes annual sediment temperature range from 13-18 °C). Organic Lake's bottom water is rich in dimethylsulfide. The marine-salinity type and the hypersaline meromictic lakes were formed by evaporation of marine waters about 10 000 years ago and have remained stable for about 4000 years [2]. They are nearly or completely perennially ice-free. The lower part of the water column remains unmixed, due to a distinct density gradient which forms a physical barrier to mixing of water. Microorganisms and other matter sink gradually through the water column. In contrast, Lake Fryxell, McMurdo Dry Valleys, is covered by perennial ice, 3-4.5 m thick. During the summer months moats are formed around the edge, which allows material to be transported into the lake from glaciers and land by streams. These streams not only discharge inorganic (PO_4^{3-} and NO_3^-) compounds but also a rich spectrum of organic material by the inflow of primary producers such as algae and cyanobacteria [8]. On a volumetric basis, Lake Fryxell is the most productive lake in the McMurdo Dry Valleys [9]. The lake contains a complex protistan community [10] which constitutes a rich reservoir of nutrients. Sediment deposited on the ice surface is migrating downwards during the Antarctic summer providing a constant input of inorganic and organic material, including microorganisms, while no reports are available on prokaryotic diversity.

The large and diverse spectrum of microorganisms detected by 16S rDNA sequences demonstrate the complexity of the microbial community in Antarctic lakes sediments and mats. As already described for other studies comparing the 16S rDNA from prokaryotic cellular organisms and uncultured representatives the match is low [11]. Most sequences fall into the radiation of higher taxa which are well represented in temperate environments analyzed so far. These are the species-rich lineages of *Proteobacteria*, Gram-positive bacteria, Cyanobacteria-Chloroplasts, Cytophaga-Flavobacterium-Bacteroides (CFB) complex and the *Verrucomicrobiales-Planctomyces-Chlamydium* (VPC) complex (Fig. 10.1). Depending on the lake sediment analyzed, different sublines are represented to a varying extent such as four of the five subclasses of the class *Proteobacteria*, low- and high G+C mol% Gram-positive bacteria (*Clostridium/Bacillus* and *Actinobacteria*, respectively), and each of the three main sublines of the CBF and VPC complexes (Table 10.1.).

Comparison of the phylogenetic diversity of uncultured organisms of the Lake Fryxell mat sample [5] with that of different sediments of hypersaline and marine-type salinity meromictic Antarctic lakes of the Vestfold Hills system [2, 3] reveals a superficial similarity in that only certain main bacterial lineages are represented. Each of the sample appear to show some specific features in community structure that may be related to the physico-chemical environment and hence may shed some lights on the functional role these bacteria might play at the sites they thrive.

Lake Fryxell is the most thoroughly investigated lake, as about 320 clone sequences have been analyzed. Gram-positive bacteria with a low GC content (clostridia and relatives) as well as flavobacteria, Bacteroides, ß-*Proteobacteria* and verrucomicrobia are dominating. Some potential species belong to gliding bacteria (*Stigmatella, Myxococcus, Cytophaga, Flavobacterium, Marinilabilia,* and *Flexibacter*). On the other hand, cyanobacteria, green non-sulfur bacteria, planctomycetes, spirochetes and haloanaerobia, sometimes found in the hypersaline and marine salinity lakes, are missing. The rich diversity of clostridia and related organisms must be viewed in light of the presence of a food web that contains anaerobic saccharolytic organisms forming CO_2, H_2 and C_1-C_4 acids and alcohols which in turn are metabolized by other anaerobic members of the community as well as by aerobes, facultative anaerobes and phototrophic organisms. The finding of some sequences being related to those of sulfur-metabolizing organisms may indicate the presence of a sulfur cycle. The occurrence of

Table 10.1. Compilation of the major bacterial groups represented in the clone library of DNA isolated from sediments and mat material of nine Antarctic lakes. Values are in percent of total clones analyzed.

Sample / Lake	*Proteo-bacteria*	Gram-positives	CFB	PVC	*Cyano-bacteria*	*Spiro-chaeta*
Fryxell	19	43	24	12	1	0
Ekho	16	36	27	1	16	3
Organic	16	0	29	0	41	0
Deep	100	0	0	0	0	0
Clear	11	27	5	8	5	0
Pendent	10	35	0	24	0	9
Scale	8	75	0	5	0	0
Ace	6	50	1	5	0	1
Burton	10	25	22	3	29	3

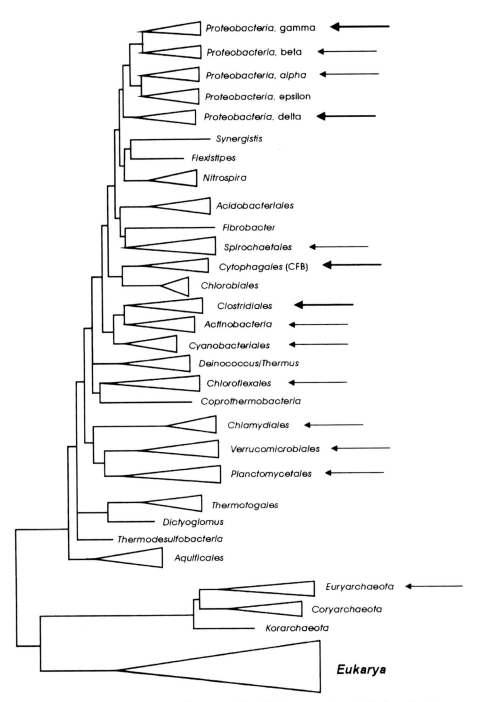

Fig. 10.1 Main 16S rDNA lines of descent (phyla, division, complexes) of cultured prokaryotes. The affiliation of clone sequences originating from DNA of sediments and mat samples of Antarctic lakes are indicated by arrows. Bold arrows: presence of members in the majority of lakes; thin arrows: members present in a few lakes only. This illustration is a dendrogram-type display of the evolution of main lineages of life forms shown in a radial fashion by Stetter, Chap. 11.

aerobic organisms [5] may be explained not only by the sedimentation effect but by the sampling site which is located in the shallow part of a moated zone; in this case, the aerobic organisms are part of the metabolically active community in a structured mat, serving both as a sink for gases, carbon and other compounds, as well as a nutrient source. The hypersaline lakes show a restricted spectrum of diversity. With respect to the main components, Ekho Lake shows a similar diversity than Lake Fryxell, but contains a high fraction of *Haloanaerobiales* (Gram-positives). In addition, it contains members of the cyanobacteria-chloroplast division. The presence of the latter organisms is even more pronounced in the shallow Organic Lake which, however, lacks the clostridia-type fermenters. Even more dramatic is the situation in the aerobic Deep Lake sediment in which bacterial representatives originate exclusively from ß- and γ-Proteobacteria.

In contrast to the hypersaline lakes, the marine-salinity-type lakes show a higher bacterial diversity than that shown in Table 10.1., e.g., members of ß- and especially δ-Proteobacteria (sulfur depended bacteria), chlamydiae, actinobacteria, planctomycetes, and green non-sulfur bacteria. Especially Burton Lake seems to be rich in *Bacteroides* and *Flavobacterium* phylotypes. However, the restricted number of clones determined for these Lakes may bias the actual degree of diversity; as shown for Lake Fryxell, even the analysis of more than 300 sequences is insufficient to explore the full spectrum of diversity.

Not unexpected, those lineages containing thermophilic and thermotolerant bacteria are missing at all such as *Thermotogales*, *Aquificales*, *Thermus*, *Deinococcus*, and the like. Though members of Proteobacteria, low G+C Gram-positive bacteria, *Cytophagales*, Bacteroides and a few *Actinobacteria* were also found in some of the marine salinity and hypersaline lakes, the phylogenetic affiliation of clones at the genus level was almost always different. In addition to the expected occurrence of anaerobic organisms in anoxigenic sediments (the lake Fryxell mat is an exception as it originated from the moated, shallow part of the lake), the analyses of clone libraries pointed toward a rich spectrum of aerobic forms as well. As discussed by Bowman et al. [3] for other Antarctic lakes, the presence of aerobic bacteria in anaerobic sediments may be due to the their accumulation followed by slow decomposition of organic carbon. Comparison of the phylogenetic affiliation of the Lake Fryxell clone sequences [5] to their nearest neighbors with those of clone sequences depicted in 16S rDNA dendrograms of Bowman et al. [2, 3] does not, except for two clones, indicate high relationship. These clones are affiliated to *Clostridium estertheticum* that thrive under different physiological conditions such as those found in Lake Pendent (2.3% salinity) and Lake Fryxell (freshwater).

The rather rich diversity in most of the lake sediment samples (which will turn out to be even larger with more clones analyzed) may be explained by the evolution of cold-adapted and halophilic forms that were present at the time of the formation of lakes. In addition, over tens of thousands of years air-born (as far as from the South American continent) and autochthonic organic matter, including microbial cells, have entered the chemically stratified meromictic and the freshwater lakes by migration through the ice cover, and discharge of creeks and streams. New microorganisms with novel biochemical potential may have enabled the system to expand the ecological diversity, allowing even eukaryotic cells to enter the niches (as derived from the rich diversity of protozoa Lake Fryxell) and from the presence chloroplasts of eukaryotic algae in the clone library of meromictic lakes. As noted by Bowman et al. [3], the upper water column of some of the meromictic hypersaline lakes contain a rich flora of novel aerobic bacteria. The same is true for the diversity of anaerobic and aerobic bacteria of the lake Fryxell mat system [12].

In general, the diversity of Archaea was found to be much more restricted than the bacterial diversity. Ninety percent of the clones of Lake Fryxell (>300 clones) could be affiliated to *Methanoculleus palmolei*, while 10% were related to an Antarctic clone sequence Ant12. Ekho Lake and Organic Lake sediments contain an insignificant number of archaeal clones while in the sediment of the aerobic and highly halophilic Deep Lake several phylogenetically different and mostly novel halobacterial lineages emerged. Sequences that were retrieved from Organic Lake were also present in Deep Lake. A significantly higher diversity of euryarchaeal clones were present in marine salinity lakes, some of which were related to clones described from salt marsh in the UK and marine sediments [2].

10.3 Conclusions

The molecular analyses of lake samples form Antarctica supports the previous notion that molecular methods are able to trace prokaryotic life in any environments of the planet Earth. The range of samples cover geysers, deep sea hydrothermal vents, deserts, alpine lakes, deep subsurface rocks, not to mention various marine and terrestrial sites, as well as microbe-eukaryotic associations. It also confirms the textbook knowledge that microbial diversity depends upon the physico-chemical status of an environmental sample - and as hardly any two niches are identical it is unlikely to find two highly similar (quality and quantity) microbial communities. In this respect, the situation of the Antarctic lakes sediments constitute vastly different environments.

Though the DNA quantities necessary for PRC amplification have not yet been investigated thoroughly, especially in mixtures of varying quantities, the power of the rDNA approach to detect life forms is at present superior to any other known method [13]. The samples analyzed in this study had the advantage of being high in microbial cell quantity, hence in extracted DNA (except for the sediment of Deep Lake). In cases of low numbers of prokaryotic cells even this method must be performed with great care as the risk of contamination with foreign DNA is high. This problem is not to be underestimated if novel environments are to be explored which have remained undisturbed for hundred thousand years (Lake Vostok [14], Polar Lake) or which may be selected for the testing of detection methods apt to search for extraterrestrial life forms (Mars, Europa) which follow the same genetically and biochemical strategies of organisms that have evolved on the planet Earth.

Acknowledgement. The Lake Fryxell part of this communication was supported by a EU grant PL 970040 (MICROMAT).

10.4 References

1   W.L. Nicholson, N. Munakata, G. Horneck, H.J. Melosh, P. Setlow, Microbiol. Mol. Biol. Rev. **64**, 548 (2000).

2   J.P. Bowman, S.M. Rea, S.A. McCammon, T.A. McMeekin, Env. Microbiol. **2**, 227 (2000).

3   J.P. Bowman, S.A. McCammon, S.M. Rea, T.A. McMeekin, FEMS Microbiol. Lett. **183**, 81-88 (2000).

4   B. Tindall, E. Brambilla, M. Steffen, R. Neumann, R. Pukall, R.M. Kroppenstedt, E. Stackebrandt, Env. Microbiol. **2**, 310 (2000).

5   E. Brambilla, H. Hippe, A. Hagelstein, B.J. Tindall, E. Stackebrandt, Env. Microbiol. (2000) in press.

6   J. Priscu, Science **280**, 2095 (1998).

7   C. Mileikowsky, F.A. Cucinotta, J.W. Wilson, B. Gladman, G. Horneck, L. Lindegren, J. Melosh, H. Rickman, M. Valtonen, J.Q. Zheng, Icarus **145**, 391 (2000).

8   D. McKnight, A. Alger, C.M. Tate, G. Shupe, S. Spaulding, in: J.C. Priscu (Ed.) *Ecosystem dynamics in a polar desert. The McMurdo Dry Valleys, Antarctica,* American Geophysical Union, Antarctic Research Series 72, Washington, DC, 1998, pp. 109.

9   M.R. James, J.A. Hall, J. Laybourn-Parry in: J.C. Priscu (Ed.) *Ecosystem dynamics in a polar desert. The McMurdo Dry Valleys, Antarctica,* American Geophysical Union, Antarctic Research Series 72, Washington, DC, 1998, pp. 255.

10   J. Laybourn-Parry, M.R. James, D. McKnight, J. Priscu, S. Spaulding, R. Shiel, Polar Biol. **17**, 54 (1997).

11   H. Rheims, A. Felske, S. Seufert, E. Stackebrandt, J. Microbiol. Meth. **36**, 65 (1998).

12   B. Tindall et al., pers. comm.

13   F. von Wintzingerode, U. Göbel, E. Stackebrandt, FEMS Microbiol. Rev. **21**, 213 (1997).

14   S.S.Abyzov, I.N. Mitskevich, M.N. Poglazova, Microbiology **67**, 451 (1998).

11 Hyperthermophilic Microorganisms

Karl O. Stetter

The first traces of life on Earth date back to the Early Archaean age. Microfossils of prokaryotes demonstrate the existence of life already 3.5 to 3.9 billion years ago [1, 2]. Although nothing is known about the original growth temperature requirements of these organisms, the Earth is generally assumed to having been much hotter at that time than today [3]. Recently, fossil-remains of thread-like microorganisms have been discovered in a 3.2 billion year old deep sea volcanogenic hydrothermal deposit, indicating the existence of hyperthermophiles already at Early Archaean times [4].

Today, most life forms known are mesophiles adapted to ambient temperatures within a range from 15 to 45 °C. Among Bacteria, thermophiles (heat-lovers) have been recognized for some time which grow optimally (fastest) between 45 and 70 °C. They thrive within sun-heated soils, self-heated waste dumps and thermal waters and are closely related to mesophiles. During the last years, hyperthermophilic Bacteria and Archaea (formerly the Archaebacteria) with unprecedented properties have been isolated mostly from areas of volcanic activity [5-7]. They grow between 80 and 113 °C and represent the organisms at the upper temperature border of life. At ambient temperatures, although unable to grow, hyperthermophiles can survive for many years in a kind of "frozen" state. Within eukaryotes, the upper temperature of growth known is about 60 °C and, therefore much lower than in Bacteria and Archaea [8]. Here, an overview will be presented about the biotopes, modes of life, and phylogeny of hyperthermophiles. Due to their likely presence already during the impact bombardment at the Early Archaean, they may have even been able to spread between the young planets of our solar system.

11.1 Biotopes of Hyperthermophiles

Hyperthermophiles are found in water-containing volcanically and geothermally heated environments situated mainly along terrestrial and submarine tectonic fracture zones where plates are colliding (subduction) or moving away from each other (spreading). Within saturated or superheated steam, life is not possible and liquid water is a fundamental prerequisite. Several hyperthermophiles exhibit growth temperatures exceeding 100 °C at an increased boiling point of water (e.g., by elevated atmospheric, hydrostatic or osmotic pressure).

Due to the presence of reducing gasses and the low solubility of oxygen at high temperatures, biotopes of hyperthermophiles are essentially anaerobic (Table 11.1.). Hyperthermophiles have been isolated from terrestrial and marine environments.

11.1.1   Terrestrial Biotopes

Terrestrial biotopes of hyperthermophiles are mainly sulfur-containing solfataric fields, named after the Solfatara Crater at Italy. Solfataric fields consist of soils, mud holes and surface waters heated by volcanic exhalations from magma chambers, a few kilometers below. Very often, solfataric fields are situated at or in the close neighborhood of active volcanoes and activity is greatly increased during eruption phases. Depending on the altitude above sea level, the maximum water temperatures are up to 100 °C. The salinity of solfataric fields is usually low. However, there are exceptions if they are situated at the beach (e.g., Faraglione, Vulcano, Italy).

The chemical composition of solfataric fields is very variable and depends on the site. Steam within the solfataric exhalations is mainly responsible for the heat transfer. CO_2 keeps the soils anaerobic and prevents penetration of oxygen into greater depths. In addition, H_2S reduces oxygen to water yielding elemental sulfur. An important gaseous energy source for hyperthermophiles is hydrogen, which may be formed either pyrolytically from water or chemically from FeS and H_2S [9]. Many solfataric fields are rich in iron minerals like ferric hydroxides, pyrite, and other ferrous sulfides.

Less usual compounds may be enriched at some sites like magnetite or arsenic minerals auripigment and realgar in Geysirnaja Valley, Kamchatka. Sometimes, solfataric fields contain silicate-rich neutral to slightly alkaline (pH 7-10) hot springs originating from the depth. Their content of sulfur compounds is usually low.

Table 11.1. Biotopes of hyperthermophiles

Characteristics	Type of thermal area	
	Terrestrial	Marine
Locations	Solfataric fields Deeply originating hot springs; Subterranean oil stratifications	Submarine hot vents and hot sediments "Black smokers"; Active sea-mounts
Temperatures	Surface: up to 100 °C *); Depth: above 100 °C	Up to about 400 °C
Salinity	Usually low (0.1 to 0.5% salt)	Usually about 3% salt
PH	0 to 10	5 to 8.5 (rarely: 3)
Major life-supporting gasses and nutrients	H_2O (steam) CO_2, CO, CH_4, H_2 H_2S, S^0, $S_2O_3^{2-}$, SO_3^{2-}, SO_4^{2-} **) NH_4^+, N_2, NO_3^-, Fe^{2+}, Fe^{3+}, O_2 (surface)	

*)   Depending on the altitude
**)   Sea water contains about 30 mmol/l of sulfate

Active volcanoes may harbor hot crater lakes which are heated by fumaroles (e.g., Askja, Iceland). Usually, those abound in sulfur and are very acidic and represent a further biotope of hyperthermophiles. Nothing is known about possible microbial life in the interior of active volcanoes. These mountains are assumed to be "hot sponges" which may contain a lot of aquifers like cracks and holes, possibly providing so far unexplored biotopes for hyperthermophiles. First evidence for the presence of communities of hyperthermophiles within geothermally heated rocks 3500 meters below the surface had been recently demonstrated [10]. Soils on the flanks of volcanoes, depending on the interior heat flow may harbor hyperthermophiles, too. For example, at the "Tramway Ridge" and southern crater on top of Mount Erebus, and on top of Mount Melbourne, both Antarctica, there are wet soils with temperatures between 60 and 65 °C and pH 5-6 at an altitude of 3500 meters. They represent "islands" of thermophilic life within a deep-frozen continent.

11.1.2 Marine Biotopes

Marine biotopes of hyperthermophiles consist of various hydrothermal systems situated at shallow to abyssal depths. Similar to ambient sea water, submarine hydrothermal systems usually contain high concentrations of NaCl and sulfate and exhibit a slightly acidic to alkaline pH (5-8.5). Otherwise, the major gasses and life-supporting mineral nutrients may be similar to that in terrestrial thermal areas. Shallow submarine hydrothermal systems are found in many parts of the world, mainly on beaches with active volcanism, like at Vulcano island, Italy with temperatures of 80 to 105 °C.

Most impressive are the deep sea "smoker" vents (Fig. 11.1) where mineral-laden hydrothermal fluids with temperatures up to about 400 °C escape into the cold (2.8 °C), surrounding deep sea water and build up huge rock chimneys. Although these hot fluids are sterile, the surrounding porous smoker rock material appears to contain very steep temperature gradients which provide zones of suitable growth temperatures for hyperthermophiles. Some smoker rocks are teeming with hyperthermophiles (for example 10^8 cells of *Methanopyrus* per gram of rock inside a Mid Atlantic Snake Pit hot vent chimney). Deep sea vents are located along submarine tectonic fracture zones for example the "TAG" and "Snake Pit" sites situated at the Mid Atlantic Ridge in a depth of about 4000 meters.

A further type of submarine high temperature environments is provided by active sea mounts. Close to Tahiti, there is a huge abyssal volcano, Macdonald Seamount (28°58.7' S, 140°15.5' W), the summit of which is situated approximately 40 meters below the sea surface. Samples taken during an active phase from the submarine eruption plume and rocks from the active crater contained high concentrations of viable hyperthermophiles [11].

11.2 Phylogeny of Hyperthermophiles

During recent years, powerful molecular techniques had been developed in order to investigate phylogenetic relationships of living organisms [12-14]. Based on the pioneering work of Carl Woese, 16S rRNA (harbored by the small ribosomal subunit) is widely used in phylogenetic studies of prokaryotes [13]. It consists of about 1500 bases and is homologous to the eukaryotic 18S rRNA. Based on 16/18S rRNA sequence comparisons, a universal phylogenetic tree is now available [15]. It exhibits a tripartite division of the living world into the bacterial (former: "eubacterial"), archaeal (former: "archaebacterial") and eukaryal (former: "eukaryotic") domains (Fig. 11.2) [16, 15]. The root was inferred from phylogenetic trees of duplicated genes of ATPase subunits and elongation factors Tu and G [17, 18]. Within the 16S rRNA phylogenetic tree, deep branches are evidence for early separation. For example, the separation of the Bacteria from the stem common to Archaea and Eucarya represents the deepest and earliest branching point. Short phylogenetic branches indicate a rather slow rate of evolution. In contrast to the Eucarya, the bacterial and archaeal domains within the universal phylogenetic tree exhibit some extremely deep and short branches. Based on the unique phylogenetic position of *Thermotoga,* a thermophilic ancestry of the Bacteria had been taken into consideration [19]. Surprisingly, hyperthermophiles are represented by all short and deep phylogenetic branches, which form a cluster around the phylogenetic root (Fig. 11.2, bold lines). The deepest and shortest phylogenetic branches are represented by *Aquifex* and *Thermotoga* within the Bacteria and *Methanopyrus, Pyrodictium,* and *Pyrolobus* within the Archaea. On the other

Fig. 11.1 Active deep sea vent chimney, a "Black Smoker" spewing superheated mineral-laden black water into the cold deep sea environment. The plume appears like black "smoke".

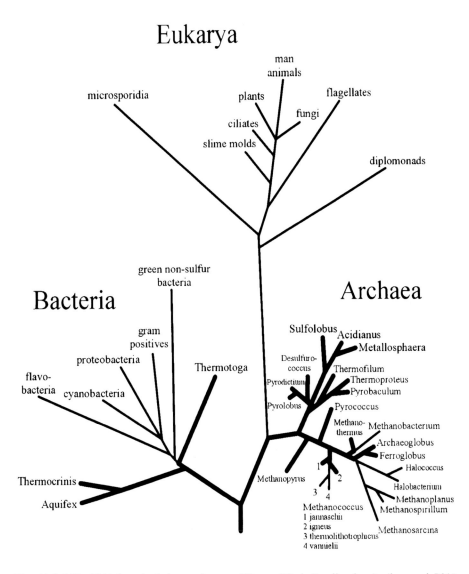

Fig. 11.2 16S rRNA-based phylogenetic tree, [6], modified. Small subunit ribosomal RNA sequences were compared according to Woese [15]. Schematic drawing. Bold lines represent hyperthermophiles.

hand, mesophilic and moderately thermophilic Bacteria and Archaea, as a rule represent long lineages within the phylogenetic tree and had a fast rate of evolution (e.g., Gram-positives; Proteobacteria; *Halobacterium; Methanosarcina;* Fig. 11.2). Total genome sequencing makes it possible to determine phylogenetic relationships of many other genes homologous at different organisms. Comparison of genes involved in gene expression confirm the 16S rRNA phylogenetic tree. Within the genes of metabolism, however, based on frequent lateral gene transfer rather a network than a tree may reflect their phylogenetic relations [20].

11.3 Taxonomy of Hyperthermophiles

So far, about 70 species of hyperthermophilic Bacteria and Archaea are known, which had been isolated from different terrestrial and marine thermal areas in the world. Hyperthermophiles are very divergent, both in terms of their phylogeny and physiological properties, and are grouped into 29 genera in 10 orders (Table 11.2.). At present, 16S rRNA sequence-based classification of prokaryotes appears to be imperative for the recognition and characterization of novel taxonomic groups. In addition, more traditional taxonomic features such as GC-contents of DNA, DNA-DNA homology, morphology, and physiological features may be used as separating characters in order to obtain high resolution of taxonomy within members of a phylogenetic lineage (e.g., description of different species and strains).

11.4 Strategies of Life
and Environmental Adaptations of Hyperthermophiles

Hyperthermophiles are adapted to distinct environmental factors including composition of minerals and gasses, pH, redox potential, salinity and temperature.

11.4.1 General Metabolic Potentialities

Most hyperthermophiles exhibit a chemolithoautotrophic mode of nutrition: inorganic redox reactions serve as energy sources (chemolithotrophic), and CO_2 is the only carbon source required to build up organic cell material (autotrophic). Therefore, these organisms are fixing CO_2 by chemosynthesis and are designated chemolithoautotrophs. The energy-yielding reactions in chemolithoautotrophic hyperthermophiles are anaerobic and aerobic types of respiration (Table 11.3.).

Molecular hydrogen serves as an important electron donor. Other electron donors are sulfide, sulfur, and ferrous iron. Like in mesophilic respiratory organisms, in some hyperthermophiles oxygen may serve as an electron acceptor. In contrast, however, oxygen-respiring hyperthermophiles are usually microaerophilic and, therefore, grow only at reduced oxygen concentrations (even as low as 10 ppm; [21]). Anaerobic respiration types are the nitrate-, sulfate-, sulfur- and carbon dioxide-respirations. While chemolithoautotrophic hyperthermophiles produce organic matter, there are some hyperthermophiles, which depend on organic material as energy- and carbon sources. They are designated as chemoorganoheterotrophs (or, shortly: heterotrophs). Several chemolithoautotrophic hyperthermophiles are facultative heterotrophs (Table 11.3.). Those are able to use organic material alternatively to inorganic nutrients whenever it is provided by the environment (e.g., by decaying cells). Heterotrophic hyperthermophiles gain energy either by aerobic or different types of anaerobic respiration, using organic material as electron donors, or by fermentation.

11.4.2   Physiological Properties

11.4.2.1   Terrestrial Hyperthermophiles

The acidic hot oxygen-exposed surface of terrestrial solfataric fields almost exclusively harbours extremely acidophilic hyperthermophiles. They consist of lobed coccoid-shaped (Fig. 11.3a) aerobes and facultative aerobes, growing between pH 1 and 5 with an optimum around pH 3 (Table 11.4.). Phylogenetically they belong to the archaeal genera *Sulfolobus, Metallosphaera* and *Acidianus*. Together with the strictly anaerobic *Stygiolobus* these genera form the *Sulfolobales* order (Fig. 11.2, Table 11.2.). Members of *Sulfolobus* are strict aerobes. During autotrophic growth, S^0, S^{2-}, and H_2 [22] may be oxidized, yielding sulfuric acid or water as end products (Table 11.3.). During heterotrophic growth, sugars, yeast extract, and peptone may serve as energy sources [8]. Members of *Acidianus* are able to grow by anaerobic and aerobic oxidation of H_2, using S^0 and O_2 as electron acceptors, respectively [23]. Alternatively, *Acidianus* is able to grow by S^0-oxidation (Table 11.3.).

Terrestrial hot springs with low salinity and the depth of solfataric fields harbour slightly acidophilic and neutrophilic hyperthermophiles, which are usually strict anaerobes. They are members of the genera *Pyrobaculum, Thermoproteus, Thermofilum, Desulfurococcus, Sulfophobococcus*, and *Thermosphaera* (Table 11.2.). Cells of *Pyrobaculum, Thermoproteus*, and *Thermofilum* are stiff, regular rods with almost rectangular edges (Fig. 11.3b). During the exponential growth phase, under the light microscope, spheres become visible at the ends ("golf clubs"). Cells of *Pyrobaculum* and *Thermoproteus* are about 0.50 μm in diameter whereas those of *Thermofilum* ("the hot thread") are only about 0.17 - 0.35 μm. *Thermoproteus tenax, Thermoproteus neutrophilus* and *Pyrobaculum islandicum* are able to gain energy by formation of H_2S from H_2 and S^0 [24] (Table 11.4.). *Thermoproteus tenax* and *Pyrobaculum islandicum* are facultative, *Pyrobaculum organotrophum, Thermoproteus uzoniensis, Thermofilum librum*, and *Thermofilum pendens* obligate heterotrophs. Members of *Desulfurococcus* are coccoid-shaped, strictly heterotrophic sulfur-respirers, while *Thermosphaera* and *Sulfophobococcus* are purely fermentative coccoid organisms. *Thermosphaera aggregans* grows in grape-shaped aggregates. Exclusively from the depth of solfataric fields in the southwest of Iceland (Kerlingarfjöll mountains), members of the genus *Methanothermus* have been isolated which appear to be endemic species of this area. They are highly oxygen-sensitive strictly chemolithoautotrophic methanogens, gaining energy by reduction of CO_2 by H_2 (Table 11.3.) and growing at temperatures between 65 and 97 °C (Table 11.4.).

Table 11.2. Taxonomy and upper growth temperatures of hyperthermophiles (Examples)

Order	Genus	Species	T max (°C)
Domain: BACTERIA			
Thermotogales	Thermotoga	T. maritima	90
		T. thermarum	84
	Thermosipho	T. africanus	77 *
	Fervidobacterium	F. nodosum	80 *
Aquificales	Aquifex	A. pyrophilus	95
		A. aeolicus	93
	Calderobacterium	C. hydrogenophilum	82 *
	Thermocrinis	T. ruber	89
Domain: ARCHAEA			
I. Kingdom: Crenarchaeota			
Sulfolobales	Sulfolobus	S. acidocaldarius	85
		S. solfataricus	87
	Metallosphaera	M. sedula	80 *
	Acidianus	A. infernus	95
	Stygiolobus	S. azoricus	89
	Sulfurococcus	S. mirabilis	86
	Sulfurisphaera	S. ohwakuensis	92
Thermoproteales	Thermoproteus	T. tenax	97
	Pyrobaculum	P. islandicum	103
		P. aerophilum	104
	Thermofilum	T. pendens	95
		T. librum	95
	Thermocladium	T. modestius	80 *
	Caldivirga	C. maquilingensis	92
Desulfurococcales	Desulfurococcus	D. mucosus	97
	Staphylothermus	S. marinus	98
	Sulfophobococcus	S. zilligii	95
	Stetteria	S. hydrogenophila	102
	Aeropyrum	A. pernix	100
	Ignicoccus	I. islandicus	100
	Thermosphaera	T. aggregans	90
	Thermodiscus	T. maritimus	98
	Pyrodictium	P. occultum	110
	Hyperthermus	H. butylicus	108
	Pyrolobus	P. fumarii	113

*) extreme thermophiles related to hyperthermophiles

Table 11.2. Continuation: Taxonomy and upper growth temperatures of hyperthermophiles (Examples)

Order	Genus	Species	T max (°C)
Domain: ARCHAEA			
II. Kingdom: Euryarchaeota			
Thermococcales	Thermococcus	T. celer	93
	Pyrococcus	P. furiosus	103
		P. abyssi	102
		P. horikoshii	102
Archaeoglobales	Archaeoglobus	A. fulgidus	92
		A. veneficus	88
	Ferroglobus	F. placidus	95
Methanobacteriales	Methanothermus	M. fervidus	97
Methanococcales	Methanococcus	M. jannaschii	86
		M. igneus	91
Methanopyrales	Methanopyrus	M. kandleri	110

*) extreme thermophiles related to hyperthermophiles

Table 11.3. Energy-yielding reactions in chemolithoautotrophic hyperthermophiles

Energy-yielding reaction	Genera
$2S^0 + 3O_2 + 2H_2O \rightarrow 2H_2SO_4$ $(2FeS_2 + 7O_2 + 2H_2O \rightarrow 2FeSO_4 + 2H_2SO_4)$ = "metal leaching"	*Sulfolobus*[a], *Acidianus*[a], *Metallosphaera*[a], *Aquifex*
$H_2 + \frac{1}{2} O_2 \rightarrow H_2O$	*Aquifex, Acidianus*[a], *Metallosphaera*[a], *Pyrobaculum*[a], *Sulfolobus*[a]
$H_2 + HNO_3 \rightarrow HNO_2 + H_2O$	*Aquifex, Pyrobaculum*[a]
$4H_2 + HNO_3 \rightarrow NH_4OH + 2H_2O$	*Pyrolobus*
$2FeCO_3 + HNO_3 + 5H_2O$ $\rightarrow 2Fe(OH)_3 + HNO_2 + 2H_2CO_3$	*Ferroglobus*
$4H_2 + H_2SO_4 \rightarrow H_2S + 4H_2O$	*Archaeoglobus*[a]
$H_2 + S^0 \rightarrow H_2S$	*Acidianus, Stygiolobus, Pyrobaculum*[a], *Ignicoccus, Pyrodictium*[a]
$H_2 + 6FeO(OH) \rightarrow 2Fe_3O_4 + 4H_2O$	*Pyrobaculum*
$4H_2 + CO_2 \rightarrow CH_4 + 2H_2O$	*Methanopyrus, Methanothermus, Methanococcus*

[a] facultatively heterotrophic

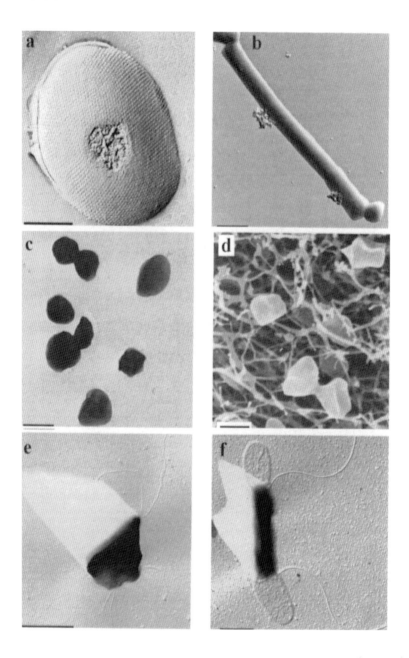

Fig. 11.3 Electron micrographs of cells of hyperthermophilic Archaea and Bacteria
    a. *Metallosphaera sedula,* freeze fracturing, bar, 0.2 μm;
    b. *Thermoproteus tenax,* Pt-shadowing, bar, 1.0 μm;
    c. *Pyrolobus fumarii,* UAc-staining, bar, 1.0 μm;
    d. *Pyrodictium abyssi,* scanning micrograph, bar, 1.0 μm;
    e. *Archaeoglobus fulgidus,* Pt-shadowing, bar, 1.0 μm;
    f. *Thermotoga maritima,* Pt-shadowing, bar, 1.0 μm.

Table 11.4. Growth conditions and morphology of hyperthermophiles

Species	Min. Temp. (°C)	Opt. Temp. (°C)	Max. Temp. (°C)	pH	Aerobic (ae) versus anaerobic (an)	Morphology
				Growth conditions		
Sulfolobus acidocaldarius	60	75	85	1-5	ae	Lobed cocci
Metallosphaera sedula	50	75	80	1-4.5	ae	Cocci
Acidianus infernus	60	88	95	1.5-5	ae/an	Lobed cocci
Stygiolobus azoricus	57	80	89	1-5.5	an	Lobed cocci
Thermoproteus tenax	70	88	97	2.5-6	an	Regular rods
Pyrobaculum islandicum	74	100	103	5-7	an	Regular rods
Pyrobaculum aerophilum	75	100	104	5.8-9	ae/an	Regular rods
Thermofilum pendens	70	88	95	4-6.5	an	Slender regular rods
Desulfurococcus mobilis	70	85	95	4.5-7	an	Cocci
Thermosphaera aggregans	67	85	90	5-7	an	Cocci in aggregates
Sulfophobococcus zilligii	70	85	95	6.5-8.5	an	Cocci
Staphylothermus marinus	65	92	98	4.5-8.5	an	Cocci in aggregates
Thermodiscus maritimus	75	88	98	5-7	an	Disks
Aeropyrum pernix	70	90	100	5-9	ae	Irregular cocci
Stetteria hydrogenophila	70	95	102	4.5-7	an	Irregular disks
Ignicoccus islandicus	65	90	100	3.9-6.3	an	Irregular cocci
Pyrodictium occultum	82	105	110	5-7	an	Disks with cannulae
Hyperthermus butylicus	80	101	108	7	an	Lobed cocci
Pyrolobus fumarii	90	106	113	4.0-6.5	ae/an	Lobed cocci
Thermococcus celer	75	87	93	4-7	an	Cocci
Pyrococcus furiosus	70	100	105	5-9	an	Cocci
Archaeoglobus fulgidus	60	83	95	5.5-7.5	an	Irregular cocci
Ferroglobus placidus	65	85	95	6-8.5	an	Irregular cocci
Methanothermus sociabilis	65	88	97	5.5-7.5	an	Rods in clusters
Methanopyrus kandleri	84	98	110	5.5-7	an	Rods in chains
Methanococcus igneus	45	88	91	5-7.5	an	Irregular cocci
Thermotoga maritima	55	80	90	5.5-9	an	Rods with sheath
Aquifex pyrophilus	67	85	95	5.4-7.5	ae	Rods

11.4.2.2   Marine Hyperthermophiles

A variety of hyperthermophiles are adapted to the high·salinity of sea water of about 3% salt. They are represented by members of the archaeal genera *Pyrolobus, Pyrodictium, Hyperthermus, Stetteria, Thermodiscus, Igneococcus, Staphylothermus, Aeropyrum, Pyrobaculum, Methanopyrus, Pyrococcus, Thermococcus, Archaeoglobus* and *Ferroglobus,* and of the bacterial genera *Aquifex* and *Thermotoga* (Table 11.2.). The organism with the highest growth temperature is *Pyrolobus fumarii*, exhibiting an upper temperature border of growth above 113 °C (Table 11.4.) [21]. *Pyrolobus* had been isolated from the walls of an active deep sea smoker chimney at the Mid Atlantic Ridge. It is so dependent on high temperatures that it is unable to grow below 90 °C. Cells of *Pyrolobus* are lobed cocci about 0.7 - 2.5 µm in diameter (Fig. 11.3c). *Pyrolobus fumarii* gains energy by chemolithoautotrophic nitrate reduction, forming ammonia as an end product. Alternatively, under microaerophilic conditions, it shows weak but significant growth on H_2 and traces of oxygen (0.05%). Close relatives to *Pyrolobus* are members of *Pyrodictium*, cells of which are disks (0.2 µm thick and up to 3 µm in diameter) which are usually connected by networks of hollow cannulae, about 25 nm in diameter (Fig. 11.3d) [25, 26]. As a rule, members of *Pyrodictium* are chemolithoautotrophs, gaining energy by reduction of S° by H_2. In addition, some isolates are able to grow by reduction of sulfite and thiosulfate [27]. A very close relative to *Pyrodictium* is *Hyperthermus*. However, in contrast to *Pyrodictium*, it is a purely fermentative hyperthermophile which does not form cannulae and does not grow at 110 °C [28]. *Stetteria hydrogenophila* represents a group of disk-shaped hyperthermophiles which gains energy on a combination of H_2 and oxidized sulfur compounds, and peptides [29]. *Ignicoccus islandicus* is a coccoid-shaped strictly chemolithoautotrophic member of the *Desulfurococcaceae* which gains energy by sulfur respiration [30]. *Pyrobaculum aerophilum* so far is the only marine member of the *Thermoproteales*. It is a rod-shaped strictly chemolithoautotrophic hyperthermophile which gains energy either anaerobically by nitrate reduction or microaerobically by reduction of O_2 (traces). Molecular hydrogen serves as electron donor. The rod-shaped *Methanopyrus kandleri* occurs the same within shallow as abyssal submarine hot vents and exhibits the highest growth temperatures in methanogens (up to 110 °C). Further marine hyperthermophilic methanogens are the coccoid-shaped *Methanococcus jannaschii* and *Methanococcus igneus* within the *Methanococcales* (Fig. 11.2) where they represent the shortest phylogenetic branch-offs. Archaeal sulfate reducers are represented by members of *Archaeoglobus* [31, 32]. *Archaeoglobus fulgidus* and *Archaeoglobus lithotrophicus* are chemolithoautotrophs able to grow by reduction of SO_4^{2-} and $S_2O_3^{2-}$ by H_2 (Table 11.3.), while *Archaeoglobus veneficus* is only able to reduce SO_3^{2-} and $S_2O_3^{2-}$. *Archaeoglobus fulgidus* is a facultative heterotroph growing on a variety of organic substrates like formate, sugars, starch, proteins, and cell extracts [33]. Sequencing of the total genome of this organism revealed the presence of genes for the fatty acid ß-oxidation pathway, indicating that its physiological properties may be even broader than initially described [33]. Cells of members of *Archaeoglobus* are coccoid to triangular-shaped (Fig. 11.3e). Like methanogens, they show a blue-green fluorescence at 420 nm under the UV-microscope. In agreement, *Archaeoglobus* possesses several coenzymes that had been thought to be unique for methanogens (e.g., F_{420}, methanopterin, tetrahydromethanopterin, methanofurane). Usually, members of *Archaeoglobus* are found in shallow and abyssal submarine hydrothermal systems. In addition, the same species have been detected in deep geothermally heated subterranean oil reservoirs and may be responsible for some of the H_2S formation there ("reservoir souring", [31, 10]). *Ferroglobus* represents a further genus within the *Archaeoglobales*. It gains energy anaerobically by reduction of NO_3^- with Fe^{2+} (Table 11.3.), H_2, and H_2S as electron donors. The unique metabolism of *Ferroglobus placidus* may offer a biological mechanism for the anaerobic formation of Fe^{3+} which may have operated during banded iron formations (BIFs) under the anoxic high-temperature conditions at the Hadean Earth [34]. Within the Bacteria domain, the deepest phylogenetic branch is represented by *Aquifex. Aquifex pyrophilus* is a motile rod-shaped strict chemolithoautotroph. It is a facultative aerobe. Under anaerobic conditions, *Aquifex pyrophilus* grows by nitrate reduction with H_2 and S^0 as electron donors. Alternatively, at very low oxygen concentrations (up to 0.5%, after adaptation) it is able to gain energy by oxidation of H_2 and S^0. Members of *Aquifex* are found in shallow submarine vents. *Aquifex pyrophilus* grows up to 95 °C, the highest growth temperature found so far within the bacterial domain (Table 11.4.).

Groups of strictly heterotrophic hyperthermophiles are thriving in submarine vents, too. *Thermodiscus maritimus* is a disk-shaped archaeal heterotroph growing by sulfur respiration on yeast extract and prokaryotic cell homogenates. In the absence of S^0, it is able to gain energy by fermentation [5]. Cells of *Staphylothermus marinus* are coccoid and arranged in grape-like aggregates. They are highly variable in diameter from 0.5 to 15 μm. *Aeropyrum pernix* is a strictly aerobic spherical-shaped marine hyperthermophile, growing optimally at neutral pH. It gains energy in metabolizing complex compounds like yeast extract and peptone [35]. Members of *Pyrococcus* and *Thermococcus* are submarine motile coccoid species, cells of which occur frequently in pairs. They gain energy by fermentation of peptides, amino acids and sugars, forming fatty acids, CO_2 and H_2. Hydrogen may inhibit growth and can be removed by gassing with N_2 [36]. Alternatively, inhibition by H_2 can be prevented by addition of S^0, whereupon H_2S is formed instead of H_2. *Pyrococcus furiosus* grows optimally at 100 °C (Table 11.4.). In addition to their marine environment, species of *Pyrococcus* and *Thermococcus* are found in geothermally heated oil reservoirs of high salinity (together with *Archaeoglobus*, see above) and are able to grow in the presence of crude oil [10]. Many submarine hydrothermal systems contain hyperthermophilic members of *Thermotoga* which represent the second deepest phylogenetic branch within the bacterial domain (Fig. 11.2). Cells of *Thermotoga* are rod-shaped, thriving together with archaeal hyperthermophiles within the same environment. They show a characteristic "toga", a sheath-like structure surrounding cells and overballooning at the ends (Fig. 11.3f). *Thermotoga* ferments various carbohydrates and proteins. As end products, H_2, CO_2, acetate and L(+)-lactate are formed. Hydrogen is inhibitory to growth. In the presence of S^0, H_2S is formed instead of H_2, which does not inhibit growth. Other genera within the *Thermotogales* are members of *Thermosipho* and *Fervidobacterium* which are strict heterotrophs, too. Similar to *Thermotoga*, cells possess a "toga". However, they are less extreme thermophilic (Table 11.2.).

11.5 Conclusions:
Hyperthermophiles in the History of Life

Although nothing is known yet about the prerequisites and mechanisms which had led to the first living cell, investigations on recent hyperthermophiles yield surprising results, enabling us to draw conclusions about possible features of the common ancestor of life. Within the (16SrRNA-based) universal phylogenetic tree of life, hyperthermophiles form a cluster around the root, occupying all the short and deep phylogenetic branches (Fig. 11.2, bold lines). This is true for both the archaeal and bacterial domains. As a rule, members of the deepest and shortest lineages exhibit the highest growth temperatures ("the shorter and deeper, the hotter"). By their 16SrRNA, these slowly evolving organisms appear to be still closest to the root (common ancestor) and, therefore the most primitive ones still existing. In general, the shortest and deepest lineages exhibit a chemolithoautotrophic mode of nutrition (e.g., *Pyrolobus*, *Pyrodictium*, *Methanopyrus*, *Aquifex*). In an ecological sense, they are primary producers of organic matter (mainly of their own cell material), which gives rise to complex hyperthermophilic microcosms. These chemolithoautotrophs are able to use various mixtures of oxidized and reduced minerals and gasses as energy sources and to assimilate carbon from CO_2. In addition, for growth they just need liquid water, trace minerals and heat. Hydrogen and sulfurous compounds are widely used inorganic energy sources which mainly occur within volcanic environments. Most hyperthermophiles do not require or even tolerate oxygen for growth. They are completely independent of sunlight. In view of a possible similarity to ancestral life, a chemolithoautotrophic hyperthermophilic common ancestor appears probable. In addition, growth conditions of recent hyperthermophiles fit well to our view of the primitive Earth about 3.9 Ga ago when life could have originated: the atmosphere was overall reducing and there was a much stronger volcanism [3]. In addition, Earth's oceans were continuously heated by heavy impacts of meteorites [37]. Therefore, within that scenario, early life had to be heat resistant to survive at all and ancestral hyperthermophiles should even have dominated in the Early Archaean age. Intense impacts, in addition, caused significant material exchange in between planets of our solar system. Based on their ability to survive for long times in the cold, even at -140 °C, hyperthermophiles could successfully have disseminated between the planets: when ejected into space by heavy impacts, insides of rocks in dormant state, they should have been able to pass long distances to other planets (see Chap. 4, Horneck et al.). After landing, they could have flourished on their new home. Since early planets like Mars are assumed to have harbored volcanic and hydrothermal activity [38], colonization by primitive hyperthermophiles appears rather probable. Concerning Mars, its surface at present is too cold and neither liquid water nor active volcanism are apparent. Therefore, hyperthermophiles could not grow there, anymore. However, the discovery of deep geothermally heated subterranean communities several kilometers below Earth's surface [10] makes it imaginable that similar active microcosms of hyper-thermophiles could still exist on Mars deep below its surface, assuming that water, heat and simple nutrients are present. In view of the early impact material exchange scenario between planets, it is uncertain if life had originated on Earth, at all. Therefore, hyperthermophiles could play a key role in search for life on Mars in the future.

Acknowledgements. I wish to thank Reinhard Rachel for electron micrographs and for establishing the phylogenetic tree. The work presented from my laboratory was supported by grants of the Deutsche Forschungsgemeinschaft, the Bundesministerium für Bildung, Wissenschaft, Forschung und Technologie, the European Commission and the Fonds der Chemischen Industrie.

11.6 References

1   J.W. Schopf, Science **260,** 640 (1993).
2   S.J. Mojzsis, G. Arrhenius, K.D. McKeegan, T.M. Harrison, A.P. Nutman, C.R.L. Friends, Nature **384,** 55 (1996).
3   W.G. Ernst, in: J.W. Schopf (Ed.) *Earth's earliest biosphere, its origin and evolution,* Princeton U.P., Princeton, N.J., 1983, pp. 41.
4   B. Rasmussen, Nature **405,** 676 (2000).
5   K.O. Stetter, in: T.D. Brock (Ed.) *Thermophiles: General, molecular and applied microbiology,* John Wiley & Sons, Inc., New York, 1986, pp. 39.
6   K.O. Stetter, in: J. and K. Trân Thanh Vân, J.C. Mounolou, J. Schneider, C. McKay (Eds.) *Frontiers of Life - Colloque Interdisciplinaire du Comité National de la Recherche Scientifique,* Proc. 3rd "Recontres de Blois", Chateau de Blois, France 1991, Editions Frontières, Gif-sur-Yvette, France, 1992, pp. 195.
7   K.O. Stetter, FEBS Letters **452,** 22 (1999).
8   T.D. Brock, Springer-Verlag, New York, 1978.
9   E. Drobner, H. Huber, G. Wächtershäuser, D. Rose, K.O. Stetter, Nature **346,** 742 (1990).
10   K.O. Stetter, R. Huber, E. Blöchl, M. Kurr, R.D. Eden, M. Fiedler, H. Cash, I. Vance, Nature **365,** 743 (1993).
11   R. Huber, P. Stoffers, J.L. Cheminee, H.H. Richnow, K.O. Stetter, Nature **345,** 179 (1990).
12   E. Zuckerk, L. Pauling, J. Theoretical Biology **8,** 357 (1965).
13   C.R. Woese, M. Sogin, D.A. Stahl, B.J. Lewis, L. Bonen, J. Molecuar Evolution **7,** 197 (1976).
14   C.R. Woese, R. Gutell, R. Gupta, H.F. Moller, Microbiol. Rev. **47,** 621 (1983).
15   C.R. Woese, O. Kandler, M.L. Wheelis, Proc. National Academy of Sciences, USA **87,** 4576 (1990).
16   C.R. Woese, G.E. Fox, Proc. National Academy of Sciences, USA **74,** 5088 (1977).
17   N. Iwabe, K. Kuma, M. Hasegawa, S. Osawa, T. Miyata, Proc. National Academy of Sciences, USA **86,** 9355 (1989).
18   J.P. Gogarten, H. Kibak, P. Dittrich, L. Taiz, E.J. Bowman, B. Bowman, M.F. Manolson, R.J. Poole, T. Date, T. Oshima, J. Konishi, K. Denda, M. Yoshida, Proc. National Academy of Sciences, USA **86,** 6661 (1989).
19   L. Achenbach-Richter, R. Gupta, K.O. Stetter, C.R. Woese, System. Appl. Microbiol. **9,** 34 (1987).
20   W.F. Doolittle, Science **284,** 2124 (1999).
21   E. Blöchl, R. Rachel, S. Burggraf, D. Hafenbradl, H.W. Jannasch, K.O. Stetter, Extremophiles **1,** 14 (1997).
22   R. Huber, T. Wilharm, D. Huber, A. Trincone, S. Burggraf, H. König, R. Rachel, I. Rockinger, H. Fricke, K.O. Stetter, Appl. Micriobiol. **15,** 340 (1992).

23 A. Segerer, A. Neuner, J. K. Kristjansson, K.O. Stetter, Int. J. Syst. Bact. **36**, 559 (1986).

24 F. Fischer, W. Zillig, K.O. Stetter, G. Schreiber, Nature **301**, 511 (1983).

25 H. König, P. Messner, K.O. Stetter, FEMS Microbiol. Letters **49**, 207 (1988).

26 G. Rieger, R. Rachel, R. Herrmann, K.O. Stetter, J. Structural Biology **115**, 78 (1995).

27 R. Huber, G. Huber, A. Segerer, J. Seger, K.O. Stetter, in: H.W. van Verseveld, J.A. Duine (Eds.) *Microbial growth on C₁ compounds*, Proc. 5ᵗʰ International Symposium, Martinus Nijhoff Publ., Dordrecht, 1987, pp. 44.

28 W. Zillig, I. Holz, D. Janekovic, H.P. Klenk, E. Imsel, J. Trent, S. Wunderl, V.H. Forjatz, R. Coutinho, T. Ferreira, J. Bacteriol. **172**, 3959 (1990).

29 B. Jochimsen, S. Peinemann-Simon, H. Völker, D. Stüben, R. Botz, P. Stoffers, P.R. Dando, M. Thomm, Extremophiles **1**, 67 (1997).

30 H. Huber, S. Burggraf, T. Mayer, I. Wyschkony, M. Biebl, R. Rachel K.O. Stetter, Intern. J. Systematic Bacteriology (2000) in press.

31 K.O. Stetter, G. Lauerer, M. Thomm, A. Neuner, Science **236**, 822 (1987).

32 K.O. Stetter, System. Appl. Microbiol. **10**, 172 (1988).

33 H.-P. Klenk, R.A. Clayton, J.-F. Tomb, O. White, K.E. Nelson, K.A. Ketchum, R.J. Dodson, M. Gwinn, E.K. Hickey, J.D. Peterson, D.L. Richardson, A.R. Kerlavage, D.E. Graham, N.C. Kyrpides, R.D. Fleischmann, J. Quackenbush, N.H. Lee, G.G. Sutton, S. Gill, E.F. Kirkness, B.A. Dougherty, K. McKenney, M.D. Adams, B. Loftus, S. Peterson, C.I. Reich, L.K. McNeil, J.H. Badger, A. Glodek, L. Zhou, R. Overbeek, J.D. Gocayne, J.F. Weidman, L. McDonald, T. Utterback, M.D. Cotton, T. Spriggs, P. Artiach, B.P. Kaine, S.M. Sykes, P.W. Sadow, K.P. D'Andrea, Ch. Bowman, C. Fujii, S.A. Garland, T.M. Mason, G.J. Olsen, C.M. Fraser, H.O. Smith, C.R. Woese, J.C. Venter, Nature **390**, 364 (1997).

34 P.W.U. Appel, Precambrian Res. **11**, 73 (1980).

35 Y. Sako, N. Nomura, A. Uchida, Y. Ishida, H. Morii, Y. Koga, T. Hoaki, T. Maruyama, Int. J. Syst. Bacteriol. **46**, No. 4, 1070 (1996).

36 G. Fiala and K.O. Stetter, Arch. Microbiol. **145**, 56 (1986).

37 P.C.W. Davies, in: G.R. Bock, J.A. Goode (Eds.) *Evolution and hydrothermal ecosystems on Earth (and Mars?)*. Proc. Ciba Foundation Symposium 202, John Wiley & Sons Ltd. Chichester, England, 1996, pp. 304.

38 M.H. Carr, in: G.R. Bock, J.A. Goode (Eds.) *Evolution and hydrothermal ecosystems on Earth (and Mars?)*, Proc. Ciba Foundation Symposium 202, John Wiley & Sons Ltd. Chichester, England, 1996, pp. 249.

12   Halophilic Microorganisms

Hans Jörg Kunte, Hans G. Trüper and Helga Stan-Lotter

Concentrated salt solutions like salt or soda lakes, coastal lagoons or man-made salterns, inhabited by only a few forms of higher life, are dominated by prokaryotic microorganisms. Global salt deposits show that evaporation of marine salt water and the development of hypersaline habitats is an ongoing process for millions of years and providing ample time for the evolution of specialized halophilic Bacteria and Archaea. Halophiles, which require more than 0.5 M NaCl for optimal growth [1], have developed two different basic mechanisms of osmoregulatory solute accumulation to cope with ionic strength and the considerable water stress. These mechanisms allow halophiles to proliferate in saturated salt solutions and to survive entrapment in salt rock. The latter was proven by the isolation of viable halophilic Archaea from several subsurface salt deposits of Permo-Triassic age. If halophilic prokaryotes on Earth can remain in viable states for long periods of time, then it is reasonable to consider, under similar extraterrestrial environments, the existence of extraterrestrial organisms. This becomes all the more plausible, considering that halite has been found in several extraterrestrial materials. Here we consider the different mechanisms of osmoadaptation, the environment of halophiles, especially of subterranean halophilic isolates, and the relevance of microbial survival in high saline environments to astrobiology.

12.1 Adaptation to Saline Environments

Adding a solute like NaCl to water will lead to changes in the characteristics of the solvent water's freezing and boiling points as well as vapor and osmotic pressures. These changes are caused by the decrease of the water's chemical potential μ_W, which can be expressed as:

$$\Delta\mu_W = \Delta H_W - T\Delta S_W \tag{12.1}$$

where ΔH is the change in enthalpy (the heat of reaction), T (K) is temperature, and ΔS is the degree of randomness (change in entropy).

According to Sweeney and Beuchat [2] the second term of the equation above ($T\Delta S_w$) is dominating and the decrease of the chemical potential largely depends on the change in entropy of the water. This is explained by the interference of salt with the ordered water structure, thereby increasing the randomness of the solvent molecules. ΔS_w, the entropy of the solvent, will therefore be positive resulting in a reduction of the water potential μ_w. A non-adapted organism exposed to a saline environment must cope with its cytoplasmic water having a higher chemical potential than the water of the surrounding environment. Water always flows from a high to low chemical potential until the potential gradient is abolished. Thus, the cytoplasm which is surrounded by a cytoplasmic membrane freely permeable to water, will lose its cytoplasmic water resulting in cell shrinkage. The reduction in cell volume is mainly caused by the loss of free water (approx. 80% of total water in a fully hydrated cell), while the bound water level remains unchanged [3]. This results in the cessation of growth, possibly due to molecular crowding, and thus, reduced diffusion rates of proteins and metabolites. In order to gain sufficient free water and to maintain an osmotic equilibrium across the membrane, the cell has to reduce the chemical potential of the cytoplasmic free water. Two principle mechanisms have evolved on Earth to lower the chemical potential of cell water, allowing an osmotic adaptation of microorganisms: "salt-in-cytoplasm mechanism" and organic-osmolyte mechanism.

1. Salt-in-cytoplasm strategy: Organisms following this strategy adapt the interior protein chemistry of the cell to high salt concentration. The thermodynamic adjustment of the cell can be achieved by raising the salt concentration in the cytoplasm.
2. Organic-osmolyte strategy: Whereby organisms keep the cytoplasm, to a large extent, free of NaCl and the design of the cell's interior remains basically unchanged. The chemical potential of the cell water is mainly reduced by an accumulation of uncharged, highly water-soluble, organic solutes.

12.1.1 Salt-in-Cytoplasm Mechanism

The "salt-in-cytoplasm mechanism", first discovered in *Halobacteria,* is considered the typical archaeal strategy of osmoadaptation. Fermenting Bacteria, acetogenic anaerobes (*Haloanaerobium, Halobacteroides, Sporohalobacter, Acetohalobium*), and sulfate reducers are now known to employ this strategy as well [4]. Despite the abundance of NaCl in the typical haloarchaeal environment, halophilic Archaea keep the cytoplasm relatively free of sodium. Instead, potassium accumulates in the cell (as shown for *Haloferax volcanii* through an energy-dependent potassium uptake system) and together with its counter ion Cl⁻, K^+ can be found in molar concentrations in the cytoplasm. Because the K^+ concentration inside the cell is 100 times higher than in the surrounding environment, a part of the proton motive force must be used to maintain the ion gradient. In this energetic respect, the situation in halophilic anaerobic Bacteria is thought to be different; there is evidence that these organisms invest as little as possible in the maintenance of ion gradients. Measurements of the ion composition of exponentially growing cells of *Haloanaerobium praevalens* show that K^+ is the dominating cation, but that Na^+ levels are also relatively high. Cells entering the stationary phase eventually replace K^+ for Na^+ [5]. Analysis of *Haloanaerobium acetoethylicum* even suggest that Na^+ could be the main cation in stationary cells as well as in exponentially growing cells [6].

The effect of the accumulation of potassium and/or sodium in the cytoplasm is that the cytoplasm is exposed to an increased ionic strength. To adapt the enzymatic machinery to an ionic cytoplasm, proteins of halophilic anaerobic Bacteria and halophilic Archaea contain an excess of acidic amino acids over basic residues [7]. This leads to a predominance of charged amino acids on the surface of enzymes and ribosomes which is thought to stabilize the hydration shell of the molecule when in high ionic surroundings. In low saline environments, the excess of negatively charged ions will destabilize the molecule's structure, due to repulsion when the shielding cations are removed [8]. This mechanism explains the fact that organisms employing the salt in cytoplasm strategy display a relatively narrow adaptation and their growth is restricted to high saline environments [9]. However, in habitats with saturated salt concentrations, halophilic Archaea outcompete organic-osmolyte producers, proving members of the "salt-in-cytoplasm mechanism" as *extreme* halophiles

12.1.2 Organic Osmolyte Mechanism

The organic osmolyte mechanism is widespread among Bacteria and Eukarya and also present in some methanogenic Archaea [10, 11]. In response to an osmotic stress, these organisms mainly accumulate organic compounds like sugars, polyols, amino acids and/or amino acid derivatives either by *de novo* synthesis or by uptake from the surrounding environment. These non-ionic, highly water-soluble compounds do not disturb the metabolism, even at high cytoplasmic concentrations and are thus aptly named *compatible solutes* [12]. Halophilic cells using compatible solutes can basically preserve the same enzymatic machinery as non-halophiles, needing only minor adjustments in their interior proteins (i.e. ribosomal proteins), which are slightly more acidic than the cytoplasmic proteins in *Escherichia coli* [13]. Halophiles employing the organic-osmolyte mechanism are more flexible than organisms employing the "salt-in-cytoplasm strategy" because even though they display wide salt tolerance, they can also grow in low salt environments.

12.1.2.1 Stress Protection by Compatible Solutes

In addition to their function of maintaining an osmotic equilibrium across the cell membrane, compatible solutes are effective stabilizers of proteins and even whole cells. They can act as protectors against heat, desiccation, freezing and thawing, and denaturants such as urea and salt [14, 15]. The reason, why these organic compounds are compatible with the metabolism and can even act as stabilizers of labile biological structures, is explained at the molecular level by the preferential exclusion model. According to this theory, compatible solutes are absent from protein surfaces due to the structural dense water bound to the protein. Compatible solutes show a preference for the less dense free water fraction in the cytoplasmic surrounding. They stabilize the two water fractions by fitting into the lattice of the free water and allowing for the formation of hydration clusters. As a consequence, unfolding and denaturation of proteins become thermodynamically unfavorable (reinforcement of hydrophobic effect). This explains, why organisms adapted to other low water-potential environments take advantage of the beneficial properties of compatible solutes. It is not surprising that cyanobacterial species found in deserts accumulate the compatible solute trehalose to compensate for the deleterious effects of desiccation [16].

12.1.2.2   Compatible Solutes of Halophiles and Their Synthetic Pathways

Halophilic Bacteria and Archaea of the organic-osmolyte mechanism synthesize *de novo* nitrogen-containing compounds as their major compatible solutes. In true halophiles analyzed so far the most predominant solutes are the amino acid-derivatives glycine-betaine and ectoine [17]. Sugars like trehalose or sucrose, which are common in a wide range of microorganisms, and necessary for osmoadaptation, rarely exceed cytoplasmic concentrations of 500 mM and are typically present in organisms of restricted salt tolerance.

In contrast, glycine-betaine and ectoine, which are energetically cheaper to synthesize, maintain suitable cell buoyancy and accumulate well above 1 M. Glycine-betaine, a typical product of halophilic phototrophic Bacteria [17], has also been found in a range of halophilic methanogenic Archaea [18, 19]. In cyanobacteria glycine-betaine is most likely synthesized via the serine/ethanol-amine pathway with choline as an intermediate [20]. In the *Ectothiorhodospira* species the biosynthesis proceeds via the direct methylation of glycine [21]. A similar pathway is proposed for methanogenic Archaea [22]. Among aerobic chemoheterotrophic Bacteria the ability to synthesize betaine is rare. Heterotrophic halophiles belonging to the Proteobacteria and Firmicutes synthesize predominantly the aspartate-derivative ectoine as their main solute [23]. The biosynthesis of ectoine proceeds via aspartic-semialdehyde, diaminobutyric acid and Nγ-acetyl-diaminobutyric acid [24]. The genes encoding the enzymes of this pathway have been isolated and sequenced [25, 26], and their regulation is under investigation [27].

12.1.2.3   Compatible Solute Transport and Osmosensing

Halophilic microorganisms do not rely entirely upon *de novo* synthesis of solutes. They are able to take up solutes or precursors from the surrounding environment, if available [28-30]. To allow for this uptake, these microorganisms are equipped with a set of different transporters, which are osmotically regulated at the level of expression and/or transport activity. These transporters facilitate a far more economical accumulation of compatible solutes. Non-halophiles, unable to synthesize nitrogen-containing compatible solutes, can switch to this energetically favorable method of osmoadaptation thus, gaining a certain degree of salt tolerance [31, 32]. In halophiles, such transport systems may have originally been intended as a means to recover compatible solutes leaking out of the cytoplasm (due to the steep solute gradient across the membrane) which would have otherwise been lost to the environment. Solute producers lacking a functional transporter would lose a significant amount of solutes to the medium [33] and thus, this explains why halophiles must also have transporters specifically for their own synthesized compatible solutes [34].

Osmoregulated compatible solute transporters have been studied mainly in *E. coli*, *Corynebacterium glutamicum*, *Bacillus subtilis* and some halotolerant microorganisms [35]. The uptake systems of these organisms are either high affinity binding protein-dependent ABC-transporter like ProU (*E. coli*), OpuA and OpuC (both *B. subtilis*) or, secondary transporters consisting of a single transmembrane protein. The ABC-systems comprise a periplasmic substrate binding domain, a transmembrane unit and a cytoplasmic protein, which fuels the transport through ATP-hydrolysis. The secondary transporters are either members of the major facilitator family (MFS; i.e. ProP (*E. coli*) and OusA *(E. chrysanthemi)*, respectively) or the sodium/solute symporter superfamily (SSSS; i.e. BetP, EctP, OpuD) [36, 37].

The only compatible solute uptake system of a halophilic bacterium, characterized at the molecular level, is the transporter for ectoine accumulation Tea from *Halomonas elongata* [34]. Tea is not related to any osmoregulated transporter known so far, but is a member of a novel type of secondary transporters called tripartite ATP-independent periplasmic transporters (TRAP-T) [38]. TRAP transporters are binding protein-dependent secondary uptake systems consisting of a substrate binding protein and two transmembrane spanning units. Tea is only the second TRAP transporter described at the molecular level and the first osmoregulated transporter of this family. The affinity of Tea for ectoine is high (K_s= 25 µM) and the transporter's design may be intended to combine this advantage with a high transport rate as known from other secondary transporters.

Since osmoregulated transporters are exposed to both the high saline environment and the cytoplasm, it was hypothesized that these systems would also function as sensors for osmotic changes. This was proven to be true for the secondary solute transporter ProP from *E. coli* [39]. Reconstitution experiments in proteoliposoms showed ProP to be a stand-alone osmosensor, able to regulate its own activity in response to osmotic stress. In the cellular background, however, ProP is also influenced by other components like the regulatory protein ProQ, which is responsible for fine-tuning the transporter's activity [40]. Successful reconstitution experiments have also been carried out using the solute uptake system BetP from *C. glutamicum* demonstrating that this transporter acts as an osmosensor as well [41]. Due to their function as transporters and sensors, systems like ProP or BetP exert an important influence on osmoadaptation of the cell e.g., osmoregulatory processes like compatible solute synthesis will be shut down, while osmoregulatory transporters are active. This implies that these systems must be integrated in signal transduction with the cell's metabolism. Whether osmoregulated transporters also take on the function as transducers in signal transmission is still to be resolved.

12.2 Saline Terrestrial Environments and Their Inhabitants

The major habitats of halophilic microorganisms are (i) salt waters (salt lakes, brines, ponds) and (ii) soils. In the latter, the matrix potential of the soil adds to the water stress caused by high salt concentrations. High saline waters originate either by seawater condensation (thalassohaline) or by evaporation of inland surface water (athalassohaline). The salt composition of thalassohaline waters resembles that of seawater with NaCl as the main constituent. Athalassohaline lakes can differ in their ion composition from seawater derived lakes. Some athalassohaline waters have a very high concentration of divalent cations (for example, the Dead Sea with the main cation Mg^{2+} instead of Na^+), while others are free of magnesium and calcium due to the presence of high levels of carbonate. Increased carbonate concentrations lead to the formation of soda lakes, which have pH-values well above 10 (for example, the Wadi

Natrun in Egypt). Microflora have been found in all of the above types of saline waters, indicating that halophilic microorganisms tolerate high salinity and can adapt to different stressors like high pH or extreme temperatures. Cold salt lakes, like the well-studied Organic Lake in the east Antarctic Vestfold region, are of interest, since they are thought to perhaps resemble the extraterrestrial environments on the Jovian moon Europa (see below).

The Organic Lake ecosystem contains salt concentrations of between 0.8 and 21% and an anoxic layer below a depth of 4 to 5 m [42]. Eukaryotic algae of the genus *Dunaliella* are found in this ecosystem, but no multicellular organisms have been detected. Procaryotes, including, moderately halophilic chemoheterotrophic Bacteria and many strains belonging to genera including *Halomonas* and *Flavobacterium* have also been isolated from the lake. Strains of *Halomonas subglaciescola*, upon further analysis, displayed a broad salt tolerance from 0.5 to 20% and were able to grow at temperatures below 0 °C.

Often overlooked and ignored saline environments are the subsurface salt deposits. These sites are of specific importance to research on extraterrestrial life, since the isolation of viable halophiles has been reported from ancient terrestrial subsurface salt environments. It has also been suggested that organisms on other planets may have survived in the planet's subsurface environment (i.e. Mars, see below). It is therefore of interest to examine in greater detail the characteristics of terrestrial subsurface salt deposits and the organisms isolated from these sites.

12.2.1   Distribution and Dating of Ancient Salt Sediments

During several periods in the Earth's history, immense sedimentation of halite and some other minerals from hypersaline seas took place. An estimated 1.3 million cubic kilometers of salt were deposited in the late Permian and early Triassic periods alone (ca. 245 to 280 million years ago; [43]). These salt sediments formed, during the existence of the supercontinent Pangaea (Permian) or the earlier Gondwanaland (Cambrian and Devonian), in large basins which were connected to the open oceans by narrow channels. Warm temperatures and an arid climate prevailed around the paleoequator, where the land masses were concentrated, causing large scale evaporation. About 100 million years ago, fragmentation of Pangaea took place, the continents were displaced to the North, and mountain ranges such as the Alps and Carpathians folded up, due to plate tectonics [44]. As a result of this shifting, the huge salt deposits, some of them up to 1200 m in thickness, are found today predominantly in the Northern regions of the continents, e.g., in Siberia, Canada (Mackenzie basin), Northern and Central Europe (Zechstein series), South-Eastern Europe (Alps and Carpathian mountains), and the Midcontinent basin in North America [43].

In contrast to other sediments, salt deposits are nearly devoid of macroscopic fossils, on which an age determination can be based. Instead, palynological and isotope studies, in addition to stratigraphical information are used. Klaus [45, 46] detected in dissolved rock salt, from numerous samples of Alpine deposits, plant spores from extinct species, which exhibited well preserved morphological features. The spores *Pityosporites*, *Gigantosporites* and others that were detected, are characteristic for the

Permian period and can be clearly distinguished from *Triadosporites* species which are found in Triassic evaporites. The formation of most of these Alpine salt sediments and the Zechstein deposits were dated to the Upper Permian period, while some Alpine deposits were dated to the Triassic period.

Sulfur isotope ratios (expressed as $\delta^{34}S$) are used for evaporites, which contain sulfates of marine origin [47]. Worldwide results from samples of doubtless stratigraphical relationships showed an extremely low $\delta^{34}S$ value for evaporites of Permian age (+8 to +12‰), and a higher $\delta^{34}S$ value (+20 to +27‰) for those of Triassic origin. Using sulfur isotope ratios from numerous anhydrite and gypsum samples, Pak and Schauberger [48] could confirm a Permian or a Triassic age for many of the alpine salt deposits ($\delta^{34}S$ values of +10 or +25/27‰, respectively). Similarly, the Zechstein series of deposits were confirmed to be of Permian origin.

12.2.2 Subterranean Halophilic Microorganisms and Their Relation to Surface Halophiles

Dombrowski [49, 50] and Reiser and Tasch [51] were the first to describe viable microorganisms which were isolated from rock salt. Other reports, one as early as 1935, demonstrated bacteria-like rods in dissolved rock salt residues and thin sections of rock salt, but recovery of these viable microorganisms was either not tried or not successful [52, 53].

Dombrowski's [49, 50] enrichments from rock salt samples of Zechstein (Northern Germany) and Precambrian (Siberia) origins, when introduced to a salt-saturated medium, yielded strains resembling *Bacillus circulans*, and another isolate, called *Pseudomonas halocrenea*. The latter was subsequently shown to be identical to *Pseudomonas aeruginosa* and thus, has to be considered a contaminant.

Tasch and coworkers [51, 54] described diplococci which were obtained from rock salt of Permian age from Kansas (Carey mine, Hutchison) and Bibo et al. [55] were able to isolate viable halophilic rods and diplococci from Zechstein salt (Neuhof near Fulda, Germany).

It is not always clear in the results of these researchers whether the isolates consisted of halophilic Archaea, since red or pink pigmentation was not mentioned by Dombrowski [49, 50] and Bibo et al. [55]. However, the papers by Tasch [54] and Nehrkorn [56], which critically evaluated Dombrowki's work, contain references to pigmentation. After 1991, publications on this topic consistently include descriptions of red or pink pigmentation of isolates, e.g., from subterranean brines of a Permian basin [57] or from rock salt of Permo-Triassic age [58, 59]. From these reports it appears that haloarchaea are the prevailing types of viable halophilic microorganisms in rock salt. To some extent, their prevalence may be due to the methods of enrichment, which favor selection of mainly aerobic, neutrophilic, heterotrophic and reasonably fast growing halophiles. It is highly likely that there exists many more haloarchaeal genera in rock salt than detected, as can be deduced from molecular analyses (see below). These other genera quite possibly have simply not been brought to growth yet. In contrast, extremely halophilic Bacteria of mostly white or yellowish appearance were recovered from brine pools, brine injection fluids, and only rarely from rock salt [58, 55].

Norton et al. [58] classified isolates from two British salt mines (Winsford, Cheshire, and Boulby, Cleveland) of Permian and Triassic age as *Haloarcula* and *Halobacterium* species on the basis of polar lipid composition. Their strain 54R was identified as a close relative of *Halobacterium* (now *Halorubrum*) *saccharovorum* by analysis of its 16S rRNA sequence [62], whole cell protein composition and presence of a V-type ATPase [60]. However, DNA:DNA hybridisation showed only 48% homology [58]. Phylogenetic analysis of 16S rRNA sequences confirmed the relationship of several isolates from British rock salt to *Haloarcula* and *Halorubrum*, and showed also that 16S rRNA genes, of which *Haloarcula* species possess at least two, did not differ significantly from those of present-day isolates [63].

Halococcus salifodinae was isolated from Permian rock salt in Austria and was described by us as a novel species [59] with strain BIp representing the type strain. Subsequently, it was found that independently isolated strains Br3 (from Cheshire, England; isolated by W. D. Grant and C. F. Norton in 1989) and BG2/2 (from Berchtesgaden, Germany; isolated by K. O. Stetter in 1988) resembled *H. salifodinae* BIp in colonial morphology and partial 16S rRNA sequences. *H. salifodinae* BIp grows in tetrads or larger clusters. This growth pattern was also observed in the strains BG2/2 and Br3 (see Fig. 12.1).

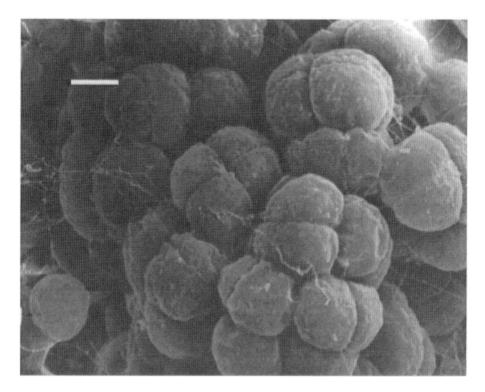

Fig.12.1 Scanning electron micrograph of *Halococcus salifodinae* Br3 (DSM 13046), grown in liquid culture medium [59]. Bar represents 1 micrometer. Photograph by Dr. G. Wanner.

Because *H. salifodinae* BIp included initially only one strain, we obtained more salt samples from the same site and recovered eight years after the original isolation several halococci, which proved to be identical to *H. salifodinae*. The detailed analysis of these isolates and the strains Br3 and BG2/2, which includes complete 16S rRNA sequences, G+C contents, sequences in a 108 base-pair insertion in the 5S rRNA gene, composition and relative abundance of polar lipids, antibiotic susceptibility, enzymatic activities and Fourier-transform infrared spectra, has been described [64]. This led to the conclusion that all isolates belong to one species, *H. salifodinae*. Thus, we demonstrated that in geographically separated halite deposits of similar geological age, identical species of halococci are present.

H. salifodinae has not yet been found in any hypersaline surface waters. While it is too early for a definite conclusion about its true native environment, it is tempting to speculate that it might be a marker organism for ancient salt deposits from certain geological periods.

12.2.3   Uncultivated Phylotypes

PCR amplification of diagnostic molecules, such as the 16S rRNA genes, and subsequent sequencing of cloned products are increasingly used for an evaluation of microbial diversity in various environments. This technique obviates culturing of microorganisms and has permitted the detection of novel and unexpected phylogenetic groups, e.g., in ocean samples [65, 66]. We used PCR with DNA prepared from dissolved rock salt (from Bad Ischl, Austria) for the amplification of 16S rRNA genes, which yielded at least two partial sequences with strong homology to haloarchaeal genes [61]. Further DNA sequences derived from about 60 cloned PCR products indicated the presence of five clusters of distinct phylotypes. Similarity values of three clusters to known 16S rRNA gene sequences were below 95%, suggesting the presence of several uncultivated novel haloarchaeal taxa [67].

In other rock salt samples (drilling cores from Altaussee, Austria, and Berchtesgaden, Germany) none, or only very few culturable cells were found, and no amplifiable DNA was detected. It is likely that these salt samples did not contain sufficient cellular material for the DNA extraction procedure modified from Benlloch et al. [68], which involves some precipitation steps. The apparent presence of uncultivated haloarchaea in only certain rock salt samples confirmed our impression that the distribution of halobacterial strains appears to vary between different strata (see below).

12.2.4   Distribution, Origin and Dispersal of Haloarchaea

The total number of colony-forming units (cfu) on complex solid medium with 20% NaCl was approximately 1.3×10^5 per kg of dry rock salt, from the salt mine at Bad Ischl, Austria [61]. Culturable strains exhibited differences in pigmentation, cellular and colonial morphology, whole cell protein patterns and other phenotypical characteristics (Fig. 12.2). The results from PCR amplification of 16S rRNA genes corroborated the presence of haloarchaeal diversity. Thus, the halophile microbial community in rock salt appears almost comparable to that of salt lakes in various parts of the world [69].

From some other rock salt samples, however, much lower numbers were obtained; for example, drilling cores from Altaussee or Berchtesgaden produced only about 10 cfu per g of dry salt [61]. Norton et al. [58] obtained on average one viable enrichment per 0.5 kg of rock salt from British salt mines. In contrast, Vreeland et al. [70] reported up to 10^4 cfu per g of rock salt from the Salado formation in New Mexico, USA.

These data suggest that the microbial content of salt sediments may vary greatly between sites. The reasons for the differences are at present unknown, but they could be due to geological (high local pressures due to tectonics, and therefore high temperatures) or perhaps biological (e.g., presence of halophilic phages or other agents, which would decimate populations) factors.

While there is no direct proof that viable microorganisms in rock salt have been entrapped since the time of their deposition, it would also be difficult to prove the opposite, namely that large numbers of extremely halophilic Archaea, or Bacteria, entered salt deposits in recent times. An influx of liquids containing microorganisms is rather improbable for the Alpine deposits in particular, since the ancient evaporites are now folded up to heights of 1000 to 2000 m, and the salt layers are located on average between 400 and 1200 m above sea level. Migration of microorganisms into the halite would require transportation by meteoric water, a source that is not known to contain extreme halophiles. In addition, layers of dolomite, limestone, marl, clay and other rocks cover alpine salt deposits. Most of these layers are water-impermeable, and have contributed to the preservation of the salt deposits during tectonics and thrust over geological times and the heavy precipitation during the ice ages.

The finding of identical strains of *H. salifodinae* in Alpine and British salt sediments could be explained by a continuous Permian hypersaline sea in Europe, which was inhabited by haloarchaea, and which gave rise to evaporites with trapped microorganisms.

H. saccharovorum was isolated from a saltern near San Francisco [71]; the similar strain 54R was obtained from Permian rock salt in England. The distance across the densely packed land masses in paleozoic times was only about 8500 km between the

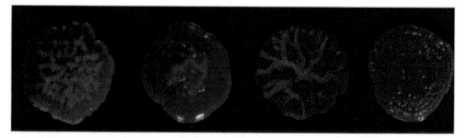

Fig.12.2 Colonies of several haloarchaeal isolates from Austrian Permo-Triassic rock salt, grown on agar plates containing nutrients and 20% NaCl.

two sites, with the huge mid-continent evaporitic basin in between [43]. It is not inconceivable that dispersal of halophiles, perhaps enclosed in crystals, did occur across the land, since wind-blown halite, together with Saharan sand, was found in England in 1984 (see [62]).

12.2.5 Long-Term Survival

Morita provides in his book [72] a history of the recognition by several microbiologists of a "cryptobiotic", "dormant", "latent" or "moribund" state of microorganisms, where no signs of a normal metabolism could be recognized, yet from which they were able to return to a physiologically active state. Such microorganisms are thought of being in a state of starvation, which could persist for long periods of time. Starvation responses of several bacterial genera have been studied in the laboratory; these include miniaturization of cells, protein degradation, reduction in endogenous respiration rate, and others (reviewed in [73]). No similar studies have yet been made with haloarchaea; thus, only suggestions for possible mechanisms of long-term survival can be considered here.

Our hypothesis is that the halophilic isolates from rock salt are the remnants of populations, which inhabited the ancient brines and were included in salt crystals upon evaporation. Whether they persisted in the dry salt sediments or in tiny fluid inclusions, which are present to various extents in rock salt [74], is not clear. At least in some Zechstein salt samples no fluid inclusions were detected in thin polished sections of 15 μm thickness [75]; microscopic inspection of several alpine salt samples has so far failed to reveal inclusions (Stan-Lotter, unpublished data). The argument for the entrapment of halophilic Archaea in fluid inclusions is based on laboratory experiments, where salt crystals with embedded Archaea were produced [76], and from which viable haloarchaea could be cultivated after at least 6 months. Similar fluid inclusions in salt sediments could contain potential energy sources, at least during the early stages of entrapment, as pointed out by McGenity et al. [62]. Later, migration of fluid inclusions could have taken place with dissolution and re-crystallization on a small scale, which could presumably transport halophiles to new energy sources.

Some further possibilities for the longevity of halophiles have been discussed by Grant et al. [63], such as the formation of resting stages or cyst-like structures, which were described for extremely halophilic Archaea from soil, for instance *Halobacterium distributum* and *Natronomonas* (formerly *Natronobacterium*) *pharaonis* [77]. Spore production has not yet been found in any halophilic Archaea, but certain thallus-like structures from coccoid haloarchaeal strains [78] are known, which apparently occur in natural environments and are lost during culturing in the laboratory. It is not sure, if such forms may be involved in long term survival. Still, these findings suggest that the potential for morphological changes, dependent on environmental conditions, appears to be present also in haloarchaea.

There is at present no method, which would allow the determination of the age of a single prokaryotic cell, since its biomass is only about 1 picogram and its composition comprises more than 3000 different molecules, putting the task beyond any procedural sensitivity obtainable to date. Most scientists, therefore, agree with the suggestions by

Kennedy et al. [79] and Grant et al. [63], who demand repeated culturing from similar environments (see also Chap. 4, Horneck et al.), and would not consider claims of single isolates [80] from ancient materials as proof for antiqueness, even if procedures were used, which are customary in the food industry for reducing the possibility of contamination [80].

12.3 Relevance to Astrobiology

The discovery of nanofossils in the carbon-containing meteorite ALH 84001 [81] was interpreted as evidence for Martian life, which is thought to resemble terrestrial microbial life. Although there is still much controversy about the true significance of the fossilized structures (e.g., [82]), possible contamination by terrestrial polycyclic aromatic hydrocarbons [83] and issues relating to the composition and inclusions of the meteorite, no evidence currently exists to completely refute the hypothesis of Martian traces of life.

In fact, several features of the discovery have strongly influenced and focussed our thoughts on what to look for, in terms of traces of life on Mars, namely: microbial life of a similar morphology of terrestrial microbes, albeit perhaps of smaller sizes, with similar building blocks [81], and an occurrence in the subsurface environment of the planet. Stevens [84] and others [85] suggested that life may have survived in the subsurface of Mars. Thus, terrestrial salt sediments of great geological age are an eminently suitable analog to perceived Martian lakebeds.

Mars and Earth share a similar geological history; they may have possessed similar early environments and thus, it is imaginable that life arose on both of these planets. If evolution followed a similar course to that on Earth, and if it proceeded at a similar rate, one could postulate that halophiles appeared early on Mars, as was suggested by Fredrickson et al. [86]. Numerous features on the Martian surface suggest the possibility of past erosion by liquid water. Recently obtained high-resolution views of the Martian surface (6 m per pixel), taken by a Mars orbiter camera [87], have led to the proposal that there is evidence for liquid water on Mars.

Traces of halite were detected in the SNC meteorites, which stem from Mars [88, 89]. Also, on the Jovian moon Europa salts were detected [90] and the Galileo spacecraft collected evidence that support the existence of a liquid ocean on Europa. Galileo's onboard magnetometer, which measures magnetic fields, detected fluctuations that are consistent with the magnetic effects of currents flowing in a salty ocean.

Recently, even macroscopic crystals of extraterrestrial halite, together with traces of sylvite (KCl) and water inclusions, were found in the Monahans meteorite, which fell in Texas in 1998 [91]. The age of this meteorite was estimated by Sr/Rb dating to 4.7 ± 0.2 billion years. It is remarkable that the contents of the liquid inclusions of this meteorite - NaCl, KCl and H_2O - are also important ingredients for growth of halophilic Archaea and Bacteria.

Lastly, the issue of forward contamination should be considered, since NASA plans to select sites on Mars which will be probed for the presence of extant or extinct life. Samples from these sites will be returned to Earth for further investigation. Recovery and identification of novel microscopic life forms should take into consideration the potential presence and extreme long-term survival of terrestrial microbes.

A deeper understanding of the practical issues, constraints and limits of prokaryotic survival under extreme conditions can be developed from the study of halophilic Bacteria and Archaea.

Acknowledgements. We thank the Deutsche Forschungsgemeinschaft (DFG) for a research grant awarded to HJK and HGT. Work in the laboratory of HSL was supported by FWF project P13995-MOB, ÖNB project 5319, and NASA Cooperative Agreement NCC2-578, while HSL was a Principal Investigator with the SETI Institute. HSL thanks C. Gruber for DNA isolation from rock salt and C. Radax for contributing sequence data prior to publication. We are grateful to Sharon Taylor for critical reading of the manuscript.

12.4 References

1   R.H. Reed, in: R.A. Herbert, G.A. Codd (Eds.) *Microbes in extreme environments*, Special publications for the Society for General Microbiology, Academic Press, London, 17, 1986, pp. 51.
2   T.E. Sweeney, C.A. Beuchat, Am. J. Physiol. **264**, R469 (1993).
3   D.S. Cayley, B.A. Lewis, H.J. Guttman, M.T. Record, J. Mol. Biol. **222**, 281, (1991).
4   A. Oren, in: A. Balows, H.G. Trüper, M. Dworkin, W. Harder, K.H. Schleifer (Eds.) *The Prokaryotes*, 2nd edn., Springer Verlag, New York, NY, 1991, pp. 1893.
5   A. Oren, M. Heldal, S. Nordland, Can. J. Microbiol. **43**, 588 (1997).
6   L.D. Mermelstein, J.G. Zeikus in: Edited by K. Horikoshi, W.D. Grant (Eds.) *Extremophiles. Microbial life in extreme environments*, Wiley-Liss, New York, 1998.
7   H. Eisenberg, E.J. Wachtel, Annu. Rev. Biophys. Chem. **16**, 69 (1987).
8   H. Eisenberg, M. Mevarech, G. Zaccai, Adv. in Protein Chemistry **43**, 1 (1992).
9   D.J. Kushner, M. Kamekura, in: F. Rodriguez-Valera (Ed.) *Halophilic Bacteria*, Vol I, CRC Press Inc, Boca Raton, 1988, pp. 109.
10 D.E. Robertson, D. Noll, M.F. Roberts, J. Biol. Chem. **267**, 14893 (1992).
11 M.C. Lai, R.P. Gunsalus, J. Bacteriol. **174**, 7474 (1992).
12 A.D. Brown, Bacteriol. Rev. **40**, 803 (1976).
13 P. Falkenberg, A.T. Matheson, C.F. Rollin, Biochim. Biophys. Acta **434**, 474 (1976).
14 K. Lippert, E.A. Galinski, Appl. Microbiol. Biotech. **37**, 61 (1992).
15 G. Malin, A. Lapidot, J. Bacteriol. **178**, 385 (1996).
16 N. Hershkovitz, A. Oren, Y. Cohen, Appl. Environ. Microbiol. **57,** 645 (1991).
17 E.A. Galinski, H.G. Trüper, FEMS Microbiol. Rev. **15**, 95 (1994).
18 D. Robertson, D. Noll, M.F. Roberts, J. Menaia, R.D. Boone, Appl. Environ. Microbiol. **56**, 563 (1990).
19 M.C. Lai, K.R. Sowers, D.E. Robertson, M.F. Roberts, R.P. Gunsalus, J. Bacteriol. **173**, 5352 (1991).
20 M.H. Sibley and J.H. Yopp, Arch. Microbiol. **149**, 43, 1987.
21 I. Tschichholz, Ph.D. Thesis, University of Bonn, 1994.
22 M.F. Roberts, M.C. Lai, R.P. Gunsalus, J. Bacteriol. **174**, 6688 (1992).
23 J. Severin, A. Wohlfarth, E.A. Galinski, J. Gen. Microbiol. **138**, 1629 (1992).

24   P. Peters, E.A. Galinski, H.G. Trüper, FEMS Microbiol. Lett. **71**, 157 (1990).
25   P. Louis, E.A. Galinski, Microbiology **143**, 1141 (1997).
26   K. Göller, A. Ofer, E.A. Galinski, FEMS Microbiol. Lett. **161**, 293 (1998).
27   K. Göller, M. Stein, E.A. Galinski, H.J. Kunte, 21st Congress of the European Society for Comparative Physiology and Biochemistry, CBP 126/A Suppl.1, 2000, pp. 61.
28   R. Regev, I. Peri, H. Gilboa, Y. Avi-Dor, Arch. Biochem. Biophys. **278**, 106 (1990).
29   P. Peters, E. Tel-Or, H.G. Trüper, J. Gen. Microbiol. **138**, 1993 (1992).
30   D. Cánovas, C. Vargas, L.N. Csonka, A. Ventosa, J.J. Nieto, J. Bacteriol. **178**, 7221 (1996).
31   M. Jebbar, R. Talibart, K. Gloux, T. Bernard, C. Blanco, J. Bacteriol. **174**, 5027 (1992).
32   H. Robert, C. LeMarrec, C. Blanco, M. Jebbar, Appl. Environ. Microbiol. **66**, 509 (2000).
33   M. Hagemann, S. Richter, S. Mikkat, J. Bacteriol. **179**, 714 (1997).
34   K. Grammann, A. Volke, H.J. Kunte, 21st Congress of the European Society for Comparative Physiology and Biochemistry, CBP 126/A Suppl.1, 2000, pp. 84.
35   B. Kempf, E. Bremer, Arch. Microbiol. **170**, 319 (1998).
36   M. Saier, Jr, Microbiol. Rev. **58**, 71 (1994).
37   H. Peter, B. Weil, A. Burkovski, R. Krämer, S. Morbach, J. Bacteriol. **180**, 6005 (1998).
38   R. Rabus, D.L. Jack, D.J. Kelly, M.H.J. Saier, Microbiology, **145**, 3431 (1999).
39   K.I. Racher, R.T. Voegele, E.V. Marshall, D.E. Culham, J.M. Wood, H. Jung, M. Bacon, M.T. Cairns, S.M. Ferguson, W.-J. Liang, P.J.F. Henderson, G. White, F.R. Hallett, Biochemistry **38**, 1676 (1999).
40   H.J. Kunte, R.A. Crane, D.E. Culham, D. Richmond, J.M. Wood, J. Bacteriol. **181**, 1537 (1999).
41   R. Rubenhagen, H. Ronsch, H. Jung, H. Krämer, S. Morbach, J. Biol. Chem. **275**, 735 (2000).
42   P.D. Franzmann, in: F. Rodriguez (Ed.) *General and Applied Aspects of Halophilic Microorganisms*, Plenum Press, New York, 1991, pp. 9.
43   M.A. Zharkov, in: *History of Paleozoic Salt Accumulation.* Springer Verlag, Berlin, 1981.
44   G. Einsele, in: *Sedimentary Basins*, Springer Verlag, Berlin, 1992.
45   W. Klaus, Z. Dtsch. Geol. Ges. **105**, 756 (1955).
46   W. Klaus, Carinthia II, **164**/Jahrg. 84, 79, Klagenfurt (1974).
47   W.T. Holser, I.R. Kaplan, Chem. Geol. **1**, 93 (1966).
48   E. Pak, O. Schauberger, Verh. Geol. B-A, Jahrg 1981, 185.
49   H.J. Dombrowski, Zbl. Bakteriol. **183**, Abt. I Originale, 173 (1961).
50   H.J. Dombrowski, Ann. N.Y. Acad. Sci. **108**, 453 (1963).
51   R. Reiser, P. Tasch, Trans. Kans. Acad. Sci. **63**, 31 (1960).
52   A. Rippel, Arch. Mikrobiol. **6**, 350 (1935).
53   E. Bien, W. Schwartz, Z. Allg. Mikrobiol. **5**, 185 (1965).
54   P. Tasch , Univ Wichita Bulletin **39**, 2 (1963).
55   F.-J. Bibo, R. Söngen, R.E. Fresenius, Kali u. Steinsalz, **8**, 36 (1983).
56   A. Nehrkorn, Arch. Hyg. Bakteriol. **150**, 232 (1967).
57   R.H. Vreeland, J.H. Huval, in: F. Rodríguez-Valera (Ed.) *General and Applied Aspects of Microorganisms*, Plenum Press, New York, 1991, pp. 53.
58   C.F. Norton, T.J. McGenity, W.D. Grant, J. Gen. Microbiol. **139**, 1077 (1993).

59 E.B.M. Denner, T.J. McGenity, H.-J. Busse, W.D. Grant, G Wanner, H. Stan-Lotter, Int. J. Syst. Bacteriol. **44**, 774 (1994).

60 H. Stan-Lotter, M. Sulzner, E. Egelseer, C. Norton, L.I. Hochstein, Orig. Life Evol. Biosph. **23**, 53 (1993).

61 H. Stan-Lotter, C. Radax, C. Gruber, T.J. McGenity, A. Legat, G. Wanner, E.B.M. Denner, in: F. Rodríguez-Valera (Ed.) *SALT 2000*, 8th World Salt Symposium 2000, Amsterdam, Elsevier Science B.V., 2000, pp. 921.

62 T.J. McGenity, R.T. Gemmell, W.D. Grant, H. Stan-Lotter, Environ. Microbiol. **2**, 243 (2000).

63 W.D. Grant, R.T. Gemmell, T.J. McGenity, Extremophiles **2**, 279 (1998).

64 H. Stan-Lotter, T.J. McGenity, A. Legat, E.B.M. Denner, K. Glaser, K.O. Stetter, G. Wanner, Microbiology **145**, 3565 (1999).

65 S.J. Giovannoni, T.B. Britschgi, C.L. Moyer, K.G. Field, Nature (London) **345**, 60 (1990).

66 E.F. DeLong, Proc. Natl. Acad .Sci. USA. **89**, 5685 (1992).

67 C. Radax, C. Gruber, N. Bresgen, H. Wieland, H. Stan-Lotter, The 3rd Internat. Congr. Extremophiles, Hamburg, Germany, 2000.

68 S. Benlloch, S.G. Acinas, A.J. Martinez-Murcia, F. Rodriguez-Valera, Hydrobiologia **329**, 19 (1996).

69 B.J. Javor, Hypersaline Environments: Microbiology and Biogeochemistry, Springer-Verlag, Berlin, 1989.

70 R.H. Vreeland, A.F. Piselli Jr, S. McDonnough, S.S. Meyers, Extremophiles **2**, 321 (1998).

71 G.A. Tomlinson, L.I. Hochstein, Can. J. Microbiol. **22**, 587 (1976).

72 R.Y. Morita (Ed.) *Bacteria in Oligotrophic Environments. Starvation-Survival-Lifestyle*, Chapman and Hall, New York, 1997.

73 H.M. Lappin-Scott, J.W. Costerton, Experientia **46**, 812 (1990).

74 E. Roedder, American Mineralogist **69**, 413 (1984).

75 H.J. Dombrowski, Natur und Museum **92**, 436 (1962).

76 C.F. Norton and W.D. Grant, J. Gen. Microbiol. **134**, 1365 (1988).

77 N.A. Kostrikina, I.S. Zvyagintseva, and V.I. Duda, Arch. Microbiol. **156**, 344 (1991).

78 A.C. Wais, Curr. Microbiol. **12**, 191 (1985).

79 M.J. Kennedy, S.L.Reader, M.L. Swiercynski, Microbiology **140**, 2513 (1994).

80 R.H.Vreeland, W.D. Rosenzweig, D.W. Powers, Nature **407**, 897 (2000).

81 D.S. McKay, E.K. Gibson Jr, K.L. Thomas-Keprta, H. Vali, C.S. Romanek, S.J. Clemett, X.D.F. Chillier, C.R. Maechling, R.N. Zare, Science **273**, 924 (1996).

82 R.A. Kerr, Science **273**, 864 (1996).

83 L. Becker, D.P. Glavin, J.L. Bada, Geochim. Cosmochim. Acta **61**, 475 (1997).

84 T.O. Stevens, in: P.S. Amy, D.L. Haldeman (Eds.) *The Microbiology of the Terrestrial Deep Subsurface*, CRC Lewis Publishers, Boca Raton, 1997, pp. 205.

85 P.J. Boston, M.V. Ivanov, C.P. McKay, Icarus **95**, 300 (1992).

86 J.K. Fredrickson, D.P. Chandler, T.C. Onstott, SPIE Proc. **3111**, 318 (1997).

87 M.C. Malin, K.S. Edgett, Science **288**, 2330 (2000).

88 J.L. Gooding, Icarus **99**, 28 (1992).

89 D.J.Sawyer, M.D. McGehee, J. Canepa, C.B. Moore, Meteoritics & Planetary Sci., **35**, 743 (2000).

90 T.B. McCord, G.B. Hansen, F.P. Fanale, R.W. Carlson, D.L. Matson, T.V. Johnson, W.D. Smythe, J.K. Crowley, P.D. Martin, A. Ocampo, C.A. Hibbitts, J.C. Granahan, the NIMS Team, Science **280**, 1242 (1998).

91 M.E. Zolensky, R.J. Bodnar, E.K. Gibson, L.E. Nyquist, Y. Reese, C.Y. Shih, H. Wiesman, Science **285**, 1377 (1999).

Part III

Electromagnetic Fields, Radiation and Life

13 Martian Atmospheric Evolution: Implications of an Ancient Intrinsic Magnetic Field

Helmut Lammer, Willibald Stumptner and Gregorio J. Molina-Cuberos

The Magnetometer / Electron Reflectometer (MAG/ER) experiment on board of Mars Global Surveyor (MGS) has detected surface magnetic anomalies of up to 1500 nT during its low aerobreaking passes, resulting from remnant crustal magnetism. These magnetic anomalies strongly indicate the existence of a strong ancient intrinsic Martian magnetic moment, which corresponded to a magnetic field strength of 10% - 100% of present Earth's.

Such an ancient intrinsic magnetic field had significant consequences for the evolution of the Martian atmosphere, especially by reducing the amount of certain atmospheric constituents lost to space. The evolution of the Martian atmosphere, with regard to water, is influenced by non-thermal atmospheric loss processes of heavy atmospheric constituents. Since Mars does not have an appreciable intrinsic magnetic field at present and a comparatively small gravitational acceleration, all known atmospheric loss processes work and several important atmospheric constituents, namely H, H_2, N, O, C, CO, O_2 and CO_2 are lost from the atmosphere. The escape rates of atmospheric constituents over time, including the loss of H_2O from Mars indicate that the red planet could have lost an atmosphere of at least 1 bar to space during the past 3.5 billion years.

The second important effect of an ancient intrinsic magnetic field and a much denser atmosphere is the shielding of the Martian surface from cosmic rays and UV radiation. Cosmic rays - which consists of charged particles - are deflected by the magnetic field depending on their energy and only high energy particles are able to penetrate and reach the surface. UV radiation is partially absorbed in planetary atmospheres, as is discussed in detail in Chaps. 14, Cockell; and 15, Rettberg and Rothschild. By investigating the surface protection during the history of the Martian atmosphere one can see that the atmospheric conditions on Mars were comparable to Earth, 3.7 billion years in the past. As life on Earth may be as old as 3.8 billion years, similar life forms may have developed on early Mars under those more favorable atmospheric conditions.

13.1 Nonthermal Atmospheric Escape Processes

Atmospheric escape occurs when atoms move upward with velocities greater than the escape velocity to an atmospheric level, where the collision probability is low (i.e. the critical level or exobase) [1]. Nearly all of the non-thermal atmospheric escape mechanisms involve ions. Many of them, such as charge exchange, dissociative recombination and sputtering processes in planetary atmospheres release energies, which appear in the form of excitation and kinetic motion of neutral products. The kinetic energies of these newly born hot atoms in some cases are in the order of electron volts. The escape fractions of a hot oxygen atom released by various non-thermal escape processes is calculated with a *Monte Carlo* technique. The hot oxygen atoms produced are assumed to become thermalized eventually by a series of elastic hard sphere collisions with the colder background gas such as CO_2, O_2 or O. Inelastic collisions are negligibly small at these low energies. On Mars all oxygen atoms at the exobase with energies greater than 2 eV are able to escape. After its release, a hot oxygen atom may collide with the neutral background gas, may change its direction, lose its energy or the atom may travel long distances in the atmosphere without collisions [2-4]. If the starting altitude is close to the exobase or above, then the escape fraction is about 1, corresponding to atmospheric levels, where the collision probability is low and thus all atoms with energies greater than the escape energy can actually escape. In Table 13.1. we have summarized exobase altitudes h_{ex} and temperatures T_{ex} (which vary with the Suns' activity cycle), escape velocities v_∞ and escape energies E for O, N and CO_2 for Venus, Earth and Mars.

13.1.1   Charge Exchange and Dissociative Recombination

Thermal ions in planetary ionospheres can be converted to fast neutral atoms with ballistic trajectories reaching very high altitudes via charge exchange interaction with neutral atomic hydrogen and oxygen (see Eqs 13.1 and 13.2).

Both reactions have been invoked to explain the existence of an extra component of suprathermal hydrogen in the exosphere of Venus [5-10]. A more important mechanism for the production of hot atomic oxygen in the Martian exosphere is the dissociative recombination of ionospheric O_2^+ ions.

Table 13.1. A summary of average exobase altitudes h_{ex} and temperatures T_{ex}, escape velocities v_∞ and escape energies E for O, N and CO_2 for the most important terrestrial planets

	h_{ex} [km]	T_{ex} [K]	v_∞ [km/s]	E_O [eV]	E_{CO2} [eV]	E_N [eV]
Venus	≈ 200	≈ 600	10.40	8.96	24.64	7.84
Earth	≈ 300	≈ 1000	11.20	10.40	28.60	9.10
Mars	≈ 200	≈ 220	5.02	2.08	5.72	1.82

$$O^+ + H \rightarrow O(hot) + H^+ \tag{13.1}$$

$$O^+ + O \rightarrow O(hot) + O^+ \tag{13.2}$$

Dissociative recombination is a photochemical process that can impart sufficient energy (about 0.125 eV/amu) to the produced atoms so they can escape Mars' gravitational field. Mostly N in the form of N_2^+, O as O_2^+ and C as CO^+, all of them Martian atmospheric constituents, are affected and the resulting neutral escape flux can be modeled. However, oxygen is by far the most important constituent for dissociative recombination on Mars. This requires knowledge of the density and composition of the Martian ionosphere and atmosphere. To understand the formation of the hot oxygen atoms [11-13], we investigate four possible channels. The oxygen atoms may be formed in the ^3P, ^1S and ^1D states, which correspond to excited oxygen atoms (one electron in a higher energetic state) with varying amounts of kinetic energy (in eV, see Eqs. 13.3 to 13.6). The upper left index means that the energy level of the exited atom belongs to a system of single-, double-, triple-, etc. levels. The letter S, P, D, etc. denotes a group of levels within this system e.g., ^3P is a triplet P level.

$$O_2^+ + e \rightarrow O\left(^3P\right) + O\left(^3P\right) + 6.96 eV \tag{13.3}$$

$$O_2^+ + e \rightarrow O\left(^3P\right) + O\left(^1D\right) + 5.00 eV \tag{13.4}$$

$$O_2^+ + e \rightarrow O\left(^1D\right) + O\left(^1D\right) + 3.02 eV \tag{13.5}$$

$$O_2^+ + e \rightarrow O\left(^1D\right) + O\left(^1S\right) + 0.8 eV \tag{13.6}$$

The branching ratios for dissociative recombination of oxygen atoms were measured for the final channels [4]: $O(^3P) + O(^3P) : O(^3P) + O(^1D) : O(^1D) + O(^1D) : O(^1D) + O(^1S) = 0.22 : 0.42 : 0.31 : 0.05$. The escape flux Φ_{esc} is obtained by integrating the energy spectrum of the cumulative hot oxygen atom flux $F(E,z_e)$ at the exobase as a function of the kinetic energy E in the energy interval between the escape energy E_{esc} and the maximum energy E_{max}.

$$\phi_{esc} = \int_{E_{esc}}^{E_{max}} F(E, z_e) \, dE \tag{13.7}$$

The escape flux for hot oxygen atoms originating from dissociative recombination (*exospheric O*) for Mars is at present about 6×10^6 cm^{-2}s^{-1} corresponding to an escape rate of 6×10^{24} s^{-1} [14, 15]. A direct result of dissociative recombination is the formation of a hot oxygen *corona* around Mars [16, 17]. Such a corona consists of atmospheric particles moving on ballistic trajectories, since their energies are higher than the energy of the background gas, but lower than the escape energy. These corona particles interact with the solar-wind and, therefore, are a source for several other escape processes like atmospheric sputtering using coronal oxygen atoms ionized by the solar-wind as sputtering agents. The solar-wind is the flux of energetic charged particles ejected by the Sun into the interplanetary medium, mainly H^+ and He^+. It has an eroding effect on atmospheres of solar system bodies which are not shielded by strong intrinsic magnetic fields.

The escape flux for the dissociative recombination of CO^+ was estimated to be at least one magnitude less than the oxygen escape [18]. Substantial quantities of nitrogen atoms have escaped from Mars through electron dissociative recombination, producing hot nitrogen atoms with energies exceeding the escape energy for ^{14}N isotopes via intermediary N_2^+ and NO^+ ions [19].

Support for this escape hypothesis was provided by the Viking mass spectrometer, which recorded an anomalous ^{15}N/^{14}N ratio. Only $N(^4S)$ and $N(^2D)$ have enough energy to escape. More ^{14}N can escape from the planet than the heavier ^{15}N isotope. The escape rate of nitrogen from Mars is to be an estimated 2.3×10^5 s^{-1} for low and 8.9×10^5 s^{-1} for high solar activity [20], as the later corresponds to higher ionospheric production rates of nitrogen ions and, therefore, an increased escape.

13.1.2   Atmospheric Sputtering

The yield for atmospheric sputtering processes Y is defined as the number of species ejected per ion incident on an exobase of a gravitationally bound gas (which is on Mars mostly O and CO_2). It includes particles ejected as a result of a direct collision with the ion and those ejected due to a cascade of collisions initiated by the ions [21]. The escape yield Y for the principal atmospheric constituent near the exobase is:

$$Y = \frac{0.5\sigma_{T \geq U}}{cos\,\theta\sigma_d} + \frac{3S_n\alpha}{\pi^2 U\sigma_d} \qquad (13.8)$$

with θ the entry pitch angle of the incident particle, T the particle energy, U the escape energy of the particle at the exobase, σ the collision cross-section for a particle receiving an energy transfer $T > U$ and σ_d the cross section for escape of a struck particle. The constants $3/\pi^2$ and α are obtained from the transport equation and S_n is the stopping cross-section [22]. Both the extreme ultraviolet (EUV) flux and the parameters of the solar-wind influence the flux of the sputtered particles. Mostly CO_2 and O are lost after being hit by reentered O^+ ions. These O^+ ions of exospheric origin are accelerated by the interaction of the solar-wind and interplanetary magnetic field (IMF) with the upper atmosphere. These ions follow helical trajectories along the interplanetary magnetic field lines draped over Mars and often re-impact the atmosphere with significant amounts of energy (upwards of 1 keV).

During the impact they can - through collisions - accelerate other particles, causing some of them to escape the planet. Most estimates do not take into account the expansion of the exosphere due to the heating caused by the ion impacts, so they may be viewed as lower limits.

There has been much debate recently about the sputtering efficiency (the number of particles ejected per incident particle) and dissociation cross section for an collision between O^+ and CO_2 [23-27]. Depending on the value chosen one gets different results in the calculation of escape rates for ancient Mars. The escape rates of sputtered O atoms from Mars are estimated at present to be about 3.0×10^{23} to 1.6×10^{24} s^{-1} and 8.0×10^{23} s^{-1} for CO_2 molecules [25].

13.1.3 Solar-Wind Interaction Processes

The massive plasma escape through the magneto-tail region on Mars, which was detected from the plasma instruments aboard the Phobos 2 spacecraft, could also be explained on the basis of ion-momentum considerations [28], since Mars at present has a negligible small intrinsic magnetic field.

A strong intrinsic magnetic field would protect the planet from most of the eroding effect of the solar-wind on the upper Martian atmosphere. Without such shielding, the solar-wind transfers momentum to atoms and ions on high ballistic trajectories and they can be swept away from the planet by the solar-wind. In Fig. 13.1 we have depicted O^+ ions following helical trajectories along magnetic field lines away from the planet.

During the Phobos 2 mission a maximum possible escape rate of about 2×10^{26} s^{-1} has been estimated if all the escaping particles are assumed to be O_2^+ ions. It is suggested that the O_2^+ ions observed from Phobos 2 owe their origin to the solar-wind erosion process in the Martian dayside ionosphere [29]. The effects of such an erosion

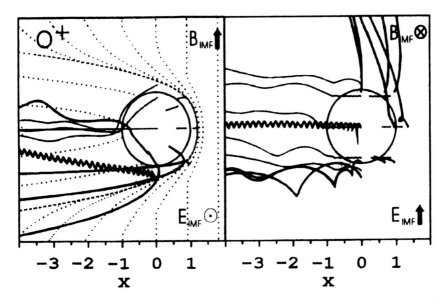

Fig. 13.1 Oxygen ions escaping Mars along interplanetary magnetic field lines on helical trajectories. BIMF and EIMF are the interplanetary magnetic and electric field. One can also see that some oxygen ions re-impact the atmosphere and act as sputtering agents.

process extend down to photochemical altitudes in the ionosphere of Mars. The estimated theoretical maximum ion escape rates from this erosion, by using Viking ionospheric data for the model, are in the range of $3\text{-}4\times10^{26}$ ions s^{-1}. As can be seen in Table 13.2., this is the most significant atmospheric escape process at present. However, during Mars' history other non-thermal escape processes e.g., sputtering have dominated.

13.1.4   Ionospheric Bubbles

A Kelvin-Helmholtz instability may cause the formation of ionospheric bubbles [30]. This instability occurs when two fluids are moving relative to each other and the boundary between them becomes unstable. This leads to a disturbance, which manifests as a travelling plasma wave in the case of Mars. This wave-like fluctuation in pressure, density and temperature grows in amplitude as long as the energy necessary to set up and maintain the instability is provided from the incoming plasma stream, i.e. the solar-wind.

The instability causes an exchange of momentum between two velocity layers, from the solar-wind plasma flow to the travelling wave to maintain its own motion. A break-up of the ionopause and the formation of a ionospheric bubble can result from these instabilities. Such an ionospheric bubble originates when a current tube is formed and ionospheric plasma is trapped in the boundary layer flow just upstream of the ionopause. The strong shear flows in this region should break up this structure and the ionospheric plasma is transported into the downstream tail region of the planet.

The formation of ionospheric bubbles happens only during a time of increased solar-wind velocity vSW (600 km/s ≤ vSW ≤ 800 km/s). About 100 bubbles can be

Table 13.2 Summary of model escape rates for oxygen, nitrogen and CO_2 for Mars at present.

Loss process	Model escape rates [s^{-1}]	Reference
Sputtered O	1.6×10^{24}	[26]
Sputtered CO_2	8.0×10^{23}	[26]
Exospheric O	5.0×10^{24}	[19]
Exospheric N	$2.0 - 9.0\times10^{5}$	[18]
Pick up O	6.0×10^{24}	[23]
Field line transport O_2^+	1.5×10^{23}	[43]
Field line transport CO_2^+	5.0×10^{22}	[43]
Ionospheric bubbles O_2^+	5.0×10^{24} / event	[31]
Erosion effect O_2^+	$3.0 - 4.0\times10^{26}$ / event	[29]

formed during one ionopause break-up, causing a loss of approximately 5×10^{24} O_2^+ ions per bubble [31]. The ASPERA ion composition experiment [32] on board of Phobos 2 has detected such an ionospheric outflow and confirmed the estimated magnitude of atmospheric loss (O_2^+, O^+ etc.), with the exception of CO_2^+ ions, where the loss cannot be explained by ionospheric bubbles, as the concentration of CO_2^+ ions near the ionopause is negligible.

All the escape rates for oxygen, nitrogen and CO_2 discussed here are summarized in Table 13.2. for easy reference.

13.2 The Early Dense Martian Atmosphere

The present thin Martian atmosphere with a surface pressure of about 7 mbar has been one of the great puzzles in our solar system. Ancient fluvial networks on the surface of Mars suggest that it was warmer and wetter three billion years ago. Surface features resembling massive outflow channels provide evidence that the Martian crust contained the equivalent of a planet wide reservoir of H_2O several hundred meters deep [33, 34]. There are two possibilities for the fate of early H_2O and CO_2 - they are either sequestered somewhere on the planet or have been lost to space.

Impact erosion [35, 36] and late impact accretion [37, 38] are other processes which could have played an important role in affecting the early Martian atmosphere. Because both the Sun [39] and the Martian atmosphere have changed over time, the importance of these evolutionary processes cannot be estimated by simply multiplying the contemporary loss rates by the age of the solar system. Models of these loss mechanisms must include the evolution of solar EUV intensity, solar-wind effects and the ancient intrinsic magnetic field barrier.

13.2.1   The Quest for Water

Calculating sputtering rates of CO_2 molecules by re-entered O^+ pick up ions for 1 EUV (present), 3 EUV (3 times present EUV intensity, 2.5 Ga ago) and 6 EUV (6 times present EUV intensity, 3.5 Ga ago) epochs of the Martian atmosphere one finds an integrated CO_2 loss equivalent to about 0.14 bar to 3 bar. Integrating the atmospheric loss rates mentioned above of O and CO_2 backward in time one gets a much denser atmosphere in the past.

Figure 13.2 shows that the early Martian atmosphere had a total surface pressure from at least 1 bar up to 5 bars, depending on the atmospheric loss models used in the study (solid line [23]; dotted line [26]). Several studies used the calculated oxygen loss rates for the estimation that Mars has lost to space an equivalent depth of 50 meters of H_2O over the last 3.5 billion years [23, 26].

A recent study [27] used a 3-dimensional (3D) Monte-Carlo model to describe the sputter interaction of the incident pick up ions with the Martian atmosphere. In this detailed analysis the extrapolated loss by sputtered O atoms from the decay of the early Mars magnetic field (≈ 3.7 Ga ago) to present time suggests lower escape rates.

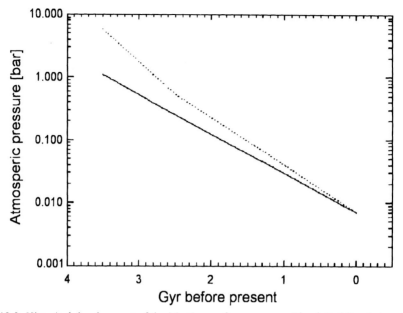

Fig. 13.2. Historical development of the Martian surface pressure. The dotted line is based on extremely high atmospheric sputtering rates [26], and the solid line on more moderate atmospheric sputtering rates [23].

In contrast to these studies we used the hydrogen loss as limiting factor for the loss of water from Mars [14], since for the early Martian atmosphere with its higher exospheric temperatures, where limiting flux conditions are likely to occur, the control of hydrogen escape by non-thermal O escape will not operate [40].

The results shown in Fig. 13.2 imply a maximum loss of H_2O to space equivalent to a depth less than 10 meters over the past 3.5 billion years. This amount is significantly lower than the early estimations, which use the non-thermal O escape rates as a limiting factor for the loss of water but is in good agreement with works based on the analysis of the D/H ratio in the Martian atmosphere [14] and in the Zagami SNC meteorite which is supposed to be of Martian origin [41]. Both isotope studies suggest that about 4 meters of H_2O were lost to space over the period considered by us, leaving today still a reservoir of crustal water, although it has been argued that the bulk of the water escape (hundreds of meters) must have happened in the first 0.5-1 Ga.

If we compare our results with the geological estimates of H_2O on Mars [33, 34] there still should be several tens of meters of water left in the form of ice and permafrost.

13.3 The Intrinsic Martian Magnetic Field

We have shown that loss of atmospheric species to space has played an important role in the evolution of the Martian atmosphere. The detection of surface magnetic anomalies by MGS, which imply the existence of an intrinsic magnetic field may have consequences for various non-thermal escape mechanisms to space in the past. Calculations by using early solar ionizing flux and solar-wind models suggest that atmospheric sputtering was more important in the past. An ancient intrinsic magnetic field can protect against the sputtering and solar-wind induced loss in several ways.

The intrinsic magnetic field deflects the solar-wind around the atmosphere by eliminating solar-wind induced ionization processes and shields ions produced in the upper atmosphere from the solar-wind. Therefore, losses of atmospheric ions and atoms by pick up or collisional sputtering are minimized during Earth-like magnetic field periods. A strong intrinsic magnetic field would be a barrier for atmospheric sputtering processes. Only loss of neutral O and N atoms originating from dissociative recombination and upward flowing ionospheric CO_2^+ and O_2^+ ions over the Martian polar caps of the magnetosphere would be important [42, 43].

Planetary intrinsic magnetic fields are generated by electrical currents circulating through the molten mantle and core of a planet. It can be explained as a magnetic dipole located in the planetary center, with elements of higher order as additional contributors. The dipole moment of Earth M_{Earth} is about 8×10^{25} Gauss cm^3. The absolute value of the magnetic field H can be calculated at a magnetic latitude ϑ and distance r with:

$$H = \frac{M}{r^3}\sqrt{1 + 3\sin^2 \vartheta} \tag{13.9}$$

This corresponds to a H $\approx$ 0,3 Gauss $\approx$ 30 000 nT at the Earth's equator (and twice that at the poles) with typical variations of the order of tens of nT.

The MAG/ER instrument on the MGS spacecraft has obtained magnetic field and plasma observations throughout the near Mars environment, from beyond the influence of Mars to just above the surface at an altitude of about 100 km. Measurements made early in the mission established that Mars does not currently possess a significant intrinsic global magnetic field, with estimated upper limits for the magnetic moment of 2×10^{21} Gauss cm^3 [44] or surface magnetic fields less than 5 nT.

During the same time the detection of strong, small scale crustal magnetic anomalies associated with the ancient, heavily cratered terrain revealed that Mars must have had an internal active dynamo in its past, which is extinct at present. The most intense magnetic crustal sources lie in the Terra Sirenum region where measured total field intensities at around 100 km altitude exceed 1500 nT [45].

The crustal magnetic anomalies in this area are sufficiently high that the magnetic fields in and above the ionosphere locally increase the total pressure that stands off and deflects the solar-wind at Mars, resulting in an asymmetric bow shock when this region rotates through the sunlit side of the planet. This configuration results in the formation of multiple small magnetospheric cusps and magnetic reconnection regions with the interplanetary magnetic field.

13.3.1 The Ancient Martian Magnetosphere: Constraints for Atmospheric Escape

One can assume two different models of core formation for Mars [46, 47]. In the first model, a solid inner core develops at approximately 1.3 Ga, which adds heat and subsequently renews the strength of the magnetic field up to present. The first measurements of MGS almost certainly ruled out this scenario. The second model is a liquid core model where postaccretionary heat drives vigorous thermal convection within the core and sustains a planetary magnetic field. The initial dynamo field can be estimated from thermal evolution models [46]. By assigning a field generating current to a toroid with a radius $(r_c + r_{ic})/2$ and a cross-sectional radius $(r_c - r_{ic})/2$, with r_c the core and r_{ic} the inner core radius, the magnetic dipole moment of Mars normalized to Earth's can be calculated from [46, 47]:

$$\frac{M_{Mars}}{M_{Earth}} = 10^9 \left(r_c - r_{ic} \right) \sqrt{P_d} \left(r_c + r_{ic} \right)^{\frac{3}{2}} \qquad (13.10)$$

where P_d is the power associated with ohmic dissipation in the liquid outer core [48]. The majority of the crustal magnetic sources detected by MGS lie in an ancient, densely cratered terrain of the Martian highlands, south of the crustal dichotomy boundary [46]. This dichotomy boundary is the geologic division between the heavily cratered highlands to the south and the relatively young, smooth plains to the north where the Martian crust is thinner.

No magnetic anomalies were detected in the major Martian volcanic areas. The large impacts that formed the Hellas and Argyre basins and that are believed to have formed about 4 Ga ago are also not associated with magnetic crustal anomalies. The absence of crustal magnetic fields in these areas implies that the Martian dynamo had already ceased to operate when these impact basins were formed. This evidence supports the magnetic field models of a hot early Mars immediately after accretion followed by rapid cooling and crust formation [46, 47].

The concentration of a light constituent like sulfur in a mainly iron core is essential for the temporal evolution of the dynamo. During the cooling process of the core, the pure iron freezes out to form a solid inner core. The boundary layer between the inner and the outer core is enriched in the light constituent, leading to gravitational instability and upward flow driving the dynamo. Model calculations show that no inner core freeze-out occurs during the first 4.5 Ga if a sulfur content >15% is assumed. If the sulfur content is much lower than 15%, a solid inner core is formed that grows rapidly on a geologic time scale once it begins to freeze out. The lack of a present intrinsic magnetic field suggests that the core has either largely frozen out or never formed. The dipole moments calculated with a sulfur content estimated from SNC meteorites ($\approx$15%) in the core are consistent with no inner core freeze-out. Rapid initial cooling of the whole planet leads to a decreased thermal convection in the core, until the Curie point is reached and the dynamo action ceases. Calculations depending on the sulfur content, P_d and core radii yield magnetic dipole moments of Mars normalized to Earth's of about 0.1 to 1.0, 3.7 Ga ago [45-48].

The first 500 to 700 million years after Mars was formed in the solar nebular its planetary magnetic field protected the atmosphere against loss by atmospheric sputtering and solar-wind interaction processes. The deflection of the solar-wind around the bulk of the atmosphere by a magnetic field limits the ion production rate in the upper atmosphere by eliminating solar-wind induced ionization processes. More important, the field shields any ions produced in the upper atmosphere (e.g., by photo-ionization) from the solar-wind magnetic field. Therefore, losses of atmospheric ions and atoms by direct sweeping or collisional sputtering, respectively, are minimized. It was found that even a small magnetosphere can significantly decrease the ion production rate by a factor of about 2 orders of magnitude [49].

One can conclude from the MAG/ER results of the MGS mission that the Martian atmosphere was unprotected by its magnetosphere since 3.5 Ga ago. After this period the non-thermal escape processes removed most of the atmosphere (see Fig. 13.2).

13.4 Shielding of Hypothetical Primitive Martian Life Forms from Energetic Cosmic Particles and Radiation

The hunt for life on Mars was reanimated after the investigation of the 1.9 kg meteorite, ALH84001. The meteorite dates from about 4.5 Ga ago, formed early in Mars' history at a depth of a few kilometers. Impacts cracked and fractured ALH84001. 3.6 Ga ago groundwater seeped through fissures and filled them with carbonate mineral. About 16 million years ago an asteroid struck the Martian surface and ejected material, including ALH84001 to escape the gravitational pull. The Martian meteorite fell to Antarctica 13 000 years ago. Three types of evidence found in the carbonate filled fractures of the meteorite back the claim for early Martian live forms: (1) Organic molecules like such produced by breakdown and geologic aging of fossilized organic matter. (2) Pancake-shaped globules, made up of minerals that on Earth can be formed by bacteria. (3) Jellybean shaped and threadlike bodies that resemble fossil microbes [50, 51].

Although there are critical arguments too [e.g., 52], the notion of life existing on Mars, now or in the past, is not implausible since the planet is like a smaller version of Earth and its atmospheric and magnetospheric conditions seem to have been similar until about 3.5 Ga ago. One of the critical issues in the discussion of environmental conditions on early Mars is its radiation environment. We discuss this in more detail in the following paragraph.

13.4.1 Cosmic Ray Particle Fluxes on the Surface of Ancient Mars

Galactic cosmic rays are isotropic radiation produced outside our solar system. They consist mainly of protons 93% and α-particles 7%, with a minor component of heavier nuclei, < 0.7 %. The integral spectrum of cosmic ray particles follows a power law dependence on kinetic energy $E^{-1.74\pm0.1}$ [53]. The main mechanisms affecting the galactic cosmic ray propagation through the solar system are deceleration by the solar-wind in the interplanetary medium and deflection by magnetic fields near the Sun or planets.

The particle flux in an atmosphere arises from the rupture of α-particles forming primary cosmic rays and from the impact of high-energy protons and neutrons on the atmospheric molecules. In such a collision other particles, such as protons and pions, are also produced:

$$\{p,n\} + N_2 \rightarrow v_p p + v_p n + v_{\pi^\pm} \pi^\pm + v_{\pi^0} \pi^0 \tag{13.11}$$

$$\pi^\pm \rightarrow \mu^\pm + neutrino$$

$$\pi^0 \rightarrow 2\gamma$$

$$\mu^\pm \rightarrow e^\pm + 2\ neutrinos$$

where v_i is the number of i-type particles resulting from the collision of a nucleon with a nucleus. The interaction of pions with the atmosphere is very complex. Charged and neutral pions decay to muons and gamma rays, respectively. Gamma rays interact with the atmosphere through the electromagnetic cascade production of electron-positron pairs, Bremsstrahlung and Compton process. Muons decay into two neutrinos and a fermion.

The atmospheric flux of the cosmic ray induced particles on the surface of ancient Mars is obtained by solving the Boltzmann equations governing the propagation of protons, neutrons, muons and pions in a 1 bar atmosphere, corresponding to more moderate non-thermal atmospheric loss rates [23]. The particle fluxes are calculated by using the same algorithm and codes [54] already used to predict the ion production rate in the atmosphere of Saturn's moon Titan [55]. Figure 13.3 shows the cosmic ray particle flux on the Martian surface 3.5 Ga ago, assuming an atmospheric pressure of about 1 bar. We neglected the magnetic field influence since it is assumed that the intrinsic Martian magnetic field vanished 3.7 Ga ago [45].

Before 3.7 Ga the ancient magnetosphere produced a radiation belt akin to the Van Allen belts of Earth, with a similar protecting effect against high energetic charged particles. Such a radiation belt reduces the flux of cosmic ray particles in the energy range of several MeV to several hundred MeV by a factor of:

$$\frac{j}{j_0} = e^{-\frac{E}{E_0}} \tag{13.12}$$

with E being the particle energy, E_0 being an energetic constant over a broad energy range, j the energy spectrum of the energetic particles in the radiation belt and j_0 being the energy spectrum of cosmic radiation in interplanetary space [56]. Only cosmic rays at the lower end of the considered energy spectrum (~10 MeV) are significantly affected by the magnetic field and the radiation belt it produced. The shielding effect of a dense early atmosphere by far dominates over the contribution of the early stronger magnetic field [57]. For comparison with the ancient Martian conditions, Fig. 13.4 shows the cosmic ray particle flux at present. Today Mars is only protected by its thin 6 mbar CO_2 atmosphere. Most of the particles reach the unprotected surface and the flux is, therefore, several orders of magnitude larger.

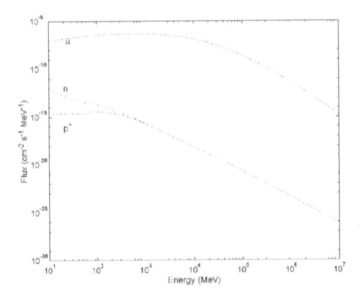

Fig. 13.3 Flux of cosmic ray particles at the surface of ancient Mars, 3.5 Ga ago shortly after the intrinsic Martian magnetic field vanished. The atmospheric pressure at this time is about 1 bar corresponding to more moderate non-thermal loss rates [23].

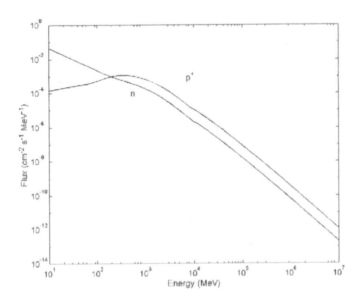

Fig. 13.4 Flux of cosmic ray particles at the present surface of Mars. At present Mars is only protected by its thin 6 mbar CO_2 atmosphere. Most of this radiation reaches the surface. The surface flux compared to the ancient atmospheric conditions is, therefore, several orders of magnitude larger.

13.5 Conclusions

The presence of an ancient strong magnetic field had significant effects on the evolution of the Martian atmosphere – especially on non-thermal escape processes like atmospheric sputtering or solar-wind pick up – and is one of the reasons for the existence of the denser early Martian atmosphere (1 bar or more) 3.7 Ga ago. The escape of water from Mars would also have been reduced by such a magnetic field, an important aspect in the discussion about extinct life on Mars. This ancient thick atmosphere was more effective in shielding the planet from harmful radiation by absorption processes than the magnetic field itself. Only cosmic rays with energy lower than ~10 MeV are significantly affected by the magnetic field, however, a dense early atmosphere is able to absorb most of the cosmic rays and reduced the flux of cosmic ray induced particles by several order of magnitude as compared to the present. The formation of a magnetosphere enabled direct shielding effects able to deflect part of the solar-wind and a Van Allen type radiation belt, which could also trap cosmic ray particles at the lower end of the cosmic ray energy spectrum.

Acknowledgments: The authors would like to thank H. Lichtenegger, Institute for Space Research, Austrian Academy of Sciences, for providing Fig. 13.1.

13.6 References

1   D.M. Hunten, Planet. Space Sci. **30**, 773 (1982).
2   F.L. Walls, G.H. Dunn, J. Geophys. Res. **79**, 1911 (1974).
3   W.B. Hanson, G.P. Mantas, J. Geophys. Res. **93**, 7538 (1988).
4   D.P. Kella, P.J. Johnson, H.B. Pedersen, V. Christensen, L.H. Andersen, Science **276**, 1530 (1997).
5   S. Kumar, D.M. Hunten, A.L. Broadfoot, Planet. Space Sci. **26**, 1063 (1978).
6   S. Kumar, D.M. Hunten, H.A. Taylor, Geophys. Res. Lett **8**, 237 (1981).
7   R.R. Hodges Jr., B.A. Tinsley, J. Geophys. Res. **86**, 7649 (1981).
8   R.R. Hodges Jr., B.A. Tinsley, Icarus **51**, 440 (1982).
9   J.M. Rodriguez, M.J. Prather, M.B. McElroy, Planet. Space Sci. **32**, 1235 (1984).
10   W.-H. Ip, Icarus **76**, 135 (1988).
11   A.F. Nagy, T.E. Cravens, J.-H. Yee, I.F. Stewart, Geophys. Res. Lett. **8**, 629 (1981).
12   J.L. Fox, A. Hac, J. Geophys. Res. **102**, 24,005 (1997).
13   J. Kim, A.F. Nagy, J.L. Fox, T.E. Cravens, J. Geophys. Res. **103**, 29,339 (1998).
14   H. Lammer, S.J. Bauer, Geophys. Res. Lett. **23**, 3353 (1996).
15   J.G. Luhmann, J. Geophys. Res. **102**, 1637 (1997).
16   H. Lammer, S.J. Bauer, J. Geophys. Res. **96**, 1819 (1991).
17   H. Lammer, W. Stumptner, S.J. Bauer, Planet. Space Sci. **48**, 1473 (2000).
18   M.B. McElroy, Science **175**, 443 (1972).
19   R.T. Brinkmann, Science **174**, 944 (1971).
20   J.L. Fox, A. Dalgarno, J. Geophys. Res. **88**, 9027 (1983).
21   E.M. Sieveka, R.E. Johnson, Astrophys. J. **287**, 418 (1984).
22   R.E. Johnson, in: *Energetic Charged-particle Interactions with Atmospheres and Surfaces*, Springer, Berlin, 1990.

23   J.G. Luhmann, R.E. Johnson, M.H.G. Zhang, Geophys. Res. Lett. **19**, 2151 (1992).
24   B.M. Jakosky, R.O. Pepin, R.E. Johnson, J.L. Fox, Icarus **111**, 271 (1994).
25   R. E. Johnson, M. Liu, Science **274**, 1932, 1996.
26   D.M. Kass, Y.L. Yung, Science **274**, 1932 (1996).
27   R.E. Johnson, F. Leblanc, Planet. Space Sci. **49**, 645 (2001).
28   R. Lundin, E.M. Dubinin, Adv. Space Res. **9**, 255 (1992).
29   J. Kar, K.K. Mahajan, R. Kohli, J. Geophs. Res. **101**, 12,747 (1996).
30   A. Miura, P.L. Prichett, J. Geophys. Res. **87**, 7431 (1982).
31   H. Lammer, M.H.G. Zhang, W. Düregger, S.J. Bauer, Internal Report No 86, Austrian Academy of Sciences, 1993.
32   R. Lundin, A. Zakharov, R. Pellinen, S.W. Barabash, H. Borg, E.M. Dubinin, B. Hulquist, H. Koskinen, I. Liede, N. Pissarenko, Geophys. Res. Lett. **17**, 873 (1990).
33   G. Neukum, R. Jaumann, E. Hauber, in *Lecture Notes in Physics*, Springer, Berlin, 2001.
34   M.H. Carr, Nature **326**, 30 (1987).
35   J.C.G. Walker, Icarus **68**, 87 (1986).
36   H.J. Melosh, A.M. Vickery, Nature **338**, 487 (1989).
37   C.F. Chyba, Nature **343**, 129 (1990).
38   G. Kargl, S.J. Bauer, in: *Sitzungsber. Abt. II (1995)*, Austrian Academy of Sciences, Springer, Vienna **131**, 1995, p. 45.
39   K.J. Zahnle, J.C.G. Walker, Rev. Geophys. **20**, 280 (1982).
40   S.C. Liu, T.M. Donahue, Icarus **28**, 231 (1976).
41   T.M. Donahue, Nature **374**, 432 (1995).
42   J. Kar, Geophys. Res. Lett. **17**, 113 (1990).
43   H. Lammer, S.J. Bauer, J. Geophys. Res. **97**, 20,925 (1992).
44   M.H. Acuña, J.E.P. Connerney, P. Wasilewski, R.P. Lin, K.A. Anderson, C.W. Carlson, J. McFadden, D.W. Curtis, D. Mitchell, H. Reme, C. Mazelle, J.A. Sauvaud, C. d'Uston, A. Cros, J.L. Medale, S.J. Bauer, P. Cloutier, M. Meyhew, D. Winterhalter, N.F. Ness, Science **279**, 1676 (1998).
45   J.E.P. Connerney, M.H. Acuña, P.J. Wasilewski, N.F. Ness, H. Rème, C. Mazelle, D. Vignes, R.P. Lin, D.L. Mitchell, P.A. Cloutier, Science **284**, 794 (1999).
46   M. Lewling, T. Spohn, Planet. Space Sci. **45**, 1389 (1997).
47   G. Schubert, T. Spohn, J. Geophys. Res. **95**, 14,095 (1990).
48   D.J. Stevenson, T. Spohn, G. Schubert, Icarus **54**, 466 (1983).
49   K.S. Hutchins, B.M. Jakosky, J. Geophys. Res. **102**, 9183 (1997).
50   D.S. McKay, E.K. Gibson Jr., K.L. Thomas-Keprta, H. Vali, C.S. Romanek, S.J. Clemett, X.D.F. Chiller, C.R. Maechling, R.N. Zare, Science **273**, 924 (1996).
51   D.S. McKay, K.L. Thomas-Keprta, C.S. Romanek, E.K. Gibson Jr., H. Vali, Science **274**, 2123 (1996).
52   J.W. Schopf, in: L.M. Celnikier, J. Trân Thanh Vân (Eds.) *Planetary Systems the long view*, Edition Frontiérs, **463**, 1997.
53   M.V. Zombeck, in: *Handbook of Astronomy and Astrophysics*, Cambridge University Press, 1990.
54   K. O'Brien, in: *REP. EML-338*, Environ. Meas. Lab., New York, 1978.
55   G.J. Molina-Cuberos, J.J. López-Moreno, R. Rodrigo, L.M. Lara, Planet. Space Sci. **47**, 1347 (1999).
56   K.J. Zahnle, J.C.G. Walker, Reviews of Geophys. and Space Phys. **20**, 280 (1982).
57   G.J. Molina-Cuberos, W. Stumptner, H. Lammer, N.I. Kömle, K. O'Brien, Icarus, 2001, in press.

14 The Ultraviolet Radiation Environment of Earth and Mars: Past and Present

Charles S. Cockell

Exactly 130 years passed between the discovery by Isaac Newton that white light was composed of colors [1] and the discovery of ultraviolet radiation by Johann Wilhelm Ritter, a German electro chemist, in 1801. We now understand that ultraviolet radiation, although representing <2% of the total number of photons that reach the surface of present-day Earth, has had an important role in the evolution of life on Earth. This is because it has a high energy, energy being proportional to the frequency of the radiation. UV radiation is damaging to a number of key macromolecules, particularly DNA. On early Earth, the lack of an ozone column probably resulted in higher biologically weighted irradiance than the surface of present-day Earth as there were no other UV absorbers in the atmosphere. This is also the case for present-day Mars and probably was for Mars in its early history.

14.1 UV Radiation on Early Earth

Ultraviolet (UV) radiation has been a ubiquitous stressor since the origin of the first microbial ecosystems during the Archean era (3.9-2.5 Ga ago). Although the UV radiation that reaches the surface of the Earth spatially and temporally depends upon many factors [2], during the history of life on Earth four distinct periods of photobiology can be recognized [3]. Firstly, the period during which UV radiation influenced chemistry on prebiotic Earth during the Hadean era (>3.9 Ga ago). This includes the beneficial involvement of UV radiation in organic complexification as well as the deleterious effects it may have had on exposed prebiotic molecules. Since this does not involve ecosystems or organisms *per se*, it is not discussed here, although discussions can be found elsewhere [3-7]. The second stage involves the role of UV radiation during the Archean, when it is supposed that the Earth lacked a significant ozone column and was therefore exposed to higher fluxes of UV-B (280-320 nm) and UV-C (200-280 nm) radiation. The third stage is the transition phase. Atmospheric oxygen partial pressures and thus ozone column abundance rose and biologically effective irradiance on the surface of the Earth was reduced. The fourth phase is the period since this transition that covers the Proterozoic and Phanerozoic (2.5 Ga ago to the present), when life has been protected by the ozone column, but subject to alterations in the UV-B radiation regime as a result of short term changes in ozone column abundance caused by stochastic alterations in the astronomical environment, such as impact events and supernovae [8] or endogenous events such as volcanism. It is the subject of this section to discuss what is known about the phase covering the Archean era (3.9-2.5 Ga ago).

14.1.1   UV Radiation During the Archean

The partial pressure of oxygen in the present-day atmosphere (~210 mb) is an imbalance caused principally by the activity of photosynthetic organisms, the burial of organic carbon and the lack of reductants from volcanic out-gassing and oceanic up-welling to mop up the oxygen so produced. A diversity of direct geologic and isotope evidence from Archean facies suggest that the Archean atmosphere was essentially anoxic. This is discussed elsewhere [e.g., 9, 10, 12-15]. The reasons for the lack of atmospheric oxygen in the Archean are still a point of discussion [e.g., 16-18]. Regardless of the mechanisms underlying the low atmospheric partial pressures of O_2 in the Archean and the arguments on the extent of oxygenic photosynthesis during this time, the photobiological consequences were identical - the early Earth lacked a significant ozone column and as a result it may have been subjected to much higher biologically effective irradiance than the present-day Earth.

In the absence of direct evidence, the effects of this photobiological environment can best be assessed using radiative transfer models that allow for the calculation of surface UV fluxes. Weighting functions can be used to calculate the biological effect of these fluxes. Because we are fairly sure that the basic structure of DNA has not changed since the Archean, action spectra for DNA damage [e.g., 19] can be useful for evaluating early Archean photobiology. Similar arguments also apply to photosystem II (PS II). The action spectra for isolated spinach chloroplasts [20] may seem an unlikely analogue for early PSs, but the experiments specifically examined the effects of UV radiation on PS II of the chloroplasts. Because PS II is similar in both chloroplasts and their non-symbiotic precursors, the cyanobacteria, this action spectrum is useful for gathering first order approximations of effects on cyanobacteria.

Once estimates of UV flux and weighted irradiance are made, then physiological responses of organisms to early environments can be assessed better.

14.1.1.1   Calculation of UV Radiation on Early Earth

The calculation of UV flux at the surface of the early Earth depends principally upon two factors - the luminosity of the early sun and the composition of the paleoatmosphere. At 3.5 Ga ago when there are unequivocal signs of life in the fossil record, the Sun was probably 25% less luminous than it is today [21, 22]. This might correspond to an approximately 25% lower flux across the UV range at wavelengths >270 nm based on the data presented by Zahnle and Walker [23] for solar fluxes at this time but with little change at lower UV-C wavelengths. These spectra are based on direct observations of young stars. The exact reductions in UV depend are model dependent [22], but ultimately, the assumptions that are made turn out to be of little consequence since the differences in weighted irradiances between early Earth and present-day Earth are overwhelmingly determined by the effect of the lack of ozone, not assumptions about whether the solar luminosity was between 0 or 35% lower. Nevertheless, our uncertainty about the UV output of the early Sun in the wavelength range of bio-logical interest is a limitation in our understanding. Early stars often emit more UV radiation at wavelengths below 200 nm [23-25]. These T-Tauri stars have been ob-served directly, and it is possible that during the formation of the Earth the Sun was emitting an intensity of UV radiation at these wavelengths 10 000 times greater than today and still four times greater at 3.5 Ga ago [24]. Because CO_2 absorbs wave-lengths of UV radiation <200 nm, it is unlikely that T-Tauri emissions reached the surface of the Archean Earth.

The composition of the Archean atmosphere is not well known, but at 3.5 Ga ago, atmospheric composition may have been approximately 1 bar CO_2 [26] with N_2 partial pressures probably similar to today (~0.8 bar). An upper limit of 10 bar CO_2 has been suggested for the very early Archean [27], but this would lead to surface temperatures ~85 °C [26]. Investigations of pCO_2 ~2.7-2.2 Ga suggest values as low as 40 mb [28]. These latter values are consistent with the lower boundary for CO_2 suggested at this time in earlier work [26]. Thus, in general it is believed that pCO_2 was high in the early Archean decreasing into the Proterozoic.

These values can be used to derive the spectral irradiance of radiation reaching the surface of the Earth. The δ-2 stream method has been described previously and is a classical approach to calculating UV radiative transfer [29, 30]. In this approach, absorption is calculated according to Beers Law and the diffuse UV flux is calculated according to a series of equations that estimate the effects of scattering in the atmos-phere. Both of these terms put together provide an approximation of the UV radiation at different wavelengths that actually reaches the ground from space. In Fig. 14.1 irradiances are shown for a zenith angle of 0° (sun overhead) for two atmospheric compositions (Early Archean at 3.5 Ga ago and late Archean at 2.7 Ga ago). Typical values for a zenith angle of 0° on present-day Earth are also shown. All cases assume cloudless skies. Clouds can have an effect on UV flux [2]. Integrated over time, com-parisons between the photobiological environment of present-day Earth and early Earth could be influenced by cloudiness. But it is unlikely that the planet would have been 100% cloudy all of the time. Therefore, the calculations presented here still pro-vide an upper boundary on instantaneous UV exposure.

The DNA weighted irradiances received at the surface of the Earth can be calcu-lated for these atmospheres. In the high pCO_2 case (1 bar) the value is 54 W/m^2 using a DNA action spectrum normalized to 300 nm. For a pCO_2 of 40 mb, then DNA weighted irradiances would increase to ~101 W/m^2. These values can be convolved with present-day UV data to postulate an ultraviolet history for Earth as is shown in Fig. 14.2 [8].

Instantaneous exposure was much higher than today, but day length was shorter be-cause of the effects of lunar tidal drag since that time. At 3-2.5 Ga ago day length may have been 14 hours [10]. The instantaneous DNA weighted irradiance would have been just over three orders of magnitude higher than today, but the daily weighted fluence would have been only 500 times greater. This would have had implications for the daily damage that a micro-organism would have had to repair and would have gone some way to off-setting the lack of an ozone column. However, it is clear that the overwhelming influence is the lack of an ozone column when comparisons are made to present-day Earth, not day length.

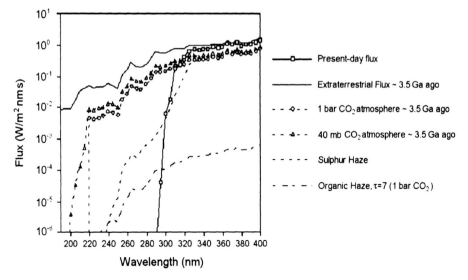

Fig. 14.1 Ultraviolet irradiance reaching the surface of Archean Earth for various assumptions about the composition of the atmosphere (see text for details).

14.1.1.2   *Atmospheric Absorbers and Effects on Archean Photobiology*

Trace quantities of other compounds could have profound consequences for UV exposure. Kasting et al. [31] investigated the surface UV effects of a sulfur haze in the early atmosphere caused by photolytic production of sulfur from SO_2 and H_2S volcanic out-gassing. At high enough temperatures ($\sim$45 °C), sulfur could have reduced the integrated UV flux by up to seven times. The photochemical arguments for this scenario are uncertain. Figure 14.1 shows the photobiological consequences of a haze with a column abundance of $\sim$1.5$\times$10^{17} cm^{-2} as they envisaged.

A plausible contaminant in the early Earth atmosphere was a CH_4-generated hydrocarbon smog, the CH_4 produced either by early methogens or nonbiological processes [32]. This is analogous to early suggestions that an organic aldehyde haze may have provided screening on early Earth [4]. At an optical depth of 7 in the UV region, (optical depth being the natural logarithm of the ratio of radiation from a source over that seen by the observer – essentially a measure of attenuation), which has been suggested for early Earth [32], DNA-weighted irradiances would have been reduced to $\sim$0.04 W/m^2, similar to exposed present-day Earth. Even modest smogs could have provided shielding for early life.

Finally, it has also been argued that appreciable levels of oxygen of 0.01-0.02 Present Atmospheric Levels (PAL) could have existed on early Earth. Numerous geological, physiological and biochemical arguments have been presented for this scenario [33, 34]. These oxygen levels which could result in ozone abundances $\sim$4$\times$10^{18} cm^{-2} would cause reductions in biologically effective irradiances by two orders of magnitude resulting in DNA weighted irradiances only two to three fold higher than typical present-day values. Although not disproven, the geologic and isotopic evidence alluded to earlier is currently more consistent with an anoxic Archean atmosphere.

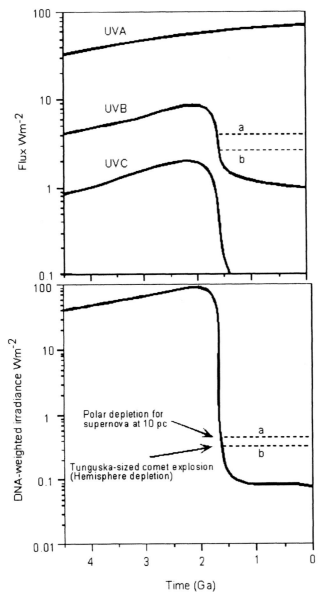

Fig. 14.2 The summarized photobiological history of Earth showing putative changes in UV-C, B and A at the surface of the Earth over time in the absence of UV absorbers in the atmosphere > 200 nm (see [8] for details). Also shown are corresponding changes in DNA-weighted biologically effective irradiances that might be expected from (a) a supernova explosion at 10 pc and (b) a Tunguska equivalent comet causing ozone depletion over a hemisphere.

14.1.2   Biological Effects of High UV Radiation Flux

The calculations shown here in an atmosphere lacking sulfur, CH_4 or ozone lead to DNA weighted irradiances two and a half to three orders of magnitude higher than on present-day Earth, similar to those presented previously [35-37]. Although a radiative transfer model was not used in [37] similar order of magnitude, differences were calculated. These differences in biologically effective irradiances have been confirmed in orbital experiments. Horneck and Rettberg [36, 38] used the extraterrestrial spectrum to observe inactivation of *Bacillus subtilis*. By measuring the change in Coomassie Blue staining, which is inversely proportional to the UV radiation received, they demonstrated that the biologically effective irradiances in Earth orbit were three orders of magnitude higher than on the surface of the Earth.

Assuming the worse case scenario (i.e. instantaneous DNA weighted irradiances were three orders of magnitude higher on Archean Earth than today), are there methods to cope with such an environment? What impact would it have on UV protection/repair and avoidance responses?

14.1.2.1   The Oceans

The oceanic water column could have been an effective screen for high UV flux. The water attenuation coefficients in the UV-C are almost an order of magnitude higher than those in the UV-B [39]. Expressed as a DNA-weighted irradiance at a depth of approximately 30 meters, irradiances could have been similar to the exposed surface of present-day Earth [8].

In the early Archean, the presence of upwelled ferrous iron could have provided additional UV attenuation. Holland suggests that ferrous iron concentrations could have been ~3 ppm [9]. With absorbance coefficients almost an order of magnitude higher than ferric iron, ferrous iron has been suggested as a potentially important UV screen [35, 40, 41]. However, if oxygenic photosynthesis existed in the early Archean, then ferrous iron could have been stripped from the photic zone. Certainly by the late Archean and early Proterozoic, when the prevalence of banded iron formations decreases [12, 13], it is likely that ferrous iron was exhausted as a screen. This could have happened before significant rises in atmospheric pO_2 [35].

Although much of the photic zone of many aquatic environments may have been clear during the early Archean and almost certainly by the late Archean, the photic zone could have been colonized by a low diversity, high UV resistant biota, which could have been numerically abundant [42]. A deep chlorophyll maximum that constitutes a deep region of high microbial abundance and quite high productivity could also have existed in the Archean as it does today [35, 42].

14.1.2.2   Intertidal and Terrestrial Habitats

Examining a range of screening methods Cockell [37] concluded that there are a variety of substrates that provide over a two order of magnitude reduction of DNA-weighted irradiances and under such substrates, Archean organisms could be exposed to weighted irradiances similar to an exposed organism on present-day Earth. Admittedly, many fully exposed single-celled organisms on present-day Earth produce UV screening compounds. Therefore, the comparison between biologically effective irradiances achieved by hiding under substrates and the exposed value on present-day Earth is not entirely accurate. However, it suffices in that we are considering order of magnitude reductions that many of these substrates provide compared to the full sky exposure.

Terrestrial habitats that would protect against UV radiation include the lithic habit (under or within rocks). In such substrates light levels are reduced to approximately 0.005% of incidence at depths, at which organisms [43] such as the cyanobacterium *Chroococcidiopsis*, are able to photosynthesize.

Reduced ferrous iron, which would have upwelled from the anoxic Archean oceans may also have protected some organisms in inter-tidal regions [35, 40]. As alluded to earlier, it could have been stripped from the water by an oxidized upper layer, but in inter-tidal regions it would have precipitated onto benthic habitats. Sediments can also provide UV protection. Garcia-Pichel and Bebout found that UV-B was reduced to 1% between 1.25 and 0.23 mm from the surface [45].

Protection of organisms may be enhanced by the matting habit, whereby the upper layer of dead organisms protect organisms underneath by virtue of their UV screening compounds. Margulis et al. [16] showed that after 3 days of continuous exposure to 254 nm radiation, a protected *Lyngbya* sp. community was still viable, although cells on the surface were killed after minutes. This matting habit is well preserved in the Archean fossil record in the form of stromatolitic layering in microbial communities [46]. Indeed, it is probably the only UV protection strategy that we can truly support with confidence based on real fossil record evidence.

14.1.2.3 Ultraviolet Radiation Screening Compounds and Repair

The evolution of UV screening compounds such as mycosporine-like amino acids (MAAs) which can screen in the UV-B and A [46, 47] as well as scytonemin, a UV-A screening compound associated with terrestrial cyanobacteria [48] would have led to important versatilities in the colonization of exposed habitats [35]. Garcia-Pichel reviews the role of these compounds in the evolution of cyanobacteria [35]. Organisms may also have been exposed to fluxes of UV-C radiation. The cyanobacterial sheath compound scytonemin has an absorbance peak at 250 nm. Experimental results showed that scytonemin can absorb UV-C radiation to an extent that is physiologically advantageous to photosynthetic carbon fixation in *Tolypothrix* sp. isolated from exposed rock surfaces [49] as well as *Calothrix* and *Chroococcidiopsis* sp. [50]. Most organics possess UV-C absorbance and are likely to have a physiologically beneficial effect against UV-C [37]. For example, plant flavonoids can provide significant protection against UV-C-induced photosynthesis inhibition in pea (*Pisum sativum* L.) [51], which certainly did not exist in the Archean! Like scytonemin, flavonoids have an absorbance peak at ~250 nm. It is likely that a diversity of organics in the upper layers of microbial mats would absorb UV-C radiation, particularly if upper layers contain a dead layer of microorganisms, from which organics would be released and subsequently photolytically degraded into smaller UV absorbing organics.

Thus, data on UV screening compounds demonstrates that biological protection against the complete UV range from 200-400 nm was probably achieved on early Earth and that despite the high energy and biological destructiveness associated with UV-C radiation, the fact that it is absorbed by most organics it passed through, probably made it one of the less challenging regions of the UV spectrum to deal with. Higher wavelengths required more specific evolutionary innovations.

Protection, either physical or biological, is never 100% efficient and the repair of DNA must also have been a key response to UV radiation that penetrated the cell.

The evidence of repair processes in the deep-branching Archaea that includes photoreactivation [52], recombination repair [53] and excision repair [52] suggest that the major pathways of repair seen in present-day organisms were developed in the Archean. Indeed, photolyase, a 310-500 nm inducible enzyme responsible for photoreactivation has been suggested to be an early photoreceptor [54]. Some of these repair processes are quite impressive. For example, the archeon-*Thermococcus stetteri* is two to three more times sensitive to gamma irradiation than *Deinococcus radiodurans* [55], but this is still a significant repair capability. *D. radiodurans* itself comes close to being able to tolerate the worse case UV environment of early Earth based on theoretical calculations. It demonstrates that repair alone might have been sufficient to deal with a high Archean UV flux [42].

In studies directed at Archean conditions, Pierson et al. demonstrated a UV tolerance in *Chloroflexus aurantiacus*, an anoxygenic photoheterotroph from the deepest branches of the eubacterial line [56]. Photoreactivation of damage caused by 254 nm UV-C radiation was also shown in the obligate anaerobe *Clostridium sporogenes*, which was suggested to be a legacy of pre-Phanerozoic evolution [57].

A further complication in developing a detailed knowledge of responses to a putatively higher Archean UV radiation regime is that the trade-off between protection and repair is quite varied. It depends upon the different energetic demands in different organisms and probably habitat. In a recent study it was found that photosynthesis in two organisms, the cyanobacterium *Lyngbya aestuarii* and a green alga *Zygogonium* sp. was affected by ambient UV-B and UV-A radiation even though these organisms do possess UV-B screening compounds and in the case of *Lyngbya*, UV-A screening scytonemin [58]. However, in the red alga *Cyanidium caldarium* which does not possess UV screening compounds, photosynthesis inhibition by UV radiation was negligible. This may be caused by different nutritional status but could also be caused by higher rates of repair.

14.1.2.4 *Photosynthesis in the Archean*

The effects of Archean UV flux on other physiological responses can be estimated. The action spectrum for PS II inhibition shows a markedly greater involvement of the UV-A region than the action spectrum for thymine dimer formation in DNA [20]. Although UV radiation can affect other parts of the photosynthetic apparatus (e.g., the photosynthetic enzyme D-ribulose 1,5-bisphosphate carboxylase-oxygenase, RuBisCO), the PS II action spectra is broadly similar to the action spectra for photosynthesis inhibition in whole organisms and so is a useful proxy for UV-induced photosynthesis inhibition in cyanobacteria and similar organisms. The UV-A contribution corresponds to the role of reactive oxygen species in PS damage. Because UV-A levels are actually higher today than in the Archean because of the more luminous Sun, this part of the spectrum would have made a lesser contribution on Archean Earth, offsetting some of the effects of greater UV-B and UV-C flux. This is why some workers have calculated that PS damage was less great in the Archean compared to DNA damage [35, 37].

14.1.3 Beneficial Effects of High UV Radiation on Archean Earth?

Insofar as UV radiation is a mutagen, then it might be expected that a two to three order of magnitude higher DNA-weighted irradiance on early Earth could lead to higher mutation rates. Then one could postulate that this might ultimately lead to higher microbial biodiversity or a greater number of mutations would have allowed faster adaptation to changing environmental conditions compared to today.

The idea is intriguing, but qualitative evidence from the Archean fossil record does not lend strong support. The microbial biodiversity of the Archean is not greater than the Proterozoic, although this may be largely a function of lack of preservation of the Archean fossil record [59]. Furthermore, cyanobacterial hyperbradytely (the extreme lack of evolutionary change in a group of organisms) is embodied in the morphological characteristics and habitat preferences of cyanobacterial stromatolites and microfossils, both modern day and their postulated Archean counterparts [59]. This evidence suggests that in fact many of the members of this phylum have not changed much, rather than being subject to great evolutionary change during the Archean as a result of UV-induced mutations. Indeed, evolutionary radiation and specialization seems to be a characteristic more of the Phanerozoic era rather than the Precambrian.

Morphometric data does not necessarily imply similar physiologies. It is plausible that higher rates of mutation may simply increase the rate of change of physiology. However, evolutionary changes and mutations normally, over time, engender morphological changes as habitat and physiology change in response to new environmental opportunities and challenges. The conservative nature of the distribution of morphologies of Archean micro-fossils in comparison to modern day cyanobacteria, particularly the Oscillatoriaceae and Chroococcaceae, as well as their similar habitats (such as inter-tidal stromatolites) might suggest physiological hypobradytely as well. Indeed, cyanobacteria are generalists [59].

UV radiation has been suggested to have other positive roles in the Archean biosphere. It has been postulated to be a trigger for the evolution of sex [16, 60]. The concept is an extension of sex as an error repair mechanism [e.g., 61]. Recombination repair, whereby new genetic material may be used to repair UV damaged DNA bears functional similarities to meiotic recombination and insofar as UV radiation causes mutations, sex has been suggested to have been stimulated by the need to repair UV-induced DNA damage [60]. Elena and Lenski [62] provide some evidence that mutations are likely to be antagonistic as much as they are synergistic and so they suggest that sex is not a good way to reduce mutational load. Nevertheless, it is clear that a population will inexorably collect mutations over time through unfaithful replication of information, the so-called "Muller's ratchet" [63] and mechanisms that allow for improved methods of transferring new information in a population might be expected to reduce mutational load [60].

14.2 The Ultraviolet History of Mars and Venus: An Exercise in Comparative Evolutionary Photobiology

The atmosphere of Mars is 95.3% CO_2 and so the radiative transfer calculations that we use to calculate the surface UV environment can essentially assume a pure CO_2 atmosphere (see [64]). Unlike the Earth, Mars does not have a significant ozone column, although some ozone build-up occurs over the poles in spring and winter. These levels, although about two orders of magnitude lower than typical terrestrial column abundances, can reduce UV-C flux reaching the ground [64, 65]. The photobiological history of the planet has been almost exclusively determined by the increase in solar luminosity over time and change in the atmospheric carbon dioxide reservoir. Haberle et al. carried out a detailed modeling study of the evolution of CO_2 on Mars over time [66]. They investigated varying initial CO_2 inventories as well as alterations in solar luminosity and the greenhouse effect. They ultimately conclude that none of the outcomes is entirely satisfactory. Large initial CO_2 inventories tend to predict Martian polar caps that are too large compared to the ones we observe today. Smaller inventories require low partial pressures of CO_2 on early Mars, which may be inconsistent with a warmer, more water rich past [67].

In view of the warmer conditions proposed for early Mars, Haberle et al. [66] propose a scenario where the initial CO_2 inventory may have been between 0.5 and 3 bar. At approximately 3.8 Ga, the CO_2 inventory may have been 0.5-1 bar. How the CO_2 atmospheric reservoir then evolved to current conditions (the present-day surface pressure is on average ~6 mb) is unknown. Either the CO_2 was slowly lost to carbonates, or the atmosphere may have collapsed. In the latter scenario the build-up of the polar ice caps results in reduced temperatures and a freeze out of CO_2. A feedback process is initiated, which leads to a collapse of the atmospheric CO_2 reservoir [66]. Because of the direct coupling between the Martian polar caps and atmospheric CO_2, the time to reach equilibrium may have been only ~200 years [68]. In Fig. 14.3, the photobiological history of Mars has been presented for an initial inventory of 2 bar declining to 1 bar in the time corresponding to the early Archean (the Martian Noachian) with an arbitrary gradual decline to present-day conditions. The rate of decline of CO_2 varies with the models used [66, 69].

The rising UV flux over time, although theoretically presenting an increasing photobiological challenge, probably does not prevent the evolution of life. The present-day DNA-weighted irradiance on the surface of Mars is similar to the weighted irradiance on the surface of Archean Earth, the biological significance of which has been discussed [37]. The photobiological deterioration of Mars could theoretically exacerbate the demise of life in synergy with other physical factors [37]. Low temperature extremes and the possible existence of peroxides in the Martian soil are two environmental stressors detrimental to life, but the lack of liquid water over most of the present-day surface is undoubtedly the worst [69].

This gradualist view of the ultraviolet history of Mars may have been different if the planet did suffer an atmospheric collapse during its history between 4.5 and 3 Ga ago [66]. A planetary atmospheric collapse has the potential to trigger an ultraviolet crisis. A reduction of the Martian atmospheric CO_2 reservoir from ~1 bar to ~6 millibar would increase DNA-weighted biologically effective irradiances by five-fold. A reduction from 0.5 bar to ~6 millibar would cause a three-fold increase.

Finally, Mars may have experienced periods of *reduced* UV radiation even since 3.5 Ga ago. Gulick et al. [70] suggest that episodic CO_2 releases of up to 2 bar resulting from catastrophic floods may have occurred. Such episodes would have resulted in an ultraviolet amelioration as illustrated in Fig. 14.1. The possibility of UV

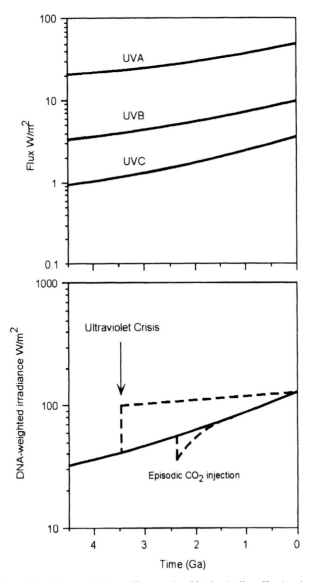

Fig. 14.3 The ultraviolet history of Mars. The graph of biologically effective irradiances also shows the theoretical photobiological consequences of an 'ultraviolet crisis' caused by an atmospheric collapse. Also shown are the effects of episodic CO_2 injections.

amelioration events should be noted, since regardless of whether there was life on Mars or not, they represent periods of increased biological potential.

Venus has a very different ultraviolet history [8]. On Venus, we suppose that the early atmosphere was thinner that it is today. As solar luminosity increased so the early water inventory was eventually lost into a runaway greenhouse effect. The water would have been photolyzed, the hydrogen disappearing into space and the oxygen lost in oxidation reactions with surface rocks. As weathering ceased in the absence of water, so there was no way for the CO_2 in the atmosphere to return to the crust, the result was the dense 9.5 MPa atmosphere we observe today. The atmosphere is in fact so dense that no UV-C or UV-B radiation reaches the surface of the planet and only very small amounts of UV-A penetrate. Thus, in a relatively short space of geological time, Venus probably went from a planet exposed to quite harsh UV radiation to one, on which the surface UV regimen is now absolutely clement. This is interesting, because the surface of Venus today is lifeless; the high (464 °C) temperature exceeds the known upper bounds for biology. Thus, Venus demonstrates nicely the unimportance of UV radiation as an evolutionary selection pressure when other factors, particularly lack of water, become limiting to life.

14.3 Conclusions

Over the past two centuries, since the discovery of UV radiation, information on the responses of micro-organisms to this agent has allowed us to develop an understanding of the role of UV radiation as an environmental stressor over geologic time periods. This perspective on Earth history is really only complete when this piece of the jigsaw is placed against others. Thus, efforts to understand the history of UV radiation on Mars can allow us to pursue an investigation of comparative evolutionary photobiology [8], elements of which were presented here. Examination of the history of UV radiation on Venus, for instance [8, 71] can extend our understanding yet further. With these perspectives in mind, we can understand better the importance, or lack of importance, of UV radiation in influencing biological evolution on planetary surfaces.

14.4 References

1   I. Newton, Trans. Royal Soc. London **6**, 3074 (1671).
2   M.A. Xenopoulus, D. Schindler, in: C.S. Cockell, A.R. Blaustein (Eds.) *Ecosystems, Evolution and UV Radiation*, Springer Verlag, 2000, in press.
3   C.S. Cockell, Knowland, J. Biol. Rev. **74**, 311 (1999).
4   C. Sagan, Journ. Theor. Biol. **39**, 195 (1973).
5   V.M. Kolb, J.P. Dworkin, S.L. Miller, J. Mol. Evol. **38**, 549 (1994).
6   H.J. Cleaves, S.L. Miller, Proc. Natl. Acad. Sci. **95**, 7260 (1998).
7   M.P. Bernstein, S.A. Sandford, L.J. Allamandola, J.S. Gillette, S.J. Clemett, R.N. Zare, Science **283**, 1135 (1999).
8   C.S. Cockell, Planet.and Space Sci. **48**, 203 (2000).

9   H.D. Holland, *The chemical evolution of the atmosphere and oceans.* Princeton University Press, Princeton, NJ., 1984, 582 pp.

10   J.C.G. Walker, C. Klein, M. Schidlowski, J.W. Schopf, D.J. Stevenson, M.R. Walter, in: J.W. Schopf (Ed.) *Earth's earliest biosphere,* Princeton University Press, Princeton, 1983, pp 260.

11   H.D. Holland, N.J. Beukes, American J. of Sci. **290**, 1 (1990).

12   H.D. Holland, in: S. Bengston (Ed.) *Early Life on Earth,* Columbia University Press, New York, 1994, pp. 237.

13   D.R. Lowe, in: S. Bengston (Ed.) *Early Life on Earth,* Columbia University Press, New York, 1994, pp. 24.

14   J.C.G. Walker, P. Brimblecombe, Precambrian Res. **28**, 205 (1985).

15   K.D. Collerson, B.S. Kamber, Science **283**, 1519 (1999).

16   L. Margulis, J.C.G. Walker, M. Rambler, Nature **264**, 620 (1976).

17   A.H. Knoll, Origin. Life Evol. Biosph. **9**, 313 (1979).

18   J.W. Schopf, J.M. Hayes, M.R. Walter, in: J.W. Schopf (Ed.).*Earth's earliest biosphere*, Princeton University Press, Princeton., 1983, pp 361.

19   A.E.S. Green, J.H. Miller, CIAP Monograph. 5, 2.60 (1975).

20   L.W. Jones, B. Kok, Plant Physiol. **41**, 1037 (1966).

21   M.J. Newman, R.T. Rood, Science **198**, 1035 (1977).

22   D.O. Gough, Solar Phys. **74**, 21 (1981).

23   K.J. Zahnle, J.C.G. Walker, Rev. Geophys. Space Phys. **20**, 280 (1982).

24   V.M. Canuto, J.S. Levine, T.R. Augustsson, C.L. Imhoff, Nature **296**, 816 (1982).

25   V.M. Canuto, J.S. Levine, T.R. Augustsson, C.L. Imhoff, M.S. Giampapa, Nature **305**, 281 (1983).

26   J.F. Kasting, Precambrian Res. **34**, 205 (1987).

27   J.C.G. Walker, Origin. Life Evol. Biosph. **16**, 117 (1986).

28   R. Rye, P.H. Kuo, H.D. Holland, Nature **378**, 603 (1995).

29   J.H. Joseph, W.J. Wiscombe, J.A. Weinman, J. Atmosph. Sci. **28**, 833 (1976).

30   R.M. Haberle, C.P. McKay, J.B. Pollack, O.E. Gwynne, D.H. Atkinson, J. Appelbaum, G.A. Landis, R.W. Zurek, D.J. Flood, in: J.S. Lewis, M.S. Mathews, M.L. Guerrieri (Eds.) *Resources of Near-Earth space,* University of Arizona Press, Tucson, 1993, pp. 845.

31   J.F. Kasting, K.J. Zahnle, J.P. Pinto, A.T. Young, Origin. Life Evol. Biosph. **19**, 95 (1989).

32   C. Sagan, C. Chyba, Science **276**, 1217 (1997).

33   K.M. Towe, Adv. Space Res. **18** (12), 7 (1996).

34   H. Ohmoto, The Geochemical News **93**, 12 (1997).

35   F. Garcia-Pichel, Origin. Life Evol. Biosph. **28**, 321 (1998).

36   P. Rettberg, G. Horneck, W. Strauch, R. Facius, G. Seckmeyer, Adv. Space Res. **22**, 335 (1998).

37   C.S. Cockell, J. Theoret. Biol. **193**, 717 (1998).

38   G. Horneck P. Rettberg, E. Rabbow, W. Strauch, G. Seckmeyer, R. Facius, G. Reitz, K. Strauch, J.U. Schott, J. Photochem. Photobiol., B: Biol. **32**, 189 (1996).

39   R.C. Smith, K.S. Baker, App. Optics **20**, 177 (1981).

40   B.K. Pierson, H.K. Mitchell, A.L. Ruff-Roberts, Origin. Life Evol. Biosph. **23**, 243 (1993).

41   C.S. Cockell, Origins Life Evol. Biosph. (2000), in press.

42   F. Garcia-Pichel, B.M. Bebout, Mar. Ecol. Prog. Ser. **131**, 257 (1996).

43   J.A. Nienow, C.P. McKay, E.I. Friedmann, Microbial Ecol. **16**, 271 (1988).

44 M.R. Walter, in: J.W. Schopf (Ed.) *Earth's earliest biosphere* Princton University Press, Princeton, 1983, pp 187.

45 F. Garcia-Pichel, B.M. Bebout, Mar. Ecol. Prog. Ser. **131**, 257 (1996).

46 D. Karentz, F.S. McEuan, M.C. Land, W.C. Dunlap, Mar. Biol. **108**, 157 (1991).

47 F. Garcia-Pichel, R.W. Castenholz, Appl. Environ. Microbiol. **59**, 163 (1993).

48 F. Garcia-Pichel, N.D. Sherry, R.W. Castenholz, Photochem. Photobiol. **59**, 17 (1992).

49 S.P. Adhikary, J.K. Sahu, J. Plant Physiol. **153**, 770 (1998).

50 J.G. Dillon, R.W. Castenholz, J. Phycol. **35**, 673 (1999).

51 K. Shimazaki, T. Igarashi, N. Kondo, Physiol. Plant **74**, 34 (1988).

52 E.R. Wood, F. Ghane, D.W. Grogan, J Bacteriol. **179**, 5693 (1997).

53 E.M. Seitz, J.P. Brockmann, S.J. Sandler, A.J. Clark, S.C. Kowalczykowski, Gen Dev **12**, 1248 (1998).

54 M.R. Walter, in: J.W. Schopf (Ed.) *Earth's earliest biosphere*, Princton University Press, Princeton, 1983, pp 187.

55 V.M. Kopylov, E.A. Bonch-Osmolovskaya, V.A. Svetlichnyi, M.L. Miroshnichenko, V.S. Skobkin, Mikrobiologikya **62**, 90 (1993).

56 B.K. Pierson, H.K. Mitchell, A.L. Ruff-Roberts, Origin. Life Evol. Biosph. **23**, 243 (1993).

57 M.B. Rambler, L. Margulis, Science **210**, 638 (1980).

58 C.S. Cockell, L.J. Rothschild, Photochem. Photobiol. **69**, 203 (1999).

59 J.W. Schopf, Proc. Natl. Acad. Sci. **91**, 6735 (1994).

60 L.J.J. Rothschild, Euk. Micro. **46**(5), 548 (1999).

61 R.E. Michod, A. Long, Thoeret. Pop. Biol. **47**, 56 (1995).

62 S.F. Elena, L. Ekunwe, N. Hajela, S.A. Oden, R.E. Lenski, Genetica **102-103**, 349 (1998).

63 D.I. Andersson, D. Hughes, Proc. Natl. Acad. Sci. **93**, 906 (1996).

64 C.S. Cockell, D.C. Catling, W.L. Davis, K. Snook, R.L. Kepner, P.C. Lee, C.P. McKay, Icarus **146**, 343 (2000).

65 W.R. Kuhn, S.K. Atreya, J. Mol. Evol. **14**, 57 (1974).

66 R.M. Haberle, D. Tyler, C.P. McKay, W.L. Davis, Icarus **109**, 102 (1994).

67 M.H. Carr, Nature **326**, 30 (1987).

68 R.B. Leighton, B.C. Murray, Science **153**, 136 (1966).

69 C.P. McKay, W.L. Davis, Icarus **90**, 214 (1991).

70 V.C. Gulick, D. Tyler, C.P. McKay, R.M. Haberle, Icarus **130**, 68 (1997).

71 C.S. Cockell, Planet. and Space Sci. **47**, 1487 (1999).

15 Ultraviolet Radiation in Planetary Atmospheres and Biological Implications

Petra Rettberg and Lynn J. Rothschild

15.1 Solar UV Radiation

The extraterrestrial solar spectrum extends far into short wavelengths of UV-C (190-280 nm) and vacuum UV (<190 nm), wavelengths that no longer reach the surface of the Earth. The intensity of solar radiation reaching the Earth's atmosphere would probably be lethal to most living organisms without the shielding afforded by the atmosphere.

Solar UV undergoes absorption and scattering as it passes through the Earth's atmosphere with the absorption by carbon dioxide, molecular oxygen and ozone being the most important processes. Carbon dioxide has a peak absorbance at 190 nm, and so attenuates radiation below 200 nm. Ozone forms a layer in the stratosphere, thinnest in the tropics (around the equator) and denser towards the poles. The amount of ozone above a point on the Earth's surface is measured in Dobson units (DU) - typically ~260 DU near the tropics and higher elsewhere, though there are large seasonal fluctuations. It is created when ultraviolet radiation strikes the stratosphere, dissociating (or "splitting") oxygen molecules (O_2) to atomic oxygen (O). The atomic oxygen quickly combines with further oxygen molecules to form ozone (O_3). Figure 15.1-A shows the absorption cross section of ozone as a function of wavelength and Fig. 15.1-B a part of the extraterrestrial solar spectrum compared to terrestrial spectra calculated for different ozone concentrations. Increasing ozone concentrations result in lower irradiances in the UV-B range of the spectrum. Surface UV-B radiation levels are highly variable because of sun angle, cloud cover, and also because of local effects including pollutants and surface reflections.

Solar UV radiation affects life on Earth today, and probably even has had a stronger impact on early evolution [1]. The composition of the Earth's atmosphere at that time differed from that of today. Although its exact composition is not known, from model calculations it can be assumed that during the Archaean era, during which the diversification of early anaerobes took place and the first anaerobic photosynthetic bacteria appeared (about 3.5 Ga ago), the amount of free oxygen in the atmosphere was significantly lower than today (see Chap. 14, Cockell)]. There was very little or no absorption of solar UV radiation by ozone. The situation on the early Mars might have been comparable (see Chap. 13, Lammer et al. and Chap. 14, Cockell). Taking

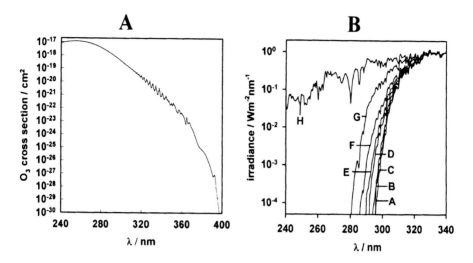

Fig. 15.1 Absorption cross section of O_3 (A) and solar irradiance calculated for different O_3 concentrations (B), A = 440 DU, B = 400 DU, C = 360 DU, D = 310 DU, E = 258 DU, F = 185 DU, G = 66 DU, H = extraterrestrial solar irradiance.

the presumed composition of the early Martian atmosphere and the lower solar luminosity into consideration radiative transfer calculations of the UV flux on the surface of Mars show a gradual increase of the irradiance including short-wavelength UV-B and UV-C over time until today. Thus, present-day solar UV irradiance on the surface of Mars may be similar to that on the surface of the Archaean Earth (see Chap. 14, Cockell).

15.2 Biological Effects of Solar UV Radiation

In biological systems, UV radiation causes photochemical reactions with different biological target molecules, the so-called chromophores. These interactions result in temporary or permanent alterations. The most important UV target in cells is the DNA because of its unique role as genetic material and its high UV sensitivity. The absorbing parts of DNA are the bases, the purine derivatives adenine and guanine, and the pyrimidine derivatives thymine and cytosine. Although the base composition of DNA is different in different genes and organisms, there are the common features of an absorption maximum in the 260 nm region and a rapid decline toward longer wavelengths (Fig. 15.2). Absorption of proteins between 240 and 300 nm is much lower than that of nucleic acids of equal concentration in weight per volume. Most proteins are present in cells in higher numbers of identical copies. Therefore, photochemical alterations in only a fraction of them do not disturb their biological function significantly. The same is true for molecules like unsaturated fatty acids, flavins, steroids, chinones, porphyrins, or carotenoids, which serve as components of the cell membrane, as coenzymes, hormones, or electron donor transport molecules.

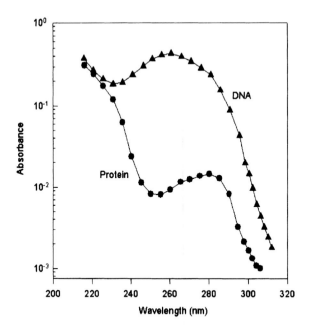

Fig. 15.2 Absorption spectra of DNA (calf thymus DNA) and a protein (bovine serum albumin) at identical concentrations.

The spectrum of UV radiation from the time that it first hits the surface of a biological object changes while it passes the outer parts of the cell or tissue to reach the sensitive targets in the cells, the chromophores. Therefore the action spectrum describing the wavelength dependence of a biological UV effect is often not identical to the absorption spectrum of a chromophore. In Fig.15.3 examples for normalized biological action spectra obtained with monochromatic radiation are given. These action spectra show a remarkable similarity of the slopes of the curves in the UV-B range. However, the curves differ from each other significantly in the UV-A range. This is caused by different photochemical reaction mechanisms. UV-B radiation is directly absorbed by the DNA molecules and causes photodamages. In contrast, UV-A radiation mainly excites so called photosensitizer molecules in the cell, which can either react with the DNA or with oxygen to give reactive oxygen species, which themselves can cause DNA damages.

Due to the wavelength specificity of biological action spectra, especially in the UV-B range, and the highly wavelength-specific absorption characteristics of components of the atmosphere like ozone, the assessment of the influence of environmental (polychromatic) UV radiation on critical biological processes requires a biological weighting of the solar UV irradiance according to the biological responses under consideration. The biological effectiveness of solar UV radiation is determined by the shape of the action spectrum of the biological endpoint and the spectral irradiance [2, 3] using equation 15.1

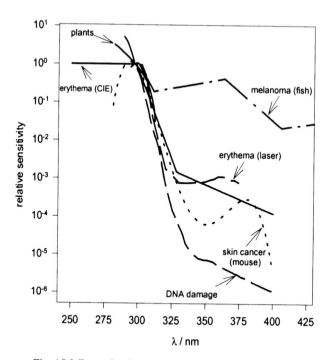

Fig. 15.3 Examples for different action spectra (from [2]).

$$E_{eff} = \int E_{\lambda}(\lambda) \cdot S_{\lambda}(\lambda) d\lambda \qquad (15.1)$$

with $E_{\lambda}(\lambda)$ =   solar spectral irradiance (W/m² nm),
      $S_{\lambda}(\lambda)$ =   action spectrum (relative units), and
      λ =   wavelength (nm).

The resulting biological effectiveness spectrum is shown exemplary for a terrestrial UV spectrum in Fig.15. 4. Integration of the biologically effective irradiance E_{eff} over time (e.g., a full day, one year) gives the biologically effective dose H_{eff} (J/m²)$_{eff}$ (e.g., daily dose, annual dose).

The significance of solar UV radiation as an environmental driving force for the early evolution of life on Earth is reflected by the development of different protection mechanisms against the deleterious biological effects of UV radiation [1]. The most important one is the development of several partly redundant enzymatic pathways for the repair (see Chap. 17, Baumstark-Khan and Facius) of UV-induced DNA damages very early in evolution [4]. Examples are (i) the photoreactivation (PHR), that is the removal of cyclobutane pyrimidine dimers and (6-4)pyrimidine-pyrimidone found in bacteria, Archaea and eukaryotes as a direct repair reaction in a single-step process, (ii) the nucleotide excision repair (NER) for the removal of bulky DNA lesions in bacteria, Archaea and eukaryotes, e.g., the UvrABCD pathway in *E. coli*, (iii) the

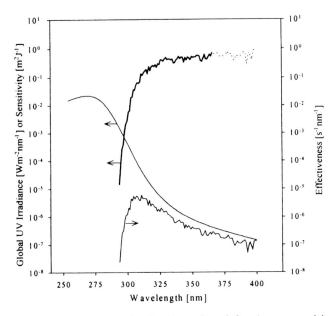

Fig. 15.4 Solar spectral irradiance at the Earth's surface (left axis, measured by A. Bais), a biological weighting function (the absolute DLR-Biofilm action spectrum, left axis) and the resulting solar effectiveness spectrum (right axis).

recently discovered alternative excision repair of UV-induced photoproducts in bacteria and eukaryotes (UV-DE pathway), (iv) the base excision repair (BER) for the removal of damaged or altered bases from the DNA backbone by DNA glycosylases in bacteria, Archaea and eukaryotes (several pathways exist for the removal of different types of oxidative DNA damages indirectly induced by UV-A, e.g., MutM in *E. coli*, Ogg 1 and 2 in yeast) and (v) recombinational repair with several pathways for homologous recombination in bacteria, Archaea and eukaryotes (for review of phylogenomic comparisons, see [4]). In addition to these essential enzymatic reactions for the protection of the genetic material against the deleterious effects of UV radiation, other mechanisms for protection against and avoidance of UV radiation have been developed, e.g., the formation of highly resistant metabolically inactive forms like spores, the synthesis of UV absorbing pigments, the trapping and binding of sediments to form microbial mats, the behavioral adaptation of motile organisms (see Chap. 16, Wynn-Williams and Edwards).

15.3 Biological UV Dosimetry

The assessment of the biological effects of a changing UV climate like that on Earth and Mars requires monitoring methods and systems, which take the strong wavelength dependence of all UV effects into account. One possibility is to use biological UV dosimeters, which consist of biological objects as UV targets and which directly weight the incident UV radiation [3, 5-7]. Up to now several different targets have been suggested and partially tested as UV dosimeters, e.g., biomolecules like provitamin D_3, uracil or DNA, different bacteriophages, whole cells like vegetative bacteria and bacterial spores, biochemical processes in eukaryotic cells like gene induction or DNA repair. Most biological UV dosimeters, such as uracil, DNA, and bacteria are simple systems based on the DNA damaging capacity of UV radiation, which is suggested as the initiating event in a variety of critical photobiological reactions in key ecological processes [2, 7, 8]. Spores of the bacterium *Bacillus subtilis* were found to be well suited for biological UV dosimetry and are applied in the UV dosimeter 'DLR-Biofilm'.

Depending on the dosimetric requirements of the individual measurement the appropriate strain of *B. subtilis* has to be chosen. DNA-repair wildtype strains are more UV-resistant than DNA-repair defective strains, which are not or only partially able to repair the different types of UV-induced DNA damages. Especially strains with mutations in the *uvr* operon show in a high UV sensitivity. They can be used in DLR-Biofilm dosimeters for short-term measurements with a high temporal resolution whereas DNA-repair wildtype strains are used in DLR-Biofilm dosimeters for longer-lasting or higher exposures. For DLR-Biofilm preparation spores of *B. subtilis* were suspended in a 0.5 % agarose solution at 65 °C and poured on the horizontally orientated surface of polyester sheets (FMC Biozym, Oldenburg, Germany) at a concentration of 5×10^5 per cm^2. After solidification at room temperature these biofilms were dried for 12 h at 60 °C without air circulation.

DLR-Biofilms are exposed in different types of exposure boxes depending on the requirements of the individual measurements. During exposure some areas on each biofilm remain unexposed and serve as dark controls for the analysis or for calibration, which is performed with each biofilm before development applying known doses of UV-C from a standard mercury low pressure lamp. For development each biofilm is incubated in nutrient medium (Tryptic Soy Broth, Difco, Augsburg, Germany) at 37 °C for 4.5 h with chaotic mixing. During this time the bacterial spores, which are not or only slightly damaged by the previous UV exposure are able to germinate and multiply inside the biofilm. After fixation and staining of the biomass formed inside the biofilm the quantitative analysis of the optical density of measurement and calibration areas on each biofilm is performed with an image analysis system using a Charge Coupled Device (CCD) camera and a data calculation unit. As result of the comparison with the constructed calibration curve biologically effective UV doses are obtained as equivalent doses of UV-C of 254 nm giving the same biological effect.

DLR-Biofilms are suited for the widespread application as wavelength- and time-integrating biological UV dosimeters [11-13]. They measure continuously even under unfavorable conditions e.g., cloudy weather. Their response is independent of temperature at least between −30 and +70 °C. DLR-Biofilms can be transported and stored in the dark at room temperature before or after exposure for more than one year without loss of sensitivity [14]. The biofilm response is additive without dose rate effects and it is very similar to an ideal cosine if no additional entrance optics is used. The normalized spectral sensitivity of the biofilm, the action spectrum for monochromatic radiation, is shown in Fig. 15.5. Below 300 nm the biofilm sensitivity increases strongly due to the increase in the absorption spectrum of DNA.

15.4 Experimental Determination of the Biological Effectiveness of Extraterrestrial Solar UV Radiation

The full unadulterated spectrum of solar UV radiation is experienced only in space without any absorption and scattering processes form the atmosphere. Therefore the biological effectiveness of pure extraterrestrial solar UV radiation can only be measured in space. In the experiment SURVIVAL II on the exposure facility BIOPAN [15, 16] the biological effectiveness of solar UV radiation was quantified for the first time with the DLR-Biofilm technique on a non-stabilized satellite (Fig. 15.6) in Earth orbit (FOTON 9 mission, June 14-30, 1994). One DLR-Biofilm was exposed to the whole solar spectrum in the space vacuum. The four parts of it were covered only by four different neutral density filters to enlarge the dynamic range of the dosimeter. The exposure was performed by the opening of a shutter system connected to a timer for 10 s. A second DLR-Biofilm mounted on BIOPAN remained unexposed as dark control to exclude the possible influence of other experimental parameters in space than UV. After post-flight calibration and development the biofilms were analyzed together with a ground control biofilm, which was exposed to the terrestrial solar spectrum in Köln on August 6, 1993, under clear sky conditions. In space, a biologically effective dose of 63 J_{eff} m^{-2} was obtained in 10 s (Fig. 15.7), whereas on Earth 1 J_{eff} m^{-2} was

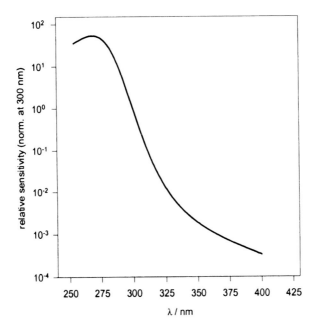

Fig. 15.5 The relative monochromatic action spectrum of the biological UV dosimeter 'DLR-Biofilm'.

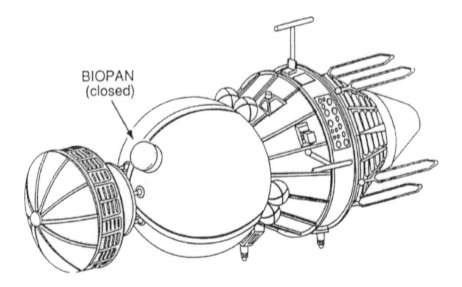

Fig. 15.6 The exposure facility BIOPAN mounted on the satellite.

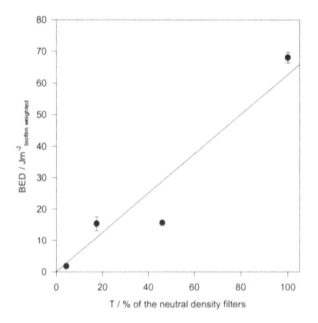

Fig. 15.7 Determination of the biologically effective dose (BED) of a 10 s exposure to extra-terrestrial solar UV radiation on BIOPAN I.

measured in 2.6 min of solar exposure. Hence it can be concluded that the extraterrestrial solar radiation has a biological effectiveness, which is about thousand higher than the actual terrestrial solar UV radiation.

15.5 Experimental Determination of the Photobiological Effects of Different Ozone Concentrations

In another space experiment different atmospheric conditions were simulated to investigate the biological effects of an increase of ozone as it took place in the terrestrial atmosphere about 3.5 Ga ago. Conditions corresponding to an atmosphere without any ozone up to ozone concentrations higher than those experienced today on Earth were chosen in the simulation. During the Spacelab mission D-2 (April 26 to May 5, 1993) in the experiment RD-UVRAD DLR-Biofilms were exposed for defined intervals to extraterrestrial solar UV radiation filtered through an optical filtering system with a shutter mechanism. The filter combinations consisted of neutral density filters to enlarge the dynamic range of the DLR-Biofilm dosimeter and of short-wavelength cut-off filters to simulate different ozone column thicknesses in the stratosphere from 66 to 440 DU. After the predefined exposure period in space the DLR-Biofilms were transported back into the laboratory, calibrated, developed and analyzed together with a ground control biofilm, which was exposed to the terrestrial solar spectrum in Köln on August 6, 1993, under clear sky conditions. The results of the biological UV dosimetry are shown in Fig. 15.8-A with the measured biologically effective irradiance as a function of ozone concentration. At simulated present ozone concentrations (about 360 DU) the measured value is in good agreement with the actual ground measurement (GC = ground control). In Fig. 15.8-B the same data are normalized at 360 DU and compared to the solar UV irradiance in physical units. The results indicate that there is a strong decrease in biological effectiveness by increasing ozone concentrations (in contrast to a small increase in irradiance measured in physical units) and confirm the results of the BIOPAN experiment, that the unfiltered extraterrestrial solar radiation has a biological effectiveness nearly three orders of magnitude higher than the actual values on the Earth's surface at normal ozone concentrations [17, 18]. These results are in accordance with model calculations of the biological effective irradiance for the early Earth and allows the characterization of the conditions, under which life has evolved on Earth and might have evolved on other planets (see Chap. 14, Cockell).

15.6 Conclusions

Solar UV radiation has been a driving force for the evolution of life on Earth. It acts as a mutagen by its DNA-damaging capacity and as a selective agent at the same time. The terrestrial organisms have developed very early several different and

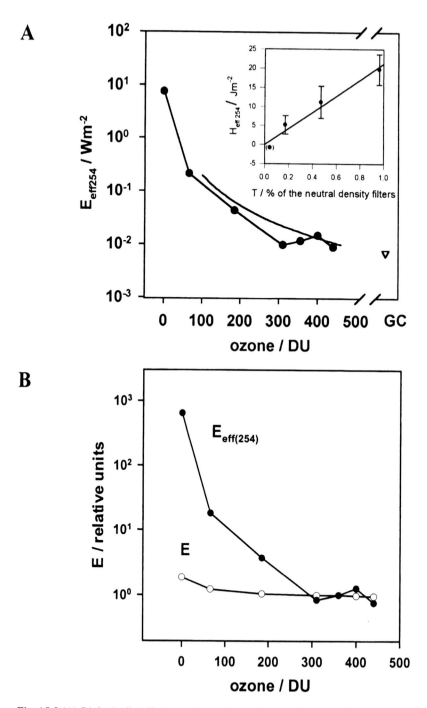

Fig. 15.8 (A) Biologically effective solar irradiance E_{eff} as a function of ozone concentrations (circles). Solid line: Calculated data under the assumption of a radiation amplification factor of 2, GC = ground control result, inset: biologically effective dose H_{eff} measured under different neutral densitiy filters; (B) The data from Fig. 15.8A are normalized at 360 DU (filled circles). Open circles: the solar irradiance in physical units, also normalized at 360 DU (from [17]).

complementary protection mechanisms against the deleterious effects of UV radiation. The most important one is the complex network of DNA repair enzyme activities. To assess the biological effectiveness of UV radiation biologically weighted measurements have been performed, which take the wavelength dependence of UV effects into account. The experimental data for the effectiveness of extraterrestrial solar UV radiation and its changes by an increasing ozone concentration in the atmosphere and the data from model calculations performed for the climatological conditions on the early Earth show a convincing agreement. This allows the characterization of the radiation conditions, under which life has successfully evolved on Earth and might have evolved also on other planets like Mars.

15.7 References

1   L.J. Rothschild, C. S. Cockell, Mut. Res 430, 281 (1999).
2   G. Horneck, Photochem. Photobiol. B: Biol. 31, 43 (1995).
3   G. Horneck, P. Rettberg, R. Facius, in: Fundamentals for the Assessment of Risks from Environmental Radiation, Kluwer Academic Publishers, 1999, pp. 451.
4   J. Eisen, P. Hanawalt, Mut. Res. 435, 171 (1999).
5   G. Horneck, P. Rettberg, R. Facius, in: Role of solar UV-B radiation on ecosystems, Ecosystem research report No. 30, European Communities, 1998, pp. 17.
6   P. Rettberg, in: Biological UV Dosimetry, A Tool for Assessing the Impacts of UV Radiation on Health and Ecosystems, Office for Official Publications of the European Communities, 1999, pp. 192.
7   G. Horneck, in: The Effects of Ozone Depletion on aquatic Ecosystems, Landes Company, 1997, pp. 119.
8   G. Horneck, Trends in Photochem. Photobiol. 4, 67 (1997).
9   G. Horneck, L.E. Quintern, P. Rettberg, U. Eschweiler, E. Görgen, in: Biological Effects of Light, W. de Gruyter, 1994, pp. 501.
10   K. Jovy, P. Rettberg, G. Horneck, Air pollution Research Report 56, 1996, pp. 663.
11   A. Bérces, A. Fekete, S. Gáspár, P. Gróf, P. Rettberg, G. Horneck, Gy. Rontó, J. Photochem. Photobiol. B: Biol. 53, 36 (1999).
12   N. Munakata, S. Kazadzis, A.F. Bais, K. Hieda, G. Rontó, P. Rettberg, G. Horneck, Photochem. Photobiol. 72, 739 (2000).
13   P. Rettberg, R. Sief, G. Horneck, in: Fundamentals for the Assessment of Risks from Environmental Radiation, Kluwer Academic Publishers, 1999, pp. 367.
14   L.E. Quintern, G. Horneck, U. Eschweiler, H. Bücker, Photochem. Photobiol. 55, 389 (1992).
15   F. Burger, ESA SP-374, 313 (1995).
16   P. Rettberg, G. Horneck, Adv. Space Res. 26, 2005 (2000).
17   G. Horneck, P. Rettberg, E. Rabbow, W. Strauch, G. Seckmeyer, R. Facius, G. Reitz, K. Strauch, J.-U. Schott, J. Photochem. Photobiol. B: Biol. 32, 189 (1996).
18   P. Rettberg, G. Horneck, W. Strauch, R. Facius, G. Seckmeyer, Adv. Space Res. 22, 335 (1998).

16 Environmental UV Radiation: Biological Strategies for Protection and Avoidance

David D. Wynn-Williams and Howell G.M. Edwards

16.1 Effects of UVR and Responses in Terrestrial Ecosystems

Any primary colonizers of surface habitats on the Earth and Mars are vulnerable to the effects of solar UV radiation (UVR, 200-400 nm) because of their dependence on photosynthetically active radiation (PAR, 400-700 nm). Inhabitants of terrestrial ecosystems show a range of strategies against potential UVR damage including screening by pigments, quenching of toxic intermediates, avoidance of UVR by movement and niche occupation (with adaptation to low levels of PAR), and damage-repair mechanisms [1]. Incident solar UV-B (280-315 nm) is attenuated by the stratospheric ozone layer, but ozone depletion over Antarctica (the ozone "hole") south of 60 °S during the austral spring can reach 80-90% [2].

During the annual spring ozone depletion in the Antarctic (and to a lesser degree in the Arctic where ozone depletion north of 60 °N is 33% less than in Antarctica [2]), elevated incident UV-B poses an increasing biological threat [3]. It has been calculated that a 65% decrease in column ozone would yield a 14-fold increase in UV-B at 305 nm [4] and that there will be a rise in biologically active UV-B of +15% per decade at 55 °S and +40% per decade at 85 °S, allowing for differential biological effectiveness of the radiation by using the annual DNA-weighted dose [5].

Antarctic terrestrial primary colonizers must obtain energy from PAR whilst minimizing UVR damage to critical molecules. DNA absorbs UV-B and UV-C radiation, especially in the region of 260 nm, which has been shown to be especially deleterious to microbes [6]. As DNA is the common factor in all prokaryotic and eukaryotic cells, UV-B has the potential to disrupt all living systems. This effect is compounded by UVR damage to proteins, and bleaching and photo-oxidation of photosynthetic systems. Different action spectra (metabolic response at different wavelengths) of biochemicals and processes must be considered when assessing damage done by UV-B radiation [7]. Structural changes to DNA interfere with RNA transcription and DNA synthesis leading to mutation or death [8].

To have survived at the surface the early Earth and early Mars before screening atmospheres with an ozone layer developed, photosynthetic microbes would have required tolerance of UV-C, UV-B and UV-A radiation [9, 10]. During recession of the water on Mars into ice-covered lakes and endolithic habitats inside rocks [11], they would also have required tolerance to low temperatures and freezing cycles, desiccation, hypersalinity and potentially to mineral limitation. Extensive research conducted on cyanobacteria in analogous deserts of Antarctica and elsewhere [12] suggests that this microbial group has the necessary attributes for the colonization of the Martian surface in diverse flowing and static water bodies [13] and may have been the pinnacle of microbial evolution. However, it is now considered unlikely that viable cyanobacteria (or other microbes) will be found on the surface of Mars because atmospheric pressure is now well below the triple point of water [14, 15]. Neither are microbes nor their residues likely to be found at shallow (<2 m) depths because of photo-oxidation in the zone of light penetration [16] and putative degradation by peroxides and superoxides such as O^{2-}, O_2^-, O_2^{2-} and H_2O_2 derivatives. These compounds result from interactions between solar radiation, the atmosphere and regolith constituents [17]. Recent imagery of the Martian surface shows channels emerging from strata in crater walls at depths ~100 m below the surrounding plains, stratified crater walls, well beneath a likely oxidized zone [11]. The water or brine probably responsible for the erosion of the channels may concurrently elute and concentrate biomolecules from any buried stromatolitic or endolithic microbial communities.

16.2 Techniques for Studying Microbes and Pigments *in situ*

16.2.1   Epifluorescence Microscopy

To demonstrate biological response to change, such as enhanced UV-B, a standardized representative natural substratum is needed. This allows replicate samples (including controls) to be taken over extended periods of time. There must also be a capacity for repeated non-destructive examination without seriously affecting the habitat. Samples must therefore contain auto-fluorescent organisms such as photosynthetic bacteria or cyanobacteria to avoid disruptive staining procedures. Mineral fines in the center of frost-sorted polygons of extreme Antarctic tundra habitats meet these requirements. Fine quartz and micaschist silt is sorted naturally to provide a relatively flat, homogeneous surface suitable for direct epifluorescence microscopy of auto-fluorescent cyanobacteria and eukaryotic algae which, as phototrophs, initiate the colonization process [18]. Their dependence on PAR restricts them to a shallow focal plane defined by the penetration of the light needed for photosynthesis. This further assists examination of the whole microbial community [19]. Similar cells can be imaged in endolithic communities within their zoned sandstone microhabitat. Quantification by image permits the monitoring of specific morphotypes. Certain strains of cyanobacteria contain not only red-fluorescing chlorophyll and their characteristic orange-red fluorescing phycocyanin but also the yellow-fluorescing accessory pigment phycoerythrin [20].

Fine silt is examined directly under a cover glass by green-light epifluorescence using a Leitz Laborlux microscope system fitted with a Ploemopak N2.1 filter block (excitation at 515-560 nm) and a ×10 objective (total ×125). Images are digitized and analyzed using Seescan Solitaire and Sonata systems (Seescan PLC, Cambridge, U.K.) to quantify auto-fluorescing microalgal colonists [21].

16.2.2   Raman Spectroscopy of Pigments and Other Compounds *in situ*

To characterize the occurrence and spatial distribution of photoprotective and photo-synthetic pigments *in situ*, a non-intrusive laser based technique such as Raman spectroscopy is required. This technique depends on the scattering of monochromatic laser radiation by its interaction with the molecular vibrations and rotations in the constituents of an untreated sample. For analysis of potentially fluorescent samples, a 1064 nm laser beam is applied to the required spot size that can be as small as 5 μm in diameter with suitable microscope optics. The relatively low energy near-IR excitation minimizes interference from auto-fluorescence. Elastic collisions between the photons and the organic and inorganic compounds that comprise the field sample result in radiation scattered mostly at the incident frequency (Rayleigh scattering). However, concurrent inelastic collisions resulting from vibrational transitions of chemical bonds in the compounds produce a small fraction of the scattered radiation with shifted frequencies. Spectral lines which are shifted to energies lower than that of the laser source (Stokes lines) are produced by ground-state molecules, whilst lines at higher frequencies (anti-Stokes lines) are due to molecules in higher excited vibrational states. The frequencies of light of different wavelengths (λ) are conventionally expressed as the equivalent wavenumber (cm^{-1}). This is the number of waves per centimeter path in a vacuum and numerically is $1/\lambda$. To present the wavelength shifts of the Raman spectrum, the Rayleigh scatter line at the excitation wavelength is standardized to a wavenumber = 0. The Raman spectrum of a given compound consists of a unique fingerprint of all its atoms, groups and bonds and their interactive effect (stretching, deformation, rotation) on each other. Diagnostic groups of corroborative peaks from this spectrum can be used to identify the compound amongst others in a mixed sample.

For studies of Antarctic samples and a Mars meteorite, spectra are recorded using a Fourier-Transform Bruker IFS66 (Bruker IR Analytische GmbH, Karlsruhe) instrument and FRA 106 Raman module attachment with 350 mW Nd/YAG laser excitation at 1064 nm and a liquid nitrogen-cooled germanium detector [22]. For spectroscopy of rock profiles and surface crusts, this is coupled via a TV camera to a Raman microscope with a x40 objective giving a spot diameter of ~40 μm at the sample. The power level is set as low as possible to minimize sample degradation whilst optimizing signal quality, and is typically 20 mW. Between 4000 and 10 000 scans (accumulated at ~2 scans per second) are needed to obtain good spectra at 4 cm^{-1} resolution with wavenumbers accurate to ±1 cm^{-1} or better. The total energy input for 4000 scans is therefore ~40 Joules, and exhaustive trials with a variety of endolithic samples showed degradation during scanning to be negligible.

For epilithic lichens encrusting the rock surface, three replicates are scanned directly without preparation. For the endolithic communities within the fabric of the rock, three replicate rock samples are fractured vertically to expose the community.

The profile (10 mm deep from the surface to the bottom of the innermost al-gal/cyanobacterial layer) is divided into up to five zones, where present, and point spectra are collected at three replicate spots per zone. Spectra of the surface crust and the zone of iron accumulation are obtained using the FT-Raman spectrometer in mac-roscopic mode with a spot diameter of 100 μm at the surface of the sample. The Ra-man spectra given in the Figures show the integrated mean of three replicate spectra per zone. The relative proportions and changes of chemical compounds detected are determined routinely from the spectra but absolute values require rigorous standardi-zation.

16.3 Screening Strategies

Terrestrial colonists in habitats exposed to high UVR produce diverse pigments with potential to protect against UV-B damage. Some are extracellular sunscreens; others absorb UV-B inside the cell before metabolically critical molecules can be damaged; some are accessory pigments with quenching properties to dissipate excess energy from UV-B, which would otherwise generate toxic single oxygen. The evidence pre-sented below suggests that there is often a combination of pigments which individually would not afford adequate protection but together minimize UVR damage either on an individual cell basis or as a mutually shading structured terrestrial community.

Cyanobacteria are susceptible to UVR damage by virtue of their need for PAR, so they have developed several protective strategies [23]. They synthesize of a variety of UV-protective pigments [24]. They also produce intracellular mycosporine-like amino acids (MAAs) which are also widespread amongst other organisms [25, 26]. Cyano-bacterial pigments associated directly or indirectly with photosynthesis are differen-tially vulnerable to photobleaching by UVR. Accessory pigments such as the acces-sory pigment phycoerythrin (seen frequently in Antarctic terrestrial cyanobacteria) bleach first [20], followed by carotenoids and finally chlorophyll [7]. UV-B is espe-cially deleterious for photobleaching of photosynthetic pigments in *Anabaena vulgaris* [27].

There is evidence for a UV screening function for phycoerythrin (major absorption peak at 555 nm with minor peaks at ~300 nm and 370 nm) in *Nostoc spongiaeforme* [28]. Its occurrence in related terrestrial Antarctic strains may therefore confer photo-protective potential against UVR as well as visible radiation. *Nostoc commune* is a typical surface colonist of frost-sorted mineral soils in the maritime Antarctic [29] and moist ground in continental Antarctica [30]. It requires extracellular screening pig-ments which transmit PAR but not potentially damaging UV-A or UV-B.

A lipid-soluble extracellular UVR protective pigment, yellow-brown scytonemin, was identified in the sheath matrix of a strain of *Nostoc commune* isolated at the pe-riphery of a fresh water pond on Bratina Island, McMurdo Ice Shelf and in diverse other cyanobacteria from habitats exposed to bright insolation [25, 31]. It absorbs 88% of incident UV-A at 370 nm *in vivo*. In mat communities, layers of maximal chlorophyll-a (chl-a) concentrations were protected from UV-A damage by a layer rich in scytonemin. Experimental removal of scytonemin from the sheath of *Chloro-gloeopsis* induces bleaching of chl-a by UV-A (and to a lesser extent, UV-B) under conditions of physiological inactivity such as desiccation [32]. Screening by scytone-min may therefore be important for the survival of *Nostoc commune* and other Ant-arctic cyanobacteria found in exposed desiccated crusts (see Raman spectrum in Fig. 16.1), especially during the elevated UV-B associated with the ozone hole [33]. The occurrence of scytonemin in species belonging to every major taxonomic group of cyanobacteria suggests that scytonemin production preceded the phylogenetic diversi-fication of cyanobacteria, presumably in the Precambrian [25].

Some cyanobacteria contain additional or alternative intracellular colorless water-soluble compounds known as MAAs [31, 34]. These compounds have a UV absorp-tion maxima in the range 310 to 360 nm. Of 22 strains of cyanobacteria from UVR-exposed habitats from diverse parts of the world, 13 contained MAAs. Of these, all but one also contained scytonemin [25].

Heterotrophic bacteria that can grow on pre-formed organic material are usually protected within the profile of the soil, but the UV-resistance of the red-pigmented terrestrial bacterium *Deinococcus radiopugnans*, which is common in ice-free valleys of Southern Victoria Land [35], may confer a selective advantage.

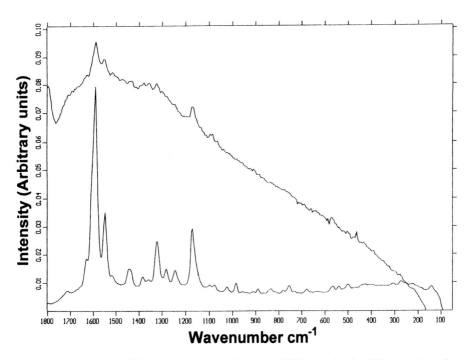

Fig. 16.1 Comparative FT-Raman spectra (excitation at 1064 nm) for the UV screening pig-ment scytonemin and *Collema*, a cyanolichen growing on a desert crust, containing the cyano-bacterium *Nostoc* which synthesizes scytonemin (lower curve). Corroborative Raman vibra-tional bands for scytonemin at 1590, 1549, 1323 and 1172 cm[-1] are detectable in the lichen (after [33]).

16.4 Quenching Strategies

Energy absorbed by chl-a from total solar radiation, especially UV-B, which is not utilized photosynthetically may be transferred to oxygen to produce highly toxic singlet oxygen. Carotenoids protect cyanobacteria by quenching this excess energy [36]. In the Antarctic, a high concentration of a carotenoid pigment with an absorption peak at 384 nm was found in a desiccated 7-year old sample of *Nostoc* commune from coastal lowland adjacent to the Ross Ice Shelf in the McMurdo Dry Valleys Region [37]. Protection from UVR damage may be especially critical in this snow-free habitat which would be fully exposed to maximum UV-B during the ozone minimum of early spring. Survival of UV-stress by *Nostoc* spp., *Gloeocapsa* spp. and other cyanobacterial species which are common in exposed Antarctic habitats may be due partly to their carotenoid content.

Lichens are symbioses (interdependent growth) of fungi and either cyanobacteria or algae. By the nature of their exposed habitats and growth forms, they are vulnerable to incident UV radiation. Carotenoids isolated from six species of Antarctic lichens [38] showed considerable variation in content throughout the growing season, requiring care to standardize conditions for inter-species and inter-site comparisons. Moreover, the pattern of occurrence of the 20 carotenoids identified differed greatly between the six species. Subsequent work showed total carotenoid contents in *Usnea* species very different from an initial study [39]. It is probable that solar radiation influences carotenoid synthesis by lichens [40]. ß-carotene is one of several pigments which can also absorb UV-C (Fig. 16.2) [41]. As this attribute is no longer required in the present atmosphere of the Earth, it may be a relic property from early evolution before full oxygenesis of the atmosphere. However, further work is required before the role of carotenoids for UV protection can be confirmed. The *para*-depside lichen pigment Atranorin has also been attributed quenching properties [42].

UV-related enzymatic DNA-repair mechanisms have been described for the cyanobacteria *Gloeocapsa alpicola* (*Synechocystis*) and *Anabaena*[43, 44]. The sum of these findings suggests that UVR damage in algae and cyanobacteria is not irrevocable.

16.5 Escape and Avoidance Strategies (Shade Adaptation)

16.5.1 Epilithic Lichens

Within the Antarctic Ozone Hole, epilithic lichens can live on the surface of rocks that are rarely covered by snow so that they are fully exposed to UV-B radiation during maximal enhanced UV-B radiation in spring. To survive this intense UV and PAR flux, they shelter their photosynthetically-active components within a structured thallus (lamina of microbial tissue) containing UV-protective pigments and compounds in the outer fungal cortex that protects an inner layer of algae and/or cyanobacteria [42].

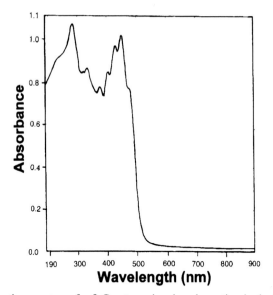

Fig. 16.2 UV absorption spectrum for ß-Carotene showing absorption in the UV-C and UV-A ranges (after [41]).

In the epilithic lichen *Xanthoria elegans* on UV-exposed rocks in Antarctica, the yellow-gray color of the underside contained a lower concentration of the UV screen parietin, an important member of the class of anthraquinone pigments. This was confirmed by very weak parietin bands at 1671 cm^{-1} and 1613 cm^{-1} in its Raman spectrum, relative to strong bands in spectra from the orange top surface. Bands for ß-carotene at 1521 cm^{-1} and 1157 cm^{-1} (Fig. 16.3) [45] were similarly weaker in the underside than in the top surface. Raman spectra therefore confirm *in situ* that the UV protective pigmentation in this lichen is concentrated at the upper cortex whilst the majority of biodegradative acids necessary to release minerals from the rock substratum accumulate in the lower layers of fungal tissue in contact with the rock [46].

The concomitant reduction in PAR is nevertheless more than sufficient for active photosynthesis. This micro-spatial information is difficult to obtain *in situ* by chemical analysis and shows the merit of the precision Raman technique. When the Raman bands of pure pigments such as ß-carotene have been identified, they can be detected amongst many other compounds as shown in *Xanthoria elegans* from a wide range of Antarctic sites (Fig. 16.4).

16.5.2   Endolithic Communities

The McMurdo Dry Valleys region is characterized by cold dry katabatic winds, minimal precipitation, and other conditions which are unfavorable for surface colonization and growth [47]. Although UV-B levels are too low to be solely responsible for the abiotic rock surface, lichens and micro-organisms adopt a stress-avoidance strategy by

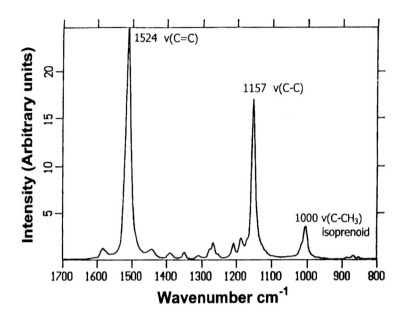

Fig. 16.3 FT-Raman spectrum (excitation at 1064 nm) of pure ß-carotene showing corroborative bands which are detectable in natural communities (after [45]).

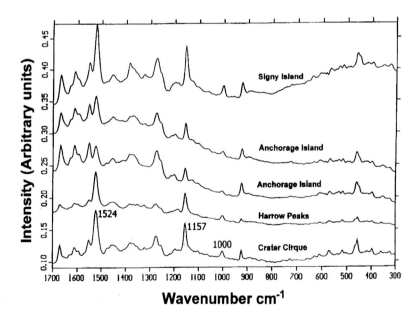

Fig. 16.4 Stack plotted FT-Raman spectra (excitation at 1064 nm) showing bands for ß-carotene in field-fresh samples of the lichen *Xanthoria elegans* from four Antarctic sites. i) from Signy Island at 60°42' S, 45°35' W, ii) and iii) two similar locations on Anchorage Island at 67°36' S, 68°14' W, iv) Harrow Peaks 74°06' S, 164°51' E, v) Crater Cirque 72°38' S 169°22' E (after [45]).

colonizing under and inside rocks [48]. This results in diverse sublithic (under stones), chasmolithic (in rock fissures) and cryptoendolithic communities (hidden within the rock fabric) [49-51]. This strategy may be critical for survival because primary production in these communities is uniquely low, with carbon turnover times as long as ~10 000 years [52]. The structure and function of endolithic microbial communities inside translucent sandstone of UV-stressed hot and cold deserts has been described in detail previously [12, 51, 53]. The habitat is exemplified by Battleship Promontory in Southern Victoria Land (Fig. 16. 5).

Fig. 16.5 Exfoliating Beacon sandstone at Battleship Promontory, Southern Victoria Land, Antarctica, which has been analyzed by Raman spectroscopy. The porous sandstone contains stratified endolithic lichen microbial communities whose algae or cyanobacteria penetrate up to 8mm into the translucent rock (after [12]).

Measurements of the spectral distribution beneath the endolithic lichen layer (3 mm depth) shows a percentage transmission of ~1.0% at 600-700 nm falling steadily to 0.1% at 400 nm [54]. Increased scattering at shorter wavelengths suggests potential protection from UVR for the innermost primary producers. Estimated annual photon fluxes calculated from field data [55] show the steepness of the light gradient penetrating the different zones [52]. Under dry conditions at temperatures above -10 °C, an annual surface receipt of 4000 mol photons m^{-2} decreases to 250 mol at the upper boundary of the pigmented fungal zone, 37 mol at its lower boundary, 2.0 mol at the upper boundary of the algal zone, and 0.16 mol at its lower boundary where cyanobacteria might be expected. More light penetrates the rock when it is wet, but even then, the photon flux at the bottom of the algal zone only reaches 0.89 mol $m^{-2}y^{-1}$. The photon flux of 0.05-1 μmol $m^{-2}s^{-1}$ (dry or even 2-10 μmol $m^{-2}s^{-1}$ when wet) is below the compensation point of most algae. However, cyanobacteria are typically capable of shade-adaptation. *Phormidium frigidum-Lyngbya martensiana* communities can fix carbon significantly (8.20 mg C m^{-2} h^{-1}) at a photon flux of only 1.5 μmol $m^{-2}s^{-1}$ in the bottom mats of ice-covered Lake Hoare in the McMurdo Dry Valleys [56].

In the Raman spectrum for endolithic cyanobacteria shown in Fig. 16.6, scytonemin is conspicuously absent, but bands associated with ß-carotene are evident, despite the cells being 8 mm inside the sandstone [57]. This suggests that there may still be a need to quench the energy of free radicals even though screening of UV-B is unnecessary at this depth. Although hot desert rocks have not been studied as intensively of those of Antarctica, the colonization potential of endolithic cyanobacteria as an avoidance strategy for survival under osmotic, thermal and UV or light stress is known to be similar to that of cold desert communities [58].

16.5.3   Stromatolites

Shade adaptation is required by bottom-dwelling cyanobacteria in ice-covered lakes, whilst enabling them to avoid UV-stress. During seasonal changes in the flow of meltwater, the thick cyanobacterial mats at the bottom of ice-covered Antarctic Dry Valley lakes become buried in silt. However, they re-grow through the mineral layer to form another mat each season, resulting in stratified stromatolites [59]. The translucency of the 3 m thick ice cover on Lake Vanda permits penetration of solar radiation through 70 m of water column to the bottom mats where the receipt of PAR is remarkably up to ~120 μmol m^{-2} s^{-1}. Even the ~15 μmol m^{-2} s^{-1} at the bottom of Lakes Hoare and Fryxell, which have thicker ice covers (about 4.5 m), is more than enough to sustain active photosynthesis by oxygenic mats [56]. This has resulted in a wide diversity of cyanobacterial mat communities in many Dry Valley lakes [59]. Analogous habitats on Mars, such as Gusev and Noachis Craters [11, 14], may provide suitable sources of fossil microbes and biomolecules such as porphyrins and cyanobacterial scytonemin [60]. Modern stromatolites are considered to be analogues of *Conophyton*, which is a columnar stromatolite abundant in Precambrian rock when UV stress would have been much greater [61].

With climatic changes on a geological scale, these lakes may dry up and their cyanobacteria become desiccated fossils [13]. This process may have helped to preserve

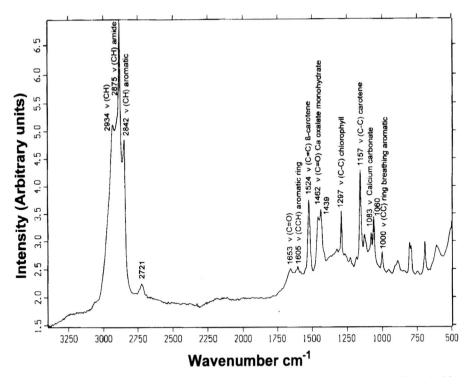

Fig. 16.6 FT-Raman spectrum (excitation at 1064 nm) of the cyanobacterial zone 8 mm inside a stratified endolithic microbial community in Beacon sandstone at Battleship Promontory in Alatna Valley, Convoy Range, Antarctica. Raman bands for ß-Carotene and chlorophyll are evident but scytonemin is not detectable (or necessary?) at this depth (after [57]).

cyanobacteria-like organisms in chert (flint) from the Australian Apex formation in Queensland for 2.5 Ga [62, 63]. Their pigments, especially early porphyrins [64] which are recalcitrant components of chlorophylls, are recognizable in oil-bearing shales [65]. A primitive photosynthetic system capable of delivering an electric charge across a membrane bi-layer could have consisted of a porphyrin (pigment), a quinone (electron donor) and a carotene (electron acceptor) to provide a chain of conjugated bonds [66]. Such molecules are valuable biomarkers on Earth [67] and may fulfil a similar function for life-detection systems on Mars.

16.6 Strategies for Extreme UV on Early Earth and Mars

Biological UV tolerance is discussed in detail in this volume (see Chap. 14, Cockell). Even with a dense atmosphere including ~1 bar CO_2 or more 3.5 Ga, substantial UV at >220 nm would have reached the surface of Mars (see Chap. 13, Lammer et al.). The ancestors of modern cyanobacteria must have been tolerant of the high UV-C and UV-B flux of the early Earth climate before their major oxygenic contribution to the atmosphere 0.5-1.5 Ga [68]. However, they were probably unable to develop large populations except in refuges such as stromatolites or endolithic habitats. UV-C is lethal and may have required additional screening strategies during the Proterozoic [10], whilst UV-B is extremely damaging to their nucleic acids, pigments and proteins [69]. Ozone depletion in Antarctica during spring enhances the receipt of present-day UV-B radiation, but there are always enough residual stratospheric constituents to absorb UV-C. This was not so on Mars [9] and UV-C would have been a hazard for any cyanobacteria developing during the putative warm wet Hesperian period, which may have had some common factors in common with conditions in the Precambrian or Proterozoic period on Earth. Although terrestrial Antarctic microbes are not exposed to UV-C, their rock or mineral habitats could afford protection from this short wavelength (190-280 nm) and would be protective in Mars analogues.

16.7 Panspermia and UV Avoidance

Cyanobacteria have a long geological history associated with past and present changes in UVR-stress [68]. Even when the cells have died under stress [70], their fossil biomolecules can be detected by FT-Raman spectroscopy [22]. Knowledge of their UV-resistance is therefore of great significance for terrestrial and extraterrestrial biology [71]. The endolithic strategy would be suitable for inter-planetary translocation of microbes within the fabric of meteorites ejected from surface communities. Their inevitably freeze-dried condition in the space environment would help to preserve the structure of key biomolecules such as UV protective pigments. Iron-doped quartz also provides a good primary UV screen before further filtering by layers of pigmented cells [72]. The ability of endolithic cyanobacteria to become enzymically reactivated after prolonged desiccation is documented [73]. Despite the damaging UV radiation environment of space, including UV-C, panspermia is now an acceptable concept [6, see Chap. 4, Horneck et al.].

16.8 Diagnosis of Key Pigments on Earth and Mars

The strategies described above show how important the spatial distribution of protective and photosynthetic pigments is for minimizing UV damage along gradients of solar radiation in habitats near the limits of life, such as hot and cold deserts [12, 74]. It is therefore critical to determine not only which pigments are present but also their relative location on the micro-community. Laser Raman spectroscopy, described in 16.2.2, is therefore valuable as a non-invasive proximal analytical technique for describing the mineral matrix and the biomolecules within it [72]. Because of its potential for remote analyses of biomolecules *in situ* and their mineral habitats, a miniature Raman spectrometer is being developed, not only for fieldwork in Antarctica, but for potential Mars lander missions [75].

Raman spectroscopy of pigments *in situ* as key biomarkers is especially valuable for understanding the significance of zonation along the solar radiation gradient when

Table 16.1. UV and visible absorption maxima for selected microbial photo-pigments together with their dominant FT-Raman spectral bands.

Eco-physiological function	Pigment	Pigment type	Absorption maxima (nm)				Raman vibrational bands (wavenumber cm^{-1})
			UVC <280	UVB 280-320	UVA 320-400	Visible light >400	
UV screening	Parietin	Anthraquinone	>257	288	330	431 422	1675 1099 551
	MAA 7437 (*Nostoc*)	Mycosporine Amino Acid		(>310)			2920 1400 820
	Scytonemin	8-ring dimer	252	300	370		1590 1549 1323 1172
Free-energy quenching	ß-Carotene	Carotenoid	>246	283	384	429 451	1524 1155
	Rhizocarpic acid	Isoprenoid	na	na	na	na	1665 1620 1596
	Atranorin	*para*-depside	<274	na	na	na	2942 1666 1303 1294 1266
Photosynthesis	Porphyrin	Tetrapyrrole ring					1453 1360
	Chlorophyll$_a$ (Cyanobacteria)	Tetrapyrrole ring				680 700	1320
	BChl$_a$ (*Rhodopseudomonas*)	Tetrapyrrole ring				850 870	na
	Chlorophyll *Chlorobium*	Tetrapyrrole ring				650 660	na
Accessory photon trapping	Phycocyanin	Phycobilin				560	
	Phycoerythrin	Phycobilin				544 620	1638 1369

* Bold face denotes strong dominant Raman bands

compared with the absorption spectra of the pigments. The summary in Table 16.1 shows not only vibrational Raman bands for selected groups of key pigments but also their absorption characteristics which include extensive potential absorption of UV-C. Although this capability is no longer necessary in current atmospheric conditions in Earth, it may have been an important asset for survival on early Earth (and early Mars?) when extensive UV-C radiation would have reached the surface before the oxic atmosphere has resulted in a protective ozone layer [10]. A combination of different pigments in strategic zones of stratified communities provides a system which enables cyanobacteria and similar photosynthetic microbes to colonize all moist illuminated habitats on the surface of the Earth, and gives them potential attributes for the same achievement on Mars, and possibly even Europa.

Acknowledgements. I am grateful to the British Antarctic Survey (1974-present), the New Zealand Antarctic Research Program (now Antarctica New Zealand) in 1982/83 and the United States Antarctic Research Program (now USAP) in 1995/96 for the opportunity to work Antarctic desert regions and collect research material.

16.9 References

1   D. Karentz, Antarct. Science **3**, 3 (1991).
2   O. Uchino, R.D. Bojkov, D.S. Balis, K. Akagi, M. Hayashi, R Kajihara, Geophys. Res. Lett. **26**, 1377 (1999).
3   R.C. Worrest, Physiol. Plant. **58**, 428 (1983).
4   J.E. Frederick, H.E. Snell, Science **241**, 438 (1988).
5   S.L. Madronich, Ergeb. Limnol. **43**, 17 (1994).
6   G. Horneck, Planet. Space Sci. **43**, 189 (1995).
7   D.-P. Häder, R.C. Worrest, Photochem. Photobiol. **53**, 717 (1991).
8   D. Karentz, J.E. Cleaver, D.L. Mitchell, Nature **350**, 28 (1991).
9   W.R. Kuhn, S.K. Atreya, J. Mol. Evol. **14**, 57 (1979).
10   C.S. Cockell, J. Theor. Biol. **193**, 717 (1998).
11   M.C. Malin, K.S. Edgett, Science **288**, 2330 (2000).
12   D.D. Wynn-Williams, in: B.A. Whitton, M. Potts (Eds.) *The Ecology of Cyanobacteria*, Kluwer, Dordrecht, 2000, pp. 341.
13   P.T. Doran, R.A.J. Wharton, D.J. Des Marais, C.P. McKay, J. Geophys. Res. Planets **103**, 28481 (1998).
14   M.H. Carr, *Water on Mars*, Oxford University Press, New York, 1996.
15   R.M. Haberle, J. Geophys. Res. Planets **103**, 28467 (1998).
16   C.R. Stoker, M.A. Bullock, J. Geophys. Res. Planets **102**, 10881 (1997).
17   M.A. Bullock, C.R. Stoker, C.P. McKay, A.P. Zent, Icarus **107**, 142 (1994).
18   M.C. Davey, P. Rothery, J. Ecol. **81**, 335 (1993).
19   D.D. Wynn-Williams, Binary **4**, 53 (1992).
20   H. Rodriguez, J. Rivas, M.G. Guerrero, M. Losada, Appl. Environ. Microbiol. **55**, 758 (1989).
21   D.D. Wynn-Williams, Polarforsch. **58**, 239 (1988).
22   N.C. Russell, H.G.M. Edwards, D.D. Wynn-Williams, Antarct. Sci. **10**, 63 (1998).
23   R.W. Castenholz, F. Garcia-Pichel, in: B.W. Whitton, M. Potts (Eds.) *The Ecology of Cyanobacteria*, Kluwer, Dordrecht, 2000, pp. 591.

24 C.S. Cockell, J. Knowland, Biol. Rev. Camb. Phil. Soc **74**, 311 (1999).

25 F. Garcia-Pichel, R.W. Castenholz, Appl. Environ. Microbiol. **59**, 163 (1993).

26 D. Karentz, J. Phycol. **32**, 24 (1996).

27 W. Nultsch, G. Agel, Arch. Microbiol. **144**, 268 (1986).

28 R. Tyagi, G. Srinwas, D. Vyas, A. Kumar, H.D. Kumar, Photochem. Photobiol. **55**, 401 (1992).

29 D.D. Wynn-Williams, Microb. Ecol. **31**, 177 (1996).

30 W.F. Vincent, *Microbial Ecosystems of Antarctica,* Studies in Polar Research, Cambridge University Press, Cambridge, 1988, pp. 304.

31 F. Garcia-Pichel, R.W. Castenholz, J. Phycol. **27**, 395 (1991).

32 F. Garcia-Pichel, N.D. Sherry, R.W. Castenholz, Photochem. Photobiol. **56**, 17 (1992).

33 D.D. Wynn Williams, H.G.M. Edwards, F. Garcia Pichel, Eur. J. Phycol. **34**, 381 (1999).

34 S. Scherer, T.W. Chen, P. Böger, Plant Physiol. **88**, 1055 (1988).

35 T.J. Counsell, R.G.E. Murray, Int. J. Syst. Bacteriol. **36**, 202 (1986).

36 H.W. Paerl, Oecologia **61**, 143 (1984).

37 M. Potts, J.J. Olie, J.S. Nickels, J. Parsons, D.C. White, Appl. Environ. Microbiol. **53**, 4 (1987).

38 B. Czeczuga, R. Gutowski, R. Czerpak, Pol. Polar Res. **7**, 295 (1986).

39 B. Czeczuga, M. Olech, Ser. Cient. Inst. Antart. Chileno **39**, 91 (1989).

40 B. Czeczuga, in: M. Galun (Ed.) *Handbook of Lichenology*, Vol. 3, CRC Press, Inc., Boca Raton, 1988, pp. 25.

41 J.M. Holder, *FT-Raman Spectroscopy of Antarctic Epilithic Lichens*, PhD Thesis, University of Bradford, Bradford, 1998.

42 L. Kappen, in: E.I. Friedmann (Ed.) *Antarctic Microbiology*, Wiley-Liss, New York, 1993, pp. 433.

43 P.A. O'Brien, J.A. Houghton, Photochem. Photobiol. **41**, 583 (1982).

44 E. Levine, T. Thiel, J. Bacteriol. **169**, 3988 (1987).

45 D.D. Wynn-Williams, J.M. Holder, H.G.M. Edwards, Bibliotheca Lichenol. **75**, 275 (2000).

46 H.G.M. Edwards, J.M. Holder, D.D. Wynn-Williams, Soil Biol. Biochem. **30**, 1947 (1998).

47 I.B. Campbell, G.G.C. Claridge, *Antarctica: Soils, Weathering Processes and Environment*, Developments in Soil Science, Vol. 16, Elsevier, Amsterdam, 1987.

48 J.A. Nienow, M.A. Meyer, Antarct. J. U. S. **21**, 222 (1986).

49 P.A. Broady, Phycologia. **20**, 259 (1981a).

50 P.A. Broady, Br. Phycol. J. **16**, 257 (1981b).

51 E.I. Friedmann, M.S. Hua, R. Ocampo-Friedmann, Polarforsch. **58**, 251 (1988).

52 J.A. Nienow, E.I. Friedmann, in: E.I. Friedmann (Ed.) *Antarctic Microbiology*, Wiley-Liss, New York, 1993, pp. 343.

53 E.I. Friedmann, Science N.Y. **215**, 1045 (1982).

54 J.A. Nienow, C.P. McKay, E.I. Friedmann, Microb. Ecol. **16**, 271 (1988).

55 E.I. Friedmann, C.P. McKay, J.A. Nienow, Polar Biol. **7**, 273 (1987).

56 B.C. Parker, R.A.J. Wharton, Arch. Hydrobiol. Alg. Stud. **38/39**, 331 (1985).

57 D.D. Wynn-Williams, H.G.M. Edwards, Planet. Space Sci., **48**, 1065 (2000).

58 J.R. Vestal, in: J. Miles, D.W.H. Walton (Eds.) *Primary Succession on Land.* Special Publication No. 12 of The British Ecological Society, Blackwell Scientific Publications, Oxford, 1993, pp. 5.

59   R.A. Wharton, in: J. Bertrand-Sarfati, C. Monty (Eds.) *Phanerozoic Stromatolites II*, Kluwer Academic Publishers, Dordrecht, 1994, pp. 53.

60   P.R. Leavitt, R.D. Vinebrooke, D.B. Donald, J.P. Smol, D.W. Schindler, Nature **388**, 457 (1997).

61   F.G. Love, G.M.J. Simmons, B.C. Parker, R.A.J. Wharton, K.G. Seaburg, Geomicrobiol. J. **3**, 33 (1983).

62   R.E. Summons, L.L. Jahnke, J.M. Hope, G.A. Logan, Nature (London) **400**, 554 (1999).

63   J.W. Schopf, in: B.A. Whitton, M. Potts (Eds.) *The Ecology of Cyanobacteria*, Kluwer, Dordrecht, 2000, pp. 13.

64   A.Y. Mulkidjanian, W. Junge, Photosyn, Res. **51**, 27 (1997).

65   B. Huseby, T. Barth, R. Ocampo, Org.Geochem. **25**, 273 (1996).

66   D.W. Deamer, Microbiol. Molec. Biol. Revs. **61**, 239 (1997).

67   J.S.S. Damste, M.P. Koopmans, Pure Appl. Chem. **69**, 2067 (1997).

68   F. Garcia-Pichel, Orig. Life Evol. Biosph. **28**, 321 (1998).

69   W.F. Vincent, A. Quesada, in: C.S. Weiler, P.A. Penhale (Eds.) *Ultraviolet radiation in Antarctica: measurements and biological effects*, American Geophysical Union. Antarctic Research Series, Vol. 62, 1994, pp. 111.

70   E.I. Friedmann, R. Weed, Science **236**, 703 (1987).

71   C.P. McKay, Orig. Life Evol. Biosph. **27**, 263 (1997).

72   D.D. Wynn-Williams, H.G.M. Edwards, Icarus **144**, 486 (2000).

73   M. Banerjee, B.A. Whitton, D.D. Wynn-Williams, Microb. Ecol. **39**, 80 (2000).

74   W.F. Vincent, in: B.A. Whitton, M. Potts (Eds.) *The Ecology of Cyanobacteria*, Kluwer, Dordrecht, 2000, pp. 321.

75   D.L. Dickensheets, D.D. Wynn-Williams, H.G.M. Edwards, C. Crowder, E.M. Newton, J. Raman Spectrosc. **31**, 633 (2000).

17 Life under Conditions of Ionizing Radiation

Christa Baumstark-Khan and Rainer Facius

Life on Earth, throughout its almost 4 billion years history, has been shaped by inter-actions of the organisms with their environment. It has developed with an ever present radiation background. As a powerful mutagen it has contributed to biological evolu-tion, however, it is potentially destructive for individual cells and organisms [1]. Ra-diation response differs extremely for different organisms and there is not necessarily a correlation between species radiation resistance and levels of exposure in the natural environment [2].

In our solar system Earth is not the only planet, which had the capacity to provide the prerequisites for sustaining life. Mars and the Jupiter moon Europa are suggested to have had the capacity to support life, at least for a certain era. The detection of meteorites on Earth that originate from Mars supports the theory of interplanetary transfer of life (see Chapter 4, Horneck et al.). Life on other planets or while it is transported through space environment encounters radiation conditions different from those on Earth.

In order to understand the likelihood of the development and the chance of persis-tence of life on Earth or other planets, following questions have to be asked:

1. What is the "normal" radiation environment under which life flourishes on Earth?
2. What is the "worst case" radiation environment that life will encounter in space?
3. How does radiation interact with living matter and how does life actively modulate radiation effects?

17.1 Space Radiation Environments

From outer space regions, which are characterized by extreme conditions involving nuclear and atomic reactions (very high temperatures, low density and high-speed subatomic particles) electromagnetic radiation (X-rays and γ-rays) is emitted which consist of soft X-rays, from about 1 to 10 nm of wavelength and of the more pene-trating hard X-rays from approximately 0.01 to 1 nm. The space ionizing radiation environment of our galaxy [3-9] is dominated by energetic, highly penetrating ions and nuclei. These particles constitute the primary radiation hazard for life in space. In the interplanetary space the primary components of the radiation field (Figure 17.1, Table 17.1) are Galactic Cosmic Rays (GCR) and Solar Cosmic Radiation (SCR).

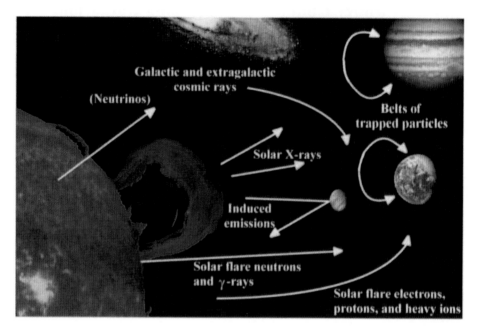

Fig. 17.1 Space radiation sources of our solar system

Table 17.1 Summary of High Energy Space Radiation Sources

Radiation Source	Influence of solar cycle	Additional modifiers	Types of orbits affected [*]
galactic cosmic ray ions	solar min: higher flux; solar max: lower flux	distance from the sun	LEO; GEO; HEO; interplanetary trajectory
solar particle events (protons, heavy ions)	large Numbers during solar max; few during solar min	distance from the sun; location of flare on sun	LEO (I>45°); GEO; HEO; interplanetary
trapped protons	solar min: higher flux; solar max: lower flux	geomagnetic field; solar flares; geomagnetic storms	LEO; HEO; transfer orbits

[*] low earth orbit (LEO); geostationary orbit (GEO); highly elliptical orbits (HEO); inclination (I);

17.1.1   Galactic Cosmic Radiation (GCR)

The spectrum of GCR is fairly well known. GCR originates from outside our solar system and consist of 98 % baryons and 2 % electrons [3]. The baryonic component is composed of 87% protons (hydrogen nuclei), 12 % alpha particles (helium nuclei), and about 1 % of heavier nuclei. (The latter are fully ionized ions of all elements from protons (atomic number Z=1) to uranium (Z=92). These particles are predominantly accelerated in the Milky Way. At the very highest energies, the particles are actually assumed to be extra-galactic; that is, originating from powerful astrophysical accelerators, such as supernova explosions, or quasars, located outside of the Milky Way.

The composition and energy spectra of GCRs change during their passage through the Galaxy, due to collisions between the GCRs and interstellar gas. The intensities and energy spectra also depend upon location in space. When GCRs enter our Solar System, they must overcome the outward-flowing solar wind. The magnetic field transported with the solar wind deflects and thereby attenuates the particle fluxes at lower energies as they penetrate into the heliospere [4]. The intensity of the solar wind varies according to the about 11-year cycle of solar activity and hence the GCR fluxes also vary with the solar cycle, an effect known as solar modulation. GCR flux is at its peak level during minimum solar activity and at its lowest level during maximal solar activity [5]. Differences between solar minimum and solar maximum are a factor of approximately five.

The energy range of GCRs extends over more than fifteen orders of magnitude, from less than 1 MeV (= 10^6 eV) to more than 10^{21} eV [3]. At peak energies of about 200-700 MeV/u, particle fluxes (flow rates) during solar minimum reach 2×10^3 protons per year per μm^2 and 0.6 Fe-ions per year per μm^2. Since dose is proportional to the square of the particle's charge the iron ions contribute nevertheless a significant part to the total radiation dose. Above about 1 GeV/u particle fluxes decrease according to a power law with an exponent of approximately 2.5.

17.1.2   Solar Particle Radiation (SPR)

SPR consists of the low energy solar wind particles that flow constantly from the sun and the so-called highly energetic solar particles events (SPEs) that originate from magnetically disturbed regions of the sun which sporadically emit bursts of energetic charged particles [6]. They are composed predominantly of protons with a minor contribution from helium ions (~10%) and an even smaller part (1%) heavy ions and electrons [7]. SPEs develop rapidly and generally last for no more than some hours, however some proton events may continue over several days. In a worst case scenario the emitted particles can reach energies up to several GeV, resulting in physically absorbed doses as high as 10 Gy to be received within a short time. The average eleven year cycle of solar activity can be divided into four inactive years with a small number of SPEs around solar minimum and seven active years with higher numbers of SPEs around solar maximum [8]. During the solar minimum phase, few significant SPE occur while during each solar maximum phase large events may occur even several times. For example, in cycle 22 (1986-1996) there were at least eight events for proton energies greater than 30 MeV [9].

17.1.3   Radiation Belts in Planetary Magnetic Fields

In the vicinity of the planets with magnetic fields, a third radiation component is present: energetic charged particles trapped by planetary magnetic fields in the so-called radiation or Van Allen belts [10]. The belts are toroidal regions of trapped protons and both inner and outer electron belts. These toroidal regions in the equatorial plane of the planetary magnetic dipole contain large fluxes of high-energy, ionized particles including electrons, protons, and some heavier ions. In each zone, the charged particles spiral around the magnetic field lines and are reflected back between the magnetic poles, acting as mirrors. Electrons in the Earth's Van Allen belts reach energies of up to 7 MeV and protons up to 600 MeV. The energy of trapped heavy ions is less than 50 MeV/u [11].

17.1.4   Modulation by Planetary Magnetic Fields, Atmospheres and Surfaces

Planetary magnetic fields, which on the one hand give rise to sources in trapped radiation belts, provide on the other hand efficient shielding against solar and cosmic particles. Incoming GCR and SPR particles are deflected by the magnetic fields, so that only regions around the magnetic poles are fully accessible to external charged particle radiation, whereas regions around the magnetic equators can only be reached by the most energetic particles [4, 12].

In contrast to shielding by magnetic fields, planetary atmospheres give rise to an initial build-up of radiation levels. This build-up is created by numerous secondary reaction products, which are generated in spallation and fragmentation reactions between the incoming primary particles and the nuclei of atmospheric constituents [13]. Only after significant amounts of matter (approx. 50 g/cm^2), total radiation levels start to decrease, so that after 1000 g/cm^2 the radiation intensity has been reduced to less than about 0.3-1 mGy/year which we normally observe at sea level on Earth.

On celestial bodies without atmospheres, the same mechanisms (spallation and fragmentation) generate significant fluxes of energetic "albedo" neutrons, which are reflected from the irradiated surface and ground [14] and thereby add to the radiation exposure from primary radiation (SCR and GCR).

17.2 Radiation Environments on Planets and Other Celestial Bodies

Natural ionizing radiation on the surface of planets arises from two sources: extraterrestrial (primary cosmic rays and secondary radiation) and terrestrial. Terrestrial radiation [15] is emitted by radioactive nuclides in rocks, soil and the hydrosphere of the planets' crusts which were formed at the birth of the planet and which persist due to their long half-lives plus the emergence of decay products (primordial radionuclides). In the atmosphere most of the radionuclides are secondary products induced by the GCRs (cosmogenic radionuclides). Some of these radionuclides can be transferred to cells and organisms and for this reason exposures may be of internal as well as of external origin.

17.2.1   Earth

On Earth, the primordial radionuclides consist mainly of the natural decay series (Table 17.2) with the exception of the already extinct neptunium series. Other examples of so-called radio-fossils are ^{40}K and ^{87}Rb. Essentially, exposure from external irradiation comes from ^{40}K and the γ-emitting intermediate products of the persisting decay series. There are considerable geographic variations of the natural radiation due to varying concentrations of radionuclides in soil and minerals. For so-called normal areas, background radiation from radionuclides lies in the range of 1 mGy per year [16]. However, there are certain areas with higher background radiation levels around the world. For an area in China dose rates of about 3-4 mGy/year have been reported [17] and on the Southwest coast of India the doses from external radiation by monazite deposits is 5-6 mGy/year with maximal doses up to 32.6 mGy/year [18]. Along certain beaches in Brazil where monazite sand deposits are found external radiation levels range up to 400 mGy/year [19].

Earth is largely protected from the cosmic radiation by the deflecting effect of the geomagnetic field and the huge shield of 1000 g/m² provided by the atmosphere. The annual dose from cosmic radiation varies somewhat with latitude and considerably with altitude, approximately doubling every 1500 m up to a few km above sea level. Typical cosmic radiation doses for Germany [20] are 0.3 mGy/year at sea level (Hamburg), 0.5 mGy/year at 1000 m (Garmisch-Partenkirchen), 1.2 mGy/year at 3000 m (Zugspitze) and up to 25 mGy/year at 15 km altitude.

17.2.2   Mars

Unlike Earth, Mars is devoid of an intrinsic magnetic field (see Chap. 13, Lammer et al.) allowing many of the free-space high-energy charged particles to reach the outer atmosphere. However, the carbon dioxide atmosphere of Mars significantly attenuates the charged particle fluxes and provides a significant amount of protection. The Mars surface environment is exposed to GCR, particles from SPEs and secondary products generated by these particles within the Martian atmosphere and surface materials [21-

Table 17.2 Natural decay series of the Earth crust

Name	Start nucleus	Final nucleus	Half-live τ of the start nucleus
Thorium series	$^{232}_{90}$Th	$^{208}_{82}$Pb	1.4×10^{10} years
Neptunium series	$^{237}_{93}$Np	$^{209}_{83}$Bi	2.2×10^{6} years
Uranium-radium series	$^{238}_{92}$U	$^{206}_{82}$Pb	4.5×10^{9} years
Uranium-actinium series	$^{235}_{92}$U	$^{207}_{82}$Pb	7.1×10^{8} years

23]. A prominent feature of the surface radiation environment is the large number of neutrons produced as secondaries. For neutron energies below 20 MeV, the backward propagating neutrons from the GCR hitting the ground are predicted to dominate those produced in the atmosphere. The backward propagating neutrons from giant SPEs are believed to be nearly equal in number to those produced in the atmosphere.

The calculation of physically absorbed radiation doses on the surface of Mars depends on the assumed density profile of the Martian atmosphere and is further limited by an as yet complete description of the physical interactions of heavy charged particles with nuclei of the Martian atmosphere and Martian surface. Table 17.3 shows calculated physically absorbed dose values for GCRs and SPRs at various Martian altitudes for different atmospheric models, which were derived from measurements during the Viking 1 and Viking 2 entries [23]. From these data the COSPAR warm, high-density atmosphere model (HDM) and the COSPAR cool, low-density atmosphere model (LDM) were developed, which result in different radiation exposures. More information will soon be brought about by the Mars Odyssey Mission 2001, which launched on April 7, 2001. It will reach the red planet in October 2001 where it will study its composition, search for water and shallow buried ice, plus measure solar and cosmic radiation. It carries a spectrometer, the Mars Radiation Environment (MARIE) experiment which will measure the accumulated dose and dose rate as a function of time, determine the radiation quality factor, determine the energy deposition spectrum from 0.1 keV/μm to 1500 keV/μm and separate the contribution of protons, neutrons and HZE particles to these quantities.

Table 17.3 Mars physically absorbed surface radiation doses at different atmospheric shielding levels. Calculations are based on the atmospheric conditions described by the COSPAR LDM and the COSPAR HDM. Data are taken from [23].

Altitude (km)		0	4	8	12
Atmospheric density (g/cm^2 CO$_2$)	LDM	16	11	7	5
	HDM	22	16	11	8
GCRs at solar minimum (mGy/a)	LDM	57	61	65	68
	HDM	53	57	61	64
GCRs at solar maximum (mGy/a)	LDM	27	27	28	28
	HDM	26	27	27	28
SPR August 1989 (mGy/event)	LDM	3	9	26	68
	HDM	1	3	9	23
SPR October 1989 (mGy/event)	LDM	32	82	189	416
	HDM	12	33	79	171

17.2.3   Jupiter's Moon Europa

Interest in Jupiter's moon Europa has intensified with exciting new findings in the last few years from NASA's Galileo mission. Perhaps the most exciting facet of Europa is that an ocean of liquid water may lie beneath its ice surface (see Chap. 7, Greenberg). The Committee on Planetary and Lunar Exploration (COMPLEX) considered Europa an exciting object for further study, with the potential for major new discoveries in planetary geology and geophysics, and the potential for studies of extraterrestrial life.

Jupiter is a rapidly rotating planet with a very strong prototypical rotationally driven magnetic field [24]. Data from the two Pioneer and the two Voyager spacecrafts indicated that its pea-shaped magnetosphere extends up to $3\text{-}7\times10^6$ km towards the Sun whereas its tail is stretched up to a distance of 7.5×10^8 km in the opposite direction. Its strong magnetic field makes Jupiter the strongest radio emitter in the Solar System after the Sun. The non-thermal radiation as a signature of Jupiter's radiation belts is a type of synchrotron radiation, and it results from high-energetic free electrons which travel nearly the speed of light in spirals shaped by the magnetic field lines.

The large Galilean satellites are embedded within Jupiter's magnetosphere and Io is known to be a source of ions and neutral particles [24]. Ions, predominantly of sulfur and oxygen, are distributed along the orbit of Io to form a large torus. Electrons and ions from Io and Jupiter's ionosphere are present throughout the Jovian magnetosphere. A substantial fraction of these particles are accelerated to extremely high energies to form intense radiation belts. The planet's inner radiation belt is analogous to Earth's Van Allen belts, but about $0.5\text{-}1\times10^4$ times more intense. The harshest radiation conditions - unmatched elsewhere in the Solar System - prevail within about 3×10^5 km off the giant planet.

Europa lies deep within the magnetosphere of Jupiter. It is located in the inner magnetosphere of Jupiter (at a radial distance of ~9.5 $R_J \cong 6.7\times10^5$ km), a region populated mainly by plasma derived from the Io plasma torus [25]. The plasma there consists of protons, oxygen and sulfur ions, and their corresponding electrons. Europa is continually bombarded by magnetically trapped, ionizing radiation. This magnetospheric particle flux is the dominant component of the radiation environment at Europa. Galactic cosmic radiation and solar particle radiation cannot access Europa because of Jupiter's magnetic field except at energies exceeding ~90 GeV, where fluxes are negligible.

The intensity of the radiation environment in the Jupiter system has been measured by several spacecrafts. Data on the radiation environment of Europa have been compiled from information gathered by the Pioneer 10 and 11, Voyager 1 and 2, and Galileo missions [24, 26]. Available measurements include electron intensities in the energy range 30 keV to >10 MeV and ion intensities from 30 keV to >100 MeV. Data on ion composition - separation of protons from helium and from ions with atomic number Z>6 - are available above about 500 keV/u. These data are input to standard models for physically absorbed radiation doses that calculate the rate of energy deposition versus depth below the target surface [27]. These results are summarized in Figure 17.2, which shows the radiation dose rate deposited at various depths in the Europan ice. Contributions from Jovian electrons and electron bremsstrahlung are

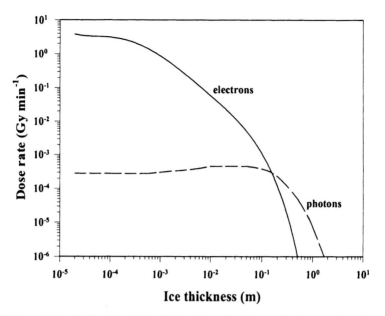

Fig. 17.2 Rates of physically absorbed radiation doses for Europa, in Gy per min of exposure below varying thicknesses of ice [modified, from 27]. The separate contributions of electrons and photons (bremsstrahlung) are shown. In addition to the theoretical uncertainties in Europa's radiation environment, natural variations of up to an order of magnitude have been observed in Jupiter's trapped-particle intensities over the 25-year span between the Pioneer and Galileo missions.

shown. The contribution from ions and electrons with energies below 30 keV may be ignored due to their restricted penetration range (10 µm).

17.3 Measures of Ionizing Radiation

17.3.1   Physical Measures

Exposure to ionizing radiation historically has been measured by use of ionizing chambers for X or γ-radiation. The measure of exposure has been the number of ion pairs generated in dry standard air. In the international system (SI), the corresponding unit is Coulomb per kilogram.

A more general measure of exposure to ionizing radiation is the amount of energy deposited per unit mass by ionizing radiation in matter. The corresponding SI-unit for the so-defined (physically) absorbed dose is the Gray (1 Gy = 1 Joule per kilogram). In radiation biology, the energy absorbed per unit mass of living tissue has to be determined.

17.3.2 Biologically Weighted Measures

The physical unit Gy, however, does not determine the biological effects of the different types of radiation, such as X-rays and γ-rays, α-particles, β-particles, neutrons and heavy ions. Instead, for a given dose in Gy, the biological effects differ with the type of radiation. A dose of energetic charged particles normally causes more damage than the same dose of energetic photons (X-rays or γ-rays). Particles with high atomic numbers and high energy (HZE particles) cause the greatest damage for a given dose.

The relative biological effectiveness (RBE) is used to express the relative amount of biological damage caused by a given dose deposited by a particular type of ionizing radiation for a specific biological endpoint. For a given test radiation, it is calculated as the dose of a reference radiation, usually X-rays, required to produce the same biological effect as with a test dose, D_T, of another radiation [28]. Thus, the value of the RBE provides a quantitative index of the effectiveness per unit of absorbed dose of any radiation. It can be experimentally determined for certain end-points and certain tissues or organisms. The RBE value for a given type of radiation usually is different for different biological systems (e.g. different organisms) or for different stages of the same system (e.g. stage in the growth cycle or other environmental factors, e.g. oxygen content). For higher organisms, such as mammalian cells, a lot of data exist, which describe RBE values as high as 20.

The primary physical feature which determines the RBE of different types of ionizing radiation is the spatial density of ionizations produced per unit absorbed dose in the irradiated tissue. Figure 17.3 gives a schematic example of this concept together with the visualization of ionization densities produced by different types of radiation in nuclear emulsions. γ-irradiation produces ionizations homogeneously within a cell, whereas densely ionizing radiation tends to produce clusters of ionization (Fig. 17.3) that are believed to cause more serious damage to the cells.

As a substitute for the spatial ionization density the Linear Energy Transfer (LET) has been introduced for its quantification. This LET [28] is defined as the average amount of energy lost per unit of particle track length, and commonly is expressed in keV μm^{-1}.

In order to quantify the RBE of different types of ionizing radiation for purposes of human radiation protection the so-called quality or radiation-weighting factor Q has been defined. The product of Q with the physically absorbed dose yields the so-called "equivalent dose", the SI unit of which is the Sievert (Sv).

17.4 Interaction of Radiation with Biological Material

Radiation, both, of terrestrial and cosmic origin, is a persistent stress factor, life has to cope with. Radiation interacts with matter primarily by ionization and excitation of the electrons in atoms and molecules. These matter-energy-interactions have ever been present during the creation and maintenance of living systems on Earth. Exposure with ionizing radiation leads to the development of reactive species in different solvents. Water (see Chap. 5, Brack) is the main constituent of all living systems. In somatic

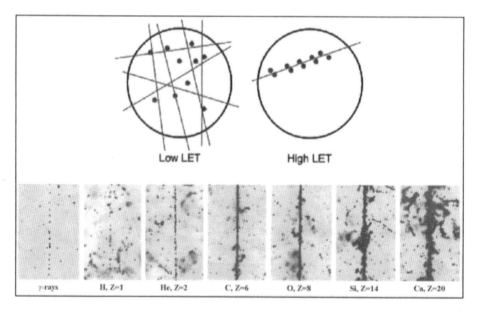

Fig. 17.3 Distribution of ionizations in equal volumes by low and high LET radiation and tracks in photo-emulsions of electrons produced by γ-rays and tracks of different nuclei of the primary cosmic radiation moving at relativistic velocities (modified from [29]). For biological radiation effects the efficiency of a radiation type increases as the ion density along the tracks increases.

and vegetative cells its fraction lies between 40 and 70 %, even in bacterial spores it amounts still around 20 %.

17.4.1   Radiation Chemistry of Water

Most of the energy of ionizing radiation is absorbed by water molecules [30-32], which are either excited or ionized (Eq. 17.1). Excitation of a water molecule is often followed by splitting of the molecule (Eq. 17.2).

$$H_2O \xrightarrow{h \cdot v} H_2O^+ + e_{aq}^- \qquad \text{(Ionization)} \qquad (17.1)$$

$$H_2O \xrightarrow{h \cdot v} H^{\bullet} + {}^{\bullet}OH \qquad \text{(Splitting)} \qquad (17.2)$$

The primary products are $H^{\bullet}$, ${}^{\bullet}OH$, H_2O^+ and electrons. All these species possess unpaired electrons, thus being highly reactive free radicals [33]. The electron is particularly reactive, it captures another water molecule thus forming a negatively charged ion (Eq. 17.3).

$$H_2O + e_{aq}^- \longrightarrow H_2O^- \qquad (17.3)$$

The ions H_2O^+ and H_2O^- are not stable, they do not persist in this form as they dissociate almost immediately (10^{-16} seconds) into H^+ ions and $^{\bullet}OH$ radicals as well as into ^-OH ions and $H^{\bullet}$ radicals (Eq. 17.4).

$$H_2O^+ \longrightarrow H^+ + {^{\bullet}OH} \tag{17.4}$$

$$H_2O^- \longrightarrow H^{\bullet} + {^-OH}$$

Commonly, there will be a number of reactions among the free radicals themselves, thereby either reconstituting water (Eq. 17.5) or forming molecular hydrogen and hydrogen peroxide (Eq. 17.6). The interactions of free radicals both among themselves and with their own reaction products is dependent primarily on how closely they have been formed. After they are formed, they must diffuse through the medium until they encounter something with which they may interact. The probability of these reactions are favored within spurs, blobs and tracks. Interactions with other solute molecules are only possible, if the primary species are able to escape these zones.

$$e_{aq}^- + H^+ \longrightarrow H_2O \tag{17.5}$$

$$\text{(Recombination)}$$

$$H^{\bullet} + {^{\bullet}OH} \longrightarrow H_2O$$

$$e_{aq}^- + H_2O^- \longrightarrow H_2 + 2\ {^-OH} \tag{17.6}$$

$${^{\bullet}OH} + {^{\bullet}OH} \longrightarrow H_2O_2$$

Densely ionizing radiations (α particles, protons, electrons) produce clusters of ions that are very close together [34]. These ions may subsequently dissociate into closely associated free radicals, which do not need to diffuse far before encountering something to interact with. Consequently, there will be a high probability of interactions between free radicals and with the products of previous radical-radical interactions.

If ionizations are brought about by sparsely ionizing radiation, the clusters of ions formed are more widely separated than those produced by densely ionizing radiation. The probability of interactions between the resultant free radicals is much smaller than that following irradiation with densely ionizing particles. Energy transfer from ionizing radiation and accordingly the number of free-radical interactions increases within a given volume. As the density of ionization increases (with increasing LET), there is also an increase in the number of changed molecules and thus an increase of radiation effects in cells.

So far only reaction in the absence of oxygen have been considered. Since this molecule plays an important role in living systems its significance in radiation biology has to be elucidated. The oxygen molecule is a reaction partner for the primary radiolytic products, the electron and the hydrogen radical (Eq. 17.7). It does not react with the $^{\bullet}OH$.

$$e_{aq}^- + O_2 \longrightarrow O_2^- \tag{17.7}$$

$$H^{\bullet} + O_2 \longrightarrow HO_2^{\bullet} \longleftrightarrow O_2^- + H^+$$

Both, $HO_2^{\bullet}$ and O_2^- are unstable and enter easily further reactions. In solutions containing oxygen, there is an increased formation of hydrogen peroxide while oxygen is partly regenerated (Eq. 17.8). There are, however, other reactions which have to be taken into account. The most important is the decomposition of hydrogen peroxide thus increasing the yield of the hydroxyl radical (Eq. 17.9).

$$O_2^- + O_2^- \longrightarrow H_2O_2 + O_2 \qquad (17.8)$$
$$HO_2^{\bullet} + HO_2^{\bullet} \longrightarrow H_2O_2 + O_2$$

$$e_{aq}^- + H_2O_2 \longrightarrow {}^{\bullet}OH + OH^- \qquad (17.9)$$

It has been shown, that highly reactive species are formed by water radiolysis. They are not only able to react with each other but also with solved molecules such as organic molecules. Thus, biological key substances such as proteins, RNA and DNA cannot only be inactivated by direct energy absorption (direct radiation effect) but also via interactions with radicals (indirect radiation effect) [35-37].

For the direct radiation effect, the mean number of inactivated molecules is directly proportional to the dose, whilst for the indirect effect the number of inactivated molecules depends on the dose and on their concentration. For more dilute solutions, there is a greater yield of damaged molecules per dose (more radicals per solute molecule) than for more concentrate solutions (less radicals per solute molecule). Only for extremely diluted solutions, this effect is reduced, due to greater diffusion distances for radicals and their higher recombination probability. The protective influence of low temperatures on biological systems (up to a factor of 100) can be attributed to reduced diffusion capabilities of radials in frozen water and thus to a perturbed reactivity with the solute molecules. As the formation of free radicals from ionized water is the major mechanism through which the indirect action of radiation proceeds, the degree of hydration of a living system is expected to influence its radiosensitivity. The greater the water content of cells is, the greater is the number of free radicals and thus the number of interactions. In dry systems, most of radiation action is thus attributable to the direct effect.

17.4.2   DNA Damage and Cellular Repair Pathways

The genetic material of all living organisms, both eukaryotes and prokaryotes as well as of most viruses, consists of deoxyribonucleic acid (DNA).

DNA is the most important molecule for living beings from two aspects. It contains all the information for structure and function (via the matrix function of DNA for the transcription step during protein biosynthesis) of each cell in a whole organism and it has the capacity to replicate accurately so that progeny cells have the same genetic information as the parental cell. For the continuous maintenance and circulation of life, the chemical stability of DNA is one of the prerequisites. In addition, the genetic material must be capable of variation, one of the bases for evolutionary change.

DNA consists of two associated polynucleotide strands that wind together in a helical fashion, called double helix. Each polynucleotide strand is a linear polymer in

which the monomers (deoxynucleotides), are linked together by means of phosphodiester bonds. These bonds link the 3' carbon in the ribose of one deoxynucleotide to the
5' carbon in the ribose of the adjacent deoxynucleotide. The four nitrogenous bases of
DNA are arranged along the sugar-phosphate backbone in a particular order (the DNA
sequence), encoding all genetic instructions for an organism. Adenine (A) pairs with
thymine (T), while cytosine (C) pairs with guanine (G). The two DNA strands are held
together by hydrogen bonds between the bases.

As a reactive chemical species DNA is the target of numerous physical and chemical agents. As a result of exposure to ionizing radiation a broad spectrum of DNA
lesions is induced in cellular DNA (Fig. 17.4, Table 17.4), depending in type and
quantity on the quality and dose of radiation. Radiochemical injuries of relevance for
biological functions induce nucleotide base damages, cross-linking, and DNA single-
and double-strand breaks. These disorders have the capability to interfere with DNA
replication as they may block the function of DNA polymerases. Nevertheless DNA is
functionally more stable than the two other cellular macromolecules, RNA and protein. This stability can be attributed to the fact that
  – the primary structure of DNA is all that is needed for transfer of information;
  – DNA carries the information in duplicate due to its double-helical structure;
  – there are molecular mechanisms of different complexity to undo the DNA
    damage thus maintaining genetic integrity [38, 39].
In response to the harmful effects of environmental radiation, life has developed a
variety of defense mechanisms, such as the increase in the production of stress proteins, and a variety of efficient repair systems for radiation-induced damages The
immense importance of repair enzymes in sustaining genetic stability of exposed cells
is obvious by a high degree of homology of particular enzymes occupied in different
repair pathways from bacteria and yeast to man.

Only some enzymes simply reverse the damaged bases to their unmodified form.
DNA repair by photolyase in the presence of light (photoreactivation) is the best
studied repair reaction of this type to remove UV-induced DNA lesions from the genome. Genes coding for photolyase activities (PHR genes) were found in many prokaryotic and eukaryotic organisms, but not in all species [40]. Another example of
direct damage reversal is repair of O^6-methyl guanine by transfer of the alkyl group
from the DNA to a cysteine in a protein, an O^6-methylguanine-DNA methyltransfer-
ase, which appears to be present in all living organisms [41]. A final example of direct
damage reversal is the sealing of a subset of nicks in DNA by DNA ligase, which can
only seal nicks having 5'-phosphates and 3'-hydroxyls.

The base excision repair system (BER) hydrolyses the 3' phosphodiester bond between the damaged base and deoxyribose moiety by a glycosylase, e.g an uracil glycosylase [42, 43]. The resulting single nucleotide gap is recognised by an AP endonucle-
ase and is filled in by DNA Pol I and then ligated. The nucleotide excision repair
pathway (NER) is an elaborate repair system which removes bulky lesions from DNA
[44-46]. In E. coli, 3 proteins, the products of the uvrA, uvrB, and uvrC genes, are
responsible for damage recognition and nicking of the DNA approximately 7 nucleotides 5' to the damage. The fourth protein, the uvrD gene product, known as DNA
helicase II, displaces the damage-containing oligonucleotide, the single stranded gap
is filled in by DNA Polymerase I and a ligase seals the newly synthesized DNA to the

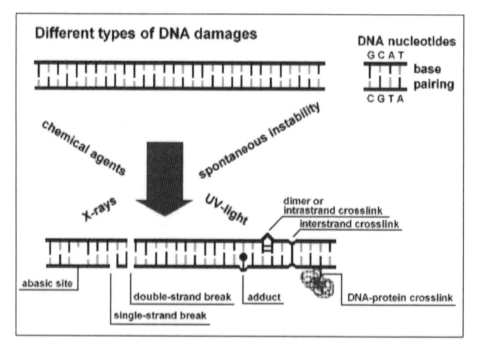

Fig. 17.4 Different types of DNA damage

Table 17.4 DNA lesions produced by different internal and external conditions and the pathways used by bacteria for their repair

DNA lesion	caused by	repair mechanism
base mismatches, loops and bubble structure	replication errors recombination	base excision and general mismatch repair
deamination/depurination	heat	base excision
oxidative damage	oxidative metabolism anoxia/hypoxia ionizing radiation	base excision
alkylation/alkyl adducts (nonbulky/bulky adducts)	nitrogen mustard polyaromatic compounds	direct repair base excision nucleotide excision
intrastrand crosslinks (Bulky adducts)	UV, cisplatin	direct repair nucleotide excision
interstrand crosslinks	psoralen, malphalan	nucleotide excision recombination
single-strand break	ionizing radiation oxidative stress	ligation
double-strand break	ionizing radiation	ligation recombination

old one. Both, BER and NER rely on the redundant information in the undamaged strand to correctly fill in the excision gap.

Of course complex DNA damages such as backbone or bulky base damage, require more complicated repair mechanisms such as multi-enzyme repair pathways.

If the genetic information in both DNA strands is damaged the recombinational repair pathway is helpful [47-49]. Non-excised lesions which persist until the onset of DNA replication lead to the formation of a gap in the newly synthesized DNA strand opposite the lesions. They can be filled by a recombinational exchange from the complementary old strand, where the respective information is available. The gaps which are now in the old strand are filled in by DNA polymerases using the newly synthesized strand as a template, followed by ligation of the newly synthesized DNA.

The inducible SOS response in *E. coli* and other bacteria is a highly mutagenic repair system which can be regarded as a driving force for evolution [50-52]. In bacteria any block to DNA replication caused by DNA damage lead to an excess of single-stranded DNA. This signal activates in *E. coli* the RecA protein, which then destroys a negatively acting regulatory protein (a repressor) that normally suppresses the transcription of the entire set of SOS response genes. Studies of mutant bacteria deficient in different parts of the SOS response indicate that the newly synthesized proteins increase cell survival as well as mutation rate by greatly rising the number of errors made in coding DNA sequences. This is most probably advantageous for long term survival because it produces a burst of genetic variability in the bacterial population and hence increases the chance that a mutant cell with increased fitness will arise.

In addition to repair systems, several other feedback controls operate to restrain the cell-cycle control in eukaryotic systems until particular conditions are satisfied. A well-characterized feedback control operates at the mitotic entry checkpoint to prevent cells with damaged DNA from entering mitosis until the damage is repaired [53, 54]. The response to DNA damage is usually studied using experimentally induced lesions in DNA, such as those created by X-rays. Many radiation-sensitive (rad) mutations have been isolated in budding yeast, and at least one, called rad 9, has been shown to code for an essential component of the feedback control mechanism. Mutants lacking rad 9 still possess the machinery for DNA repair, but they fail to delay in G2-phase of the mitotic cell cycle when they have been irradiated. As a result, they proceed into mitosis with damaged chromosomes; not surprisingly, they are killed by doses of radiation that normal cells would survive.

17.5 Cellular Radiation Responses

Two cellular functions are most important for the development and maintenance of life on Earth or other planets. These functions are cellular reproducibility and conservation of genetic stability. Both tasks can be disturbed by ionizing radiation with the result of cellular inactivation and mutation induction.

17.5.1   Radiation Sensitivity of Organisms

Cell survival, in radiobiological terms, is understood as the ability for indefinite re-production. Its impairment by radiation is fundamental for the assessment of radiation hazards. The colony forming ability (CFA) assay has long been used for unicellular microorganisms like bacteria, yeast and algae. CFA is of such central importance, that its loss is often equalized with cell death [55], which is strictly not correct since many other cell functions may still be undisturbed.

The surviving fraction of irradiated cells relative to that of non-irradiated cells is plotted on the ordinate log-arithmetically versus dose on a linear abscissa scale. With increasing dose the number of survivors decreases. The corresponding dose-effect curve declines continuously and can be simply characterized by two parameters. One is the D_0 which is defined as the dose necessary to reduce survival to e^{-1} (=0.37), it can be calculated from the slope of the terminal straight part of the curve (−1/slope). The other parameter is the extrapolation number (n) which is calculated from the backward extrapolation of the straight proportion of the effect curve. Radiation sensitivity of different organisms can be compared on the basis of these parameters (Fig. 17.5, modified from [56], Table 17.5).

Radiation sensitivity of various organisms is generally related to the amount of ge-netic material [57]. The most resistant organisms are exclusively single-stranded vi-ruses, followed by double-stranded viruses, bacteria, algae and yeast. For simple eukaryotes, it could be shown that haploid cells are about twice as sensitive as diploid cells [58]. Bacterial sensitivity to radiation is correlated with DNA base composition [59].

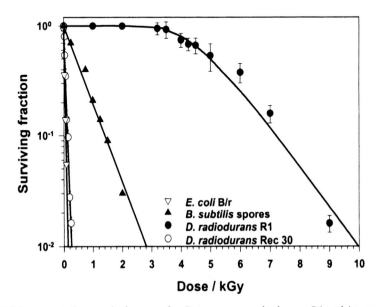

Fig. 17.5 Representative survival curves for *Deinococcus radiodurans R1* and its recombina-tion deficient mutant Rec 30, compared to survival curves for spores of *Bacillus subtilis* and *Escherichia coli B/r* following exposure to X-rays. (modified from [56]).

Table 17.5 D_0-values and cell DNA content for prokaryotic and eukaryotic organisms

Name	Species	DNA content (g/cell)	D_0 (Gy)
T1-phage	Virus	8.1×10^{-17}	2600
Escherichia. coli B/r	Bacteria	1.9×10^{-14}	30
Bacillus subtilis cells	Bacilla	7.7×10^{-15}	33
Saccharomyces cerevisiae	Yeast	2.8×10^{-14}	150
Chlamydomonas	Algae	1.4×10^{-13}	24
Human	Human	5.4×10^{-13}	1.4

The variety of cellular processes involved in the repair of damaged DNA have to be considered to be primarily responsible for radiation resistance. For bacterial cells as well as for yeasts and higher eukaryotes D_0-values from repair proficient and repair deficient cells obtained from the same species differ more than a factor of 100 from each other [60].

The most radiation resistant bacterium known is *Deinococcus radiodurans*. It was originally isolated from samples of canned meat that were thought to be sterilized by high doses of γ-radiation [61]. Typically, it is found in locations where most other bacteria have died from extreme conditions, ranging from the shielding pond of a radioactive cesium source to the surfaces of Arctic rocks. Researchers examined the bacterium's cellular repair genes and discovered that, while *D. radiodurans* contained the usual complement of repair genes found in other radiation-sensitive bacteria, it has an unusually large redundancy of repair functions [62]. Furthermore, the regulation of its repair activities is highly coordinated with the cell cycle. In Figure 17.5 the radiation sensitivity of the highly radiation resistant bacterium *D. radiodurans* R1 is compared with that of a mutant *D. radiodurans* Rec 30, that is deficient in DNA repair and two other bacterial representatives: spores of *B. subtilis* and cells of *E. coli*. The D_{37} dose for the *E. coli* culture is 30 Gy, which is approximately 200 times lower than that of *D. radiodurans*. *D. radiodurans* has a characteristic shoulder of resistance to approximately 3×10^3 Gy; at doses below that value there is no loss of viability although up to about 100 double strand breaks have been induced in the chromosome of each cell [56]. Above 5×10^3 Gy, there is an exponential decline in viability and a D_{37} dose of between $6-7 \times 10^3$ Gy for cultures in exponential phase. It is remarkable, that in *D. radiodurans* Rec 30 that has lost its repair capacities the radiation sensitivity reaches similar values as in *E. coli* [56].

17.5.2 Mutation Induction

Mutations, alterations of the genetic material, result in a permanent change in the expression of genes. This is known as a naturally occurring phenomenon, which occurs spontaneously with a rare but gene-dependent fixed frequency, thereby being the driver of biological evolution.

Certain agents are able to increase the rate of mutation, ionizing radiation being most effective [63, 64]. Radiation acts by ionizing the bases in the DNA chain, this is in particular especially important during DNA synthesis. Ionization of one ore more of the bases or interactions of bases with free radicals may alter base structures in that way that they no longer possess the characteristics they had before. Under such circumstances, a mismatch base pairing might occur which in turn might result in an altered sequence of bases in the newly synthesized chain. Deletions occur, if sufficient numbers of ionizations or interactions with free radicals take place. Mutational frequency increases linearly with radiation dose. It increases with increasing LET reaching a peak and diminishing thereafter. Except for very rare cases, which give rise to evolutionary advancements, most mutations are detrimental to cells as they bring about sharp deviations from the status quo. Such mutational changes are negative mutations, resulting in a cellular loss of function impairing viability by a shortened life-span (years, days, minutes) of the organism [65]. Some genes are so important, that their changes bring death essentially immediately.

17.5.3   Factors Influencing Cellular Radiation Effects

It could be deduced from the aforementioned facts, that the biological effects of ionizing radiation are not rigid and precise. There are variations in the degree of response of organisms, even among populations of individuals of the same species. Some of the factors influencing the response to radiation are the subject matter of this part.

17.5.3.1   *Dose Rate and Dose Fractionation*

For most biological effects the effectiveness of a given dose will be related to the rate at which the dose is delivered. Generally speaking, the effectiveness of a given dose of sparsely ionizing radiation is reduced as the time to deliver that dose is increased [66]. The effectiveness of any given radiation is not only reduced, when the dose-rate is reduced. There will be an alteration in the magnitude of the effect when the same radiation is delivered at interrupted intervals, in fractions with some time between each fraction [67]. The difference in the magnitude of effects observed as a function of dose rate or of fractionating the dose are usually explained by recovery phenomena which are based upon cellular repair of DNA damages [39, 59].

17.5.3.2   *Oxygen Pressure*

The tension of oxygen in cells at the time of irradiation is an important determinant of the degree of severity of radiation damage, as in presence of oxygen, the amount of oxidizing radicals increases, due to the fact that the oxygen molecule is a reaction partner for the primary radiolytic products (see 17.4, Eq. 17.7). Oxygen always enhances the action of radiation [68]. For comparing radiation effects on organisms with different metabolic conditions (anoxia, hypoxia, hyperoxia), the oxygen enhancement ratio describes the surplus of effect in presence of oxygen ($+O_2$) over the effect in absence of oxygen ($-O_2$).

17.5.3.3   Chemicals as Radioprotectors or Radiosensitizers

A major influence on cellular radiation sensitivity in terms of protective effects comes from cellular constituents which serve as radical scavengers [69]. Examples for this kind of reaction are alcohols, e.g. ethyl alcohol which react with radicals according to Eq. 17.10. The resultant alcohol radical is much less reactive than the ^+OH so that the net effect is reduced [70].

$$HO^\bullet + CH_3CH_2OH \longrightarrow CH_3C^\bullet HOH + H_2O \qquad (17.10)$$

Many other substances react as radioprotectors by radical scavenging in living organisms, e.g. cysteine, cysteamine and dimetyl sulfoxid [71].

Other substances have the capacity to worsen the cellular radiation effect. Among them are oxygen mimetic substances like N-acetyl-maleimide and stable free radical-type sensitizes like triacetonamine-N-oxyl [72]. Furthermore chemicals which interact with cellular repair systems, such as novobiocin and 9-ß-D-arabinofuranosyladenine (ara A) sensitize exposed cells towards the detrimental effects of ionizing radiation [73].

17.5.3.4   Metabolic Rate and Temperature

The development of radiation injury is dependent on metabolic rate. Accordingly, varying the metabolic rate will bring about variations in the rate of development of radiation effects. If metabolism is speeded up the lethal effects of radiation are enhanced, while retarding the metabolic rate results in less radiation effects [74-76]. Metabolic active and metabolic resting cells of the same species differ in their radiosensitivity by a factor of about 10. Irradiation at reduced temperatures can result in opposite effects. Due to a metabolic slow-down, less cellular radiation effects are to be expected for temperatures down to 4 °C. However, due to reduced activities of repair enzymes, the radiation response can be worsened. If temperatures drop well below 0 °C during exposure to ionizing radiation the reduced diffusion ability of radials leads to a protective effect. In frozen water the indirect action of radiation is reduced and the damaging power towards DNA is attributed mainly to the direct effect [77].

17.5.3.5   Proliferation Rate

Cell populations that are rapidly proliferating are more sensitive to radiation than populations having lower generation turnovers. Radiation sensitivity depends on the cell cycle stage in which the cells are exposed. The mitotic process of eukaryotic cells is a relatively radiosensitive phase in a cell's life [78, 79]. This is well documented for sparsely ionizing radiation. Radiation also slows down the progression of cells through the cycle [80]. Sensitivity variations as a function of cell cycle stage are not only found for colony formation but also for chromosome aberrations or mutation induction. The highest yield of mutations is found for cells exposed during G1-phase with the S-phase being quite resistant.

17.6 The Chances of Life
Surviving Space Radiation Conditions

17.6.1   Interplanetary Transfer of Life

Te question of the chances of interplanetary transfer of life has been elucidated by Horneck et al. in Chapter 4 of this volume. From the radiation biology point of view, the chances for life to survive space radiation conditions are not too bad.

As shown by space experiments and ground experiments with accelerated heavy ions [81-85], B. subtilis spores can survive even a central hit of a heavy ion of cosmic radiation, which are speculated to be the ultimate limit on the survival of spores in space because they penetrate even thick shielding. Due to the comparative low fluxes of heavy ions in our galaxy a spore can escape from a hit by a HZE particle (e.g. iron of LET >100 keV/μm) for hundred thousands up to 1 million of years. This time span meets the estimates which are calculated for medium-sized rocks to travel from one planet of our solar system to another, e.g. from Mars to Earth [86]. For transfer of microscopically small particles only a few months have been estimated to be sufficient for an interplanetary passage [87].

As already mentioned, the component of the space radiation environment that presents the largest uncertainty in predictions are the SPEs accompanying solar flares, which develops rapidly but generally last for no more than some hours. Concerning shielding against radiation in space, less than 0.5 g/cm^2 is required against the diffuse X-rays, but the high energy particles in SPEs and the GCR particles are difficult to shield due their high penetration depth. Experiments with B. subtilis suggest, however, an estimated density of 1.8 g/cm² of meteorite material of 2 to 3 m thickness to shield the spores sufficiently. Such calculations show that yet after 25 million years in space, a substantial fraction of a bacterial spore population (10^{-6}) would survive the exposure to cosmic radiation if shielded under comparable conditions.

17.6.2   Putative Habitats on Other Planets

The unusual extent of radiation resistance, which is displayed by the bacterium D. radiodurans has already been mentioned. Because there are no known radioactive environments on Earth that can explain the evolution of D. radiodurans's resistance to radiation, there is general agreement that this organism's resistance to radiation is a secondary characteristic developed in response to some other environmental stress. The consensus view is that the mechanisms that evolved to permit survival in very dry environments additionally confer resistance to radiation [88]. It is possible that other desiccation-resistant micro-organisms, not yet described as radiation-resistant, could exist in planetary biospheres. Despite the discrediting of the often repeated claim that live bacteria were recovered from Surveyor 3's camera after surviving on the Moon's surface from 1967-1969, experiments conducted aboard a variety of spacecraft including the European Retrievable Carrier (EURECA) and the Long Duration Exposure Facility (LDEF) indicate that a variety of common terrestrial bacteria are able to withstand the free space environment for periods as long as 6 years [89, 90]. Since the radiation-resistance characteristics of many common organisms (and most extremophiles) are unknown, it is conceivable that many bacteria classified as desiccation- and/or radiation-resistant have the capacity to survive also in harsh radiation environments.

Although the radiation dose rate on the surface of Mars is about 100 times higher than on Earth (see 17.2, Table 17.3), it cannot be considered as a limiting factor for microbial life on Mars, as discussed by Gilichinsky in Chapter 8 of this volume. As already described, *D. radiodurans* tolerates radiation doses up to 3×10^3 Gy without any significant inactivation. The annual dose on Mars is $3-4 \times 10^4$ times lower, a fact which would not impair growth of such microorganisms if they possess potent DNA repair systems. Even vegetative cells of *B. subtilis* or *E. coli* cells with a D_0 of about 30 Gy would be able to survive radiation exposure under Martian conditions for extended periods of time.

The nowadays harsh conditions for life on the Marian surface (low atmospheric pressure, no oxygen, low temperature) would require life to persevere in stages of reduced metabolic activity or even in resting stages (reduced or absent proliferation rate, reduced water content, cells in frozen state). This in turn are conditions which modify radiation effects considerably (reduced diffusion capacity of radiolysis products, reduced indirect radiation effects, less DNA damage per dose) resulting in increased cellular radiation resistance. From the radiobiology point of view, simple life forms in dormant states are ideal candidates to survive even harsh radiation conditions for extended time periods. For the Martian radiation environment of about yearly 90 mGy (calculating 57 mGy/a of GCR and one SPR event of October 1989 size of 32 mGy per year) it would last 6600 years, to accumulate a dose of 600 Gy (resulting in a surviving fraction of 0.37) in spores of *Bacillus subtilis*. Even after about 1×10^6 years (accumulated dose of 8 kGy), 1 out of 10^6 irradiated spores can be expected to survive Martian surface radiation conditions. For these calculations, the modification of cellular radiation response based on the factors described in 17.5.3 has not been considered.

Other than for the planet Mars, the Jupiter moon Europa displays an extremely harsh radiation field near its surface (see 17.2.3, Fig. 17.2), which is a key factor governing the viability of organisms. For a dose rate of 3-4 Gy/min, *E. coli* cells and vegetative cells of *B. subtilis* would be inactivated within some 10 minutes. Under these conditions it can be calculated, that only 1 out of 10^6 cells would survive an acute irradiation of 2 hrs. Even spores of *B. subtilis* will be expected to yield a deadly dose of ionizing radiation (survival fraction of 10^{-6}) within 40 hrs.

On Europa, life-sustaining, near-surface environments may exist within or under regions of water ice (see Chap. 7, Greenberg), since ice will provide microbes with some degree of radiation protection. A psychrophilic or psychrotolerant microbial species (see Chap. 9, Wagner et al., Chap. 10, Stackebrandt and Brambilla) with the radioresistance of *D. radiodurans* would grow continuously without any effect on its growth rate in 1 mm depth of Europa's surface ice (Fig. 17.2), where the dose rate is reduced to about 1 Gy/min. At 10 cm depth the radiation environment in the Europan ice has been reduced by a factor of 5×10^3 to ~50 Gy per month (1×10^{-3} Gy/min), a radiation dose, which can be tolerated by most bacteria. For microorganisms living below a shallow depth at Europa – at most a few tens of meters below the surface – radiation itself is no longer a significant environmental factor. At greater depths, the radiation environment continues to decrease, reaching values similar to those in Earth's biosphere below depths of 20 to 40 m.

17.7 References

1   D.E. Lea (Ed.) *Action of radiation on living cells*, Cambridge University Press, London & New York, 1955, 416 pp.

2   A. Nasim, A.P. James, in: DJ Kushner (Ed.) *Microbial Life in Extreme Environments*, Academic Press, London New York San Francisco, 1978, pp. 409.

3   J.A. Simpson, Ann. Rev. Nucl. Part. Sci. **33**, 323 (1983).

4   W. Heinrich, J. Beer, Adv. Space Res. **4**(10), 133 (1984).

5   G.D. Badhwar, F.A. Cucinotta, P.M. O'Neill, Radiat. Res. **138**, 201 (1994).

6   S.W. Kahler, N.R. Sheeley Jr., R.A. Howard, M.L. Koomen, D.J. Michels, R.E. McGuire, T.T. von Rosenvinge, D.F. Reames, J. Geophys. Res. **89**, 9683 (1984).

7   G. Gloeckler, Rev. of Geophys. **17**, 569 (1979).

8   M.A. Shea, D.F. Smart, J. Geomag. Geoelect. **42**, 1107 (1990).

9   J. Feynman, T.P. Armstrong, L. Dao-Gibner, S. Silverman, J. Spacecraft **27**, 403 (1990).

10   J.A. van Allen, L.A. Frank, Nature, **183**, 430 (1959).

11   G.D. Badhwar, D.E. Robbins, Adv. Space Res. **17**(2), 151 (1996).

12   K. O'Brien, W. Friedberg, H.H. Sauer, D.F. Smart, Environ. Int. 2 Suppl. **1**, S9 (1996).

13   W. Schimmerling, J.W. Wilson, F.E. Nealy, S.A. Thibeault, F.A. Cucinotta, J.L. Shinn, M. Kim, R. Kiefer, Adv Space Res. **17**(2), 31 (1996).

14   J. Miller, Adv Space Res. **14**(10), 831 (1994).

15   United Nations Scientific Committee on the Effects of Atomic Radiation, *Sources and Effects of Ionizing Radiation*, Volume I: Sources, UNSCEAR 2000 Report to the General Assembly, with Scientific Annexes, United Nations Sales Publication, Sales No. E.00.IX.3, 2000.

16   J.A. Sorenson, Semin. Nucl. Med. **6**, 158 (1986).

17   H. Morishima, T. Koga, K. Tatsumi, S. Nakai, T. Sugahara, Y. Yuan, L. Wie, J. Radiat. Res. (Tokyo) **41**, 9 (2000).

18   A.R. Gopal-Ayengar, G.G. Nayar, K.P. George, K.B. Mistry, Indian J. Exp. Biol. **8**, 313 (1970).

19   A. Malanca, V. Pessina, G. Dallara, Health Phys. **65**, 298 (1993).

20   Bayerisches Staatsministerium für Landesentwicklung und Umweltfragen. Strahlenschutz - Radioaktivität und Gesundheit. München, 1991.

21   E.J. Conway, L.W. Townsend, in: E.B. Pritchard (Ed.) *Mars: past, present, and future*, American Institute of Aeronautics and Astronautics, Washington, DC, 1992, pp. 239 (Progress in astronautics and aeronautics , vol. 145).

22 L.C. Simonsen, J.E. Nealy, L.W. Townsend, J.W. Wilson, J. Spacecr. Rockets. **27**, 353 (1990).

23 L.C. Simonsen, J.E. Nealy, NASA-TP 3300 (1993)

24 H.M. Fischer, E. Pehlke, G. Wibberenz, L.J. Lanzerotti, J.D. Mihalov, Science **272**, 856 (1996).

25 M.A. McGrath, Science **278**, 268 (1997).

26 Committee on Planetary and Lunar Exploration, Space Studies Board, Commission on Physical Sciences, Mathematics, and Applications. A Science Strategy for the Exploration of Europa. National Academy Press, Washington, DC, USA, 1999.

27 Task Group on the Forward Contamination of Europa, Space Studies Board, Commission on Physical Sciences, Mathematics, and Applications, National Research Council. Preventing the Forward Contamination of Europa. National Academy Press, Washington, DC, USA, 2000.

28 D.T. Goodhead, J. Radiat. Res. **40** Suppl. 1 (1999).

29 C.F. Powell, P.H. Fowler, D.H. Perkins (Eds.) *The study of elementary Particles by the Photographic Method*, Pergamon Press, London New York Paris Los Angeles, 1959, pp. 669.

30 S.M. Pimblott, J.A. LaVerne, Radiat. Res. **129**, 265 (1992).

31 C.D. Jonah, Radiat Res **144**, 141 (1995).

32 J.A. LaVerne, Radiat. Res. **153**, 196 (2000).

33 A. Mozumder, Radiat. Res. Suppl. **8**, S33 (1985).

34 K.F. Baverstock, W.G. Burns, Radiat Res **86**, 20 (1981).

35 F.C. Steward, R.D. Holsten, M. Sugii, Nature **213**, 178 (1967).

36 H.B. Michaels, J.W. Hunt, Radiat Res **74**, 23 (1978).

37 G.D. Jones, T.V. Boswell, J. Lee, J.R. Milligan, J.F. Ward, M. Weinfeld, Int. J. Radiat. Biol. **66**, 441 (1994).

38 W. Vermeulen, J. Hoeijmakers, D. Bootsma, HELIX, Amgens Magazine of Biotechnology **5**, 22 (1996).

39 C. Baumstark-Khan, in: C. Baumstark-Khan, S. Kozubek, G. Horneck (Eds.) *Fundamentals for the Assessment of Risks from Environmental Radiation*. Kluwer Academic Publisher, Dordrecht Boston London, 1999, pp. 103.

40 D.J. Davies, S.A. Tyler, R.B. Webb, Photochem. Photobiol. **6**, 371 (1970).

41 T.H. Bestor, G.L. Verdine, Current Opinion in Cell Biology **6**, 380 (1994).

42 O.D. Scharer, J. Jiricny, Bioessays **23**, 270 (2001).

43 S.S. Parikh, C.D. Mol, J.A. Tainer, Structure **5**, 1543 (1997).

44 A.S. Balajee, V.A. Bohr, Gene **250**, 15 (2000).

45 A. Sancar, M.S. Tang, Photochem. Photobiol. **57**, 905 (1993).

46 S. Hoare, Y. Zou, V. Purohit, R. Krishnasamy, B. VanHouten, N.E. Geacintov, A.K. Basu, Biochemistry **39**, 12252 (2000).

47 T. Lindahl, R.D. Wood, Curr. Opin. Cell Biol. **1**, 475 (1989).

48 A. Kuzminov, Microbil. Mol. Biol. Rev. **63**, 751 (1999).

49 P. Baumann, S.C. West, Trends Biochem. Sci. **23**, 247 (1998).

50 M. Radman, Basic Life Sci. **5A**, 355 (1975).

51 F.W. Perrino, D.C. Rein, A.M. Bobst, R.R. Meyer, Mol. Gen. Genet. **209**, 612 (1987).

52 A. Bouyoub, G. Barbier, J. Querellou, P. Forterre, Gene **167**, 147 (1995).

53 T.A. Weinert, L.H. Hartwell, Science **241**, 317 (1988).

54 D. Lydall, T. Weinert, Science **270**, 1488 (1995).

55 G.W. Barendsen, BJR Suppl. **24**, 53 (1992).

56   H. Zimmermann, *Wirkung locker- und dichtionisierender Strahlung auf Zellen von Deinococcus radiodurans*. DLR-FB 94-14, DLR, Köln, Germany, 1994.

57   H.S. Kaplan, L.E. Moses, Science **145**, 21 (1964).

58   Y. Takamori, E.R. Lochmann, W. Laskowski, Z. Naturforsch. **B 21**, 960 (1966).

59   R.H. Haynes, Radiat. Res. Suppl. **6**, 1 (1966).

60   J. Kiefer, E Wagner, Radiat. Res. **63**, 336 (1975).

61   J.R. Battista, A.M. Earl, M.J. Park, Trends in Microbiol. **7**, 362 (1999).

62   O. White, J.A. Eisen, J.F. Heidelberg, E.K. Hickey, J.D. Peterson, R.J. Dodson, D.H. Haft, M.L. Gwinn, W.C. Nelson, D.L. Richardson, K.S. Moffat, H. Qin, L. Jiang, W. Pamphile, M. Crosby, M. Shen, J.J. Vamathevan, P. Lam, L. McDonald, T. Utterback, C. Zalewski, K.S. Makarova, L. Aravind, M.J. Daly, K.W. Minton, R.D. Fleischmann, K.A. Ketchum, K.E. Nelson, S. Salzberg, H.O. Smith, J.C. Venter, C.M. Fraser, Science **286**, 1571 (1999).

63   R.F. Kimball, Mutat. Res. **55**, 85 (1978).

64   B.A. Bridges, Int. J. Radiat. Biol. Relat. Stud. Phys. Chem. Med. **37**, 93 (1980).

65   M.B. Baird, A.M. Clark, Exp Gerontol **6**, 1 (1971).

66   E.J. Hall, Int. J. Radiat. Biol. **59**, 595 (1991).

67   J.H. Hendry, R.I. Mackay, S.A. Roberts, N.J. Slevin, Int. J. Radiat. Biol. **73**, 383 (1998).

68   M. Quintiliani, Int. J. Radiat. Oncol. Biol. Phys. **5**, 1069 (1979).

69   H. Tomita, M. Kai, T. Kusama, A. Ito, Radiat. Environ. Biophys. **36**, 105 (1997).

70   S.M. Afzal, P.C. Kesavan, Int. J. Radiat. Biol. Relat. Stud. Phys. Chem. Med. **35**, 287 (1979).

71   D.R. Singh, J.M. Mahajan, D. Krishnan, Mutat. Res. **37**, 193 (1976).

72   E.L. Powers, Int. J. Radiat. Biol. Relat. Stud. Phys. Chem. Med. **42**, 629 (1982).

73   A. Collins, Int. J. Radiat. Biol. Relat. Stud. Phys. Chem. Med. **51**, 971 (1987).

74   B.S. Rao, M.S. Murthy, N.M. Reddy, P. Subrahmanyam, U. Madhvanath, Mutat. Res. **28**, 183 (1975).

75   S. Matsumoto, Can. J. Microbiol. **17**, 179 (1971).

76   B.A. Bridges, M.J. Ashwood-Smith, R.J. Munson, Biochem. Biophys. Res. Commun. **35**, 193 (1969).

77   J. Hüttermann, M. Lange, J. Ohlmann, Radiat. Res. **131**, 18 (1992).

78   R. Fingerhut, J. Kiefer, F. Otto, Mol. Gen. Genet. **193**, 192 (1984).

79   N.L. Oleinick, Radiat. Res. **51**, 638 (1972).

80   A. Maity, W.G. McKenna, R.J. Muschel, Radiother. Oncol. **31**, 1 (1994).

81   H. Bücker, G. Horneck, in: O.F. Nygaard, H.I. Adler, W.K. Sinclair (Eds.) *Radiation Research*, Academic Press, New York, 1975, pp. 1138.

82   K. Baltschukat, G. Horneck, Radiat. Environ. Biophys. **30**, 87 (1991).

83   U. Weisbrod, H. Bücker, G. Horneck, G. Kraft, Radiat. Res. **129**, 250 (1992).

84   M. Schäfer, Int. J. Radiat. Biol. **69**, 459 (1996).

85   H. Zimmermann, M. Schäfer, C. Schmitz, H. Bücker., Adv. Space Res. **14**(10), 213 (1994).

86   H.J. Melosh, Nature **332**, 687 (1988).

87   M.A. Moreno, Nature **336**, 209 (1988).

88   V. Mattimore, J.R. Battista, J. Bacteriol. **177**,5232, (1996).

89   R.L. Mancinelli, M.R. White, L.J. Rothschild, Adv. Space Res. **22**(3), 327, (1998).

90   G. Horneck, Adv. Space Res. **22**(3), 317, (1998).

Part IV

Gravity and Life

18 Graviperception and Graviresponse at the Cellular Level

Richard Bräucker, Augusto Cogoli and Ruth Hemmersbach

The evolution of life on Earth occurred under the persistent influence of gravity. Even protists (unicellular organisms such as flagellates and ciliates) had to find and to stay in environments whose chemical and physical parameters fit their needs. Consequently, already unicellular organisms developed organelles for active (oriented) movement (cilia, flagella) and sensors for diverse stimuli. Among environmental parameters, gravitational acceleration is a most reliable reference for orientation, because it is virtually constant in its magnitude and direction. Consequently, graviorientation can be already found on very early, unicellular, stages of development [1-3]. As protists are heavier than water, they had to develop mechanisms to compensate sedimentation. Without graviorientation, a population of *Paramecium,* for instance, would sink to the ground of a 1m depth pond within 3½ hours.

The mode of gravibehavior in protozoa seems to depend on the preferred living conditions. *Paramecium* which feeds on aerobe bacteria, shows a negative gravitaxis, i.e. the cell population swims mainly upwards, against the gravity vector, thus reaching oxygen saturated layers. In contrast, *Loxodes* which prefers low oxygen concentrations, shows a positive or a bimodal gravitaxis [4]: alternately moving upwards or downwards increases the chance of finding optimal conditions of light and O_2 concentrations which change with the depth of the water column in an aquatic ecosystem (Fig. 18.1).

In addition to gravitaxis (the oriented movement with respect to gravity), some species show a gravity related active regulation of swimming rate, called gravikinesis. This results in an increased swimming rate during upward swimming, and a decreased swimming rate during downward swimming.

18.1 Protists

18.1.1 Perception of Varied Acceleration

Changing the magnitude of a stimulus is a common method to study its influence on a living system. Experiments under microgravity conditions as in space flights, sounding rockets and drop towers are necessary to confirm that gravitational acceleration is the reason of the phenomena examined [5-7]. Furthermore, it was demonstrated that a step transition from 1 ×g to microgravity in a drop tower led to an unexpected slow transient loss of graviresponses, which gives some hint on the nature of the gravireceptor in ciliates.

Using a centrifuge in microgravity environment gives the chance of investigating the behavior of cells under hypogravity conditions and during application of a distinct acceleration force for threshold studies. In search for the threshold of graviperception, graviresponses were documented in ciliates and flagellates at 0.1 to 0.3 ×g. Long-term cultivation for up to 14 days in µg did not show dramatic changes in morphology of the protists although the proliferation rate increased [8]. Behavioral changes due to the lack in gravity (Fig. 18.1B) were restored after return to 1 ×g conditions [9].

Hypergravity conditions which can be achieved on earth with comparatively low expense, showed no saturation of gravireactions in ciliates up to 5 ×g conditions [10-11].

From these experiments it may be speculated that perceiving of and reacting to gravitational forces is not the limiting factor for survival of ciliates on other planets (Table 18.1.), whereas most remaining physical or chemical parameters (temperature, pressure, atmosphere, radiation) will probably be past endurance of protists.

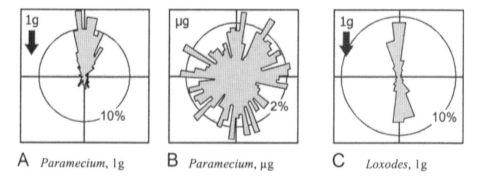

A *Paramecium*, 1g   B *Paramecium*, µg   C   *Loxodes*, 1g

Fig. 18.1 Orientation polarograms of populations of *Paramecium* and *Loxodes* swimming in a vertical plane (g = direction of acceleration). Under normal gravity conditions (A) *Paramecium* shows a clear orientation against the gravity vector which disappears in microgravity (B). C: *Loxodes* showing a bimodal orientation at 1g. Scales give the percentage of cells swimming into either direction. Width of directional classes 5.5° in A and B; 10° in C.

Table 18.1. Equatorial escape velocity and equalorial surface gravity of several celestial bodies

Celestial body	Moon	Mars	Venus	Earth	Jupiter
Escape velocity [km s⁻¹]	2.38	5.02	10.36	11.18	59.56
Surface gravity [m s⁻²]	1.62	3.72	8.87	9.78	22.88
Equatorial surface gravity [g]	0.166	0.380	0.879	1	2.339

Understanding the mechanisms of graviperception in protists would possibly help to also answer questions on the role of gravity in smaller cell systems such as mammalian cells.

18.1.2   Models of Graviperception

Since the detailed description of the negative gravitaxis (orientation against the direction of gravity) of the ciliate *Paramecium* at the end of the 19th century [12], several hypotheses tried to explain this behavior, either assuming physical or physiological mechanisms (for review see [1-3]).

While "static" hypotheses try to explain graviorientation as a result of unequal mass distributions inside the cytoplasm (buoy-like mechanism) [12-14], "hydrodynamic" hypotheses suppose that the shape of the cell causes a gravitaxis via different sinking speeds of the anterior and posterior hemisphere of the cell [15].

Since ciliates are actively swimming organisms, some authors presumed the difference between a cellular center of gravity and a center of propulsion to be the reason of graviorientation [16].

Major arguments against a pure physical mechanism of gravitaxis in protists are the facts that immobilized cells sediment in variable orientation of their longitudinal axis [17] and that the direction of gravitaxis switches under special circumstances without changes of the cell form [18, 19].

Alternative models of graviperception assume that the cell is capable of active *perceiving* the direction of the gravity vector and orients itself *actively*.

It has been proposed that a *Paramecium* either senses the hydrostatic pressure difference between its upper and lower hemisphere (hydrostatic hypothesis [20]), or that the cell swims into the direction of increased resistance (i.e. upwards) by measuring its energy consumption [21]. Both latter assumptions presume storage and temporal evaluation of measured values, which is unlikely in a unicellular organism.

Alternatively, other authors [22-25] suggested a statocyst-like mechanism, assuming that heavy compartments of the cell act as statolith inducing a selective stimulation of the locomotion apparatus. This mechanism would be similar to graviperception in plants (see Chap. 19, Schnabl). Due to the lack of a morphological distinct statocyst-like organelle in most ciliates, it was proposed that in *Paramecium* the mass of the whole cytoplasm, whose density exceeds the density of the surrounding medium by 4%, causes a mechanical load on the lower cell membrane, thus stimulating gravisensors ("statocyst hypothesis"). Detailed knowledge about the connection of cellular membrane potential and swimming behavior (electromotor coupling, [26]) and the cognition of distributions of mechanically sensitive ion channels [27] allowed proving this hypothesis in behavioral experiments.

In *Paramecium*, the ciliary activity and thus the swimming behavior is under the rigid control of the membrane potential. Hyperpolarization (shift of the negative membrane potential to more negative values) increases the swimming rate, whereas depolarization (shift of the membrane potential to more positive values) leads to decreased swimming rates or even backward swimming. Mechano-sensitive ion channels are located in a characteristic, bipolar distribution in the plasma membrane: mechano-sensitive K-channels mainly posteriorly and mechano-sensitive Ca-channels mainly anteriorly (Fig. 18.2). Stimulation of these channels leads to either depolarization (Ca-channels) or hyperpolarization (K-channels) of the membrane.

In an upward swimming cell, a gravistimulation of the posterior hemisphere hyperpolarizes the membrane potential, resulting in an increased swimming rate. In a downward swimming cell, stimulation of the anterior cell pole leads to decreased swimming rate. The consequence is gravikinesis which has been found in several

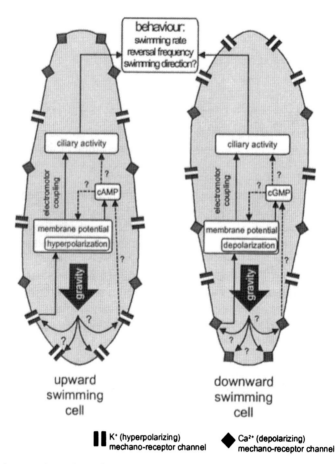

Fig. 18.2 Scheme of gravitransduction in *Paramecium*. Two populations of mechano-(=gravi?) receptor channels are distributed in the cell membrane, forming two spatial gradients. The mass of the cytoplasm generates a mechanical load on then lower plasma membrane. This mechanical stimulus is possibly mediated or reinforced by auxiliary structures, such as the filament system. Dependent on the orientation of the cell [with respect to the gravity vector], different populations of gravisensors are activated which cause either depolarization or hyperpolarization of the membrane potential. The shift in membrane potential results in a change in ciliary activity, which alters the swimming behavior of the cell. Possibly, second messengers are involved, which may affect ciliary activity or the membrane potential.

ciliate species [1, 28, 6]. From electrophysiological studies, we have first evidences of a gravireceptor potential in *Paramecium* depending on the orientation of the cell with respect to the gravity vector [29].

Break-down of an existing calcium gradient by means of the ionophore calcimycin (A23187), as well as manipulating the membrane potential by means of the lipophilic cation triphenylmethylphosphonium (TPMP$^+$) [30] resulted in a loss of gravitaxis in the flagellate *Euglena*. This also confirms that changes in the membrane potential of the cell are involved in graviperception.

Further support of the statocyst hypothesis came from density experiments. Increasing density of the medium (Ficoll or Percoll) impaired gravitaxis of *Euglena* and *Paramecium*. Orientation was completely disturbed at 1.04 to 1.05 g/cm^3, which corresponds to the mean cell density. Higher densities than the density of the cells reversed the direction of movement [31, 32]. These results indicate that intracellular organelles are not involved in graviperception of *Euglena* and *Paramecium*. In contrast, the ciliate *Loxodes* maintained its gravitaxis independently of the density of the surrounding medium, thus demonstrating the existence of an additional intracellular gravireceptor [32-34] (compare Chap. 20, Anken et al). These so-called Müller organelles or Müller vesicles are specialized vacuoles containing a heavy body of BaSO$_4$ which is fixed to a modified cilium.

Destruction of the Müller organelles by means of a laser beam leads to a loss of the gravitactic orientation in *Loxodes* [32], demonstrating its role in gravity sensing. It can be concluded that different graviperception mechanisms have been evolved in several systems in parallel, but is there a limit concerning cell size for gravisensation?

18.1.3 Energy Considerations

Gravikinesis and gravitaxis have been demonstrated in several protists of various sizes. Especially regarding the smaller species the question arises, whether the mechanical load of the cytoplasm is sufficient for signal transduction via mechanosensitive (= gravisensitive?) ion channels in the lower cell membrane. Calculations have given the values listed in Table 18.2.

Compared to the thermal noise level ($2 \cdot 10^{-21}$ J) the signal to noise ratio is sufficient for *Bursaria*, *Paramecium* and *Tetrahymena*, but may be critical in *Euglena* and even smaller cell systems such as lymphocytes. These energetic considerations, slow re-

Table 18.2. Activation energies calculated for small ciliates

Species	Volume [μm^3]	Gating energy [J]	Reference
Bursaria truncatella	3.0×10^7	1.2×10^{-16}	[28]
Paramecium caudatum	3.3×10^5	4.6×10^{-19}	[35]
Tetrahymena pyriformis	2.2×10^4	6.5×10^{-20}	[36]
Euglena gracilis	2.6×10^3	4.0×10^{-21}	[36]

sponse times (10 s to 60 s [5]), as well as threshold values in the range of 0.1 ×g to 0.3 ×g (determined by means of a centrifuge microscope in space) [37, 38, 9] demand the involvement of supporting structures such as the microfilament system. This assumption is supported by results from electric field experiments [39].

18.1.4 Gravisensory Channels

The gravisensory channels in protists have not yet been identified. They may be stretch activated or linear activated channels [40, 31]. The use of channel blockers provided controversial results so far, obviously due to their lack of selectivity. Gadolinium has been used in the study of graviperception, as it blocks the gravitropism in plants [41] and the gravitaxis in *Euglena* [31, 42]. Therefore, it was postulated that stretch-activated channels (SACs) in the cell membrane are stimulated by the sedimenting cell mass. However, similar results could not be achieved in case of *Paramecium,* neither by behavioral nor by electrophysiological studies [43], indicating that gadolinium is not specific for mechanosensitive channels in this species. Nagel and Machemer postulated that the mechanoreceptors in *Paramecium* are not activated by stretch, but by a pressure acting perpendicular to the cell membrane [43] (see also Chap. 20, Anken et al.). Energetical considerations support this hypothesis: The threshold value for the tension of stretch activated channels has been suggested 10^{-4} N/m [44] possibly being too large for graviperception in small cells. However, linear activated mechanoreceptors in the hair cells of the inner ear of vertebrates [45] require much lower activation energies, better fitting the data calculated for small ciliates such as *Tetrahymena* (Table 18.2.) [36].

Recent studies concentrate on the potential role of second messengers in the gravitransduction chain [42]. In ciliates, second messengers such as cyclic adenosinemonophosphate (cAMP) and cyclic guanosinemonophosphate (cGMP) are involved in behavioral responses, though there are still questions concerning the time course of events in the signal transduction chain. Due to current knowledge, cAMP is coupled to hyperpolarization events and cGMP to depolarization events in *Paramecium* [46]. Biochemical assays of cells which had been exposed to varied acceleration levels (by means of centrifuge, sounding rocket flights) are currently under investigation and will help to elucidate the role of second messengers in graviperception.

18.2 Mammalian Cells

18.2.1 Gravity Effects and Consequences

Studies under varied acceleration conditions demonstrated the capability of free living cells to adapt quickly to different environments, but what is the situation in higher specialized cells such as mammalian cells? The answer of this question is not only essential for understanding the complex machinery of cells but also with respect to manned space missions.

In an experiment carried out in 1983 in Spacelab 1 it was discovered that mitogenic activation of T lymphocytes was nearly completely inhibited at µg [47]. A follow-on experiment in Spacelab D-1 revealed that lymphocytes were highly damaged under microgravity conditions, whereas cells cultured at 1 ×g in flight on a 1 ×g–reference centrifuge undergo mitosis and blastogenesis (Cogoli et al. in [48]). Ultrastructural changes observed by electron microscopy suggested that programmed cell death (apoptosis) is increased in microgravity. This hypothesis was later supported by biochemical and microscopical studies [49, 50].

It was also shown that white blood cells are capable of autonomous movements, of cell-cell contacts and of the formation of aggregates in microgravity [51]. This again was a surprising and unpredictable finding as it was thought that mammalian cells which lack special locomotion organelles, can move only on a substratum and gravity is somehow driving the motion. It was also seen that the cytoskeleton which plays an important role in signal transduction, undergoes structural changes few seconds after exposure to µg [52]. A major break-through was achieved in studies on the self-assembly of microtubules at µg [53]. Microtubules self organize *in vitro* in a reaction-diffusion process, which appears to be gravity-dependent. In fact, in a 13 minutes flight on a sounding rocket, microtubules did almost show no self-organization in the absence of gravity compared to samples in an in-flight 1g-centrifuge [53]. Such findings are of primary importance because they point to a direct effect of gravity on basic structural elements of the cell.

Signal transduction is an extremely complicated process involving membrane receptors, G-proteins, the cytoskeleton, several protein kinases, transcription factors and oncogenes.

Limouse et al. [54] and de Groot et al. [55] were the firsts to investigate intermediate steps of signal transduction in space. They reported that protein kinase C (PKC) is inhibited µg [54], and furthermore that the intracellular distribution of this important enzyme is changed in microgravity [56]. In addition, the genetic expression of the early oncogenes *c-fos, c-myc, c-jun* was found to be significantly depressed in µg. Genetic expression was remarkably changed also in T-lymphocytes exposed to simulated microgravity in the random positioning simulator, also called 3-D clinostat [57]. Finally, using DNA microarrays, it was shown recently that the expression of 1632 genes was changed in human renal cortical cells cultured for 8 d in space [58].

In summary, data from experiments with different mammalian cell systems show that gravity may influence signal transduction pathways by altering the function of the cytoskeleton (probably via protein G), of important enzymes such as PKC, and of the expression of several genes, in particular of oncogenes and cytokines. As the signal transduction paths are essentially similar in all animal cells, it can be assumed that some of the effects seen in lymphocytes may be detected also in other cells. It is important to notice that gravitational effects are seen mainly in cells undergoing differentiation rather than in quiescent or non-differentiating cells. If differentiation is the response to the reception of a specific signal and consists of new expression of specific genes, we can expect that cancer cells which are dividing spontaneously (i.e. without the intervention of specific signal), are less or not at all dependent on gravitational changes.

Several tests were also conducted on lymphocytes from space crewmembers *in vivo*, and in parallel to the investigations *in vitro* (reviewed in [18]). Thereby, it was seen that the conditions of space flight (launch, in flight and landing activities) cause in more than 50% of the subjects significant depression of the cellular immune response, which is known to be T cell dependent. Such effect is induced by the psychological and physical stress of space flight on the neuroendocrine/neuroimmune system rather than by a direct effect of weightlessness *per se*. Nevertheless, the depression of the immune system may become a serious health issue on prolonged space missions such as a flight to and from Mars.

18.3 Conclusions

Studies under varied acceleration conditions demonstrated that free living cells such as protists are able to perceive changes of the acceleration conditions. Recent studies favorite the hypothesis that in these systems gravity is perceived either by intracellular receptors (statocyst-like organelles), heavy cell organelles (such as nucleus) and/or by sensing the cell mass by means of ion channels located in the cell membrane.

Mammalian cells in microgravity were profoundly influenced. Alteration in the cellular mechanisms and structures in mammalian cells like signal transduction and the cytoskeleton were detected. It can be speculated that the depression of the immune system may become a serious health issue on flights to and from Mars.

Acknowledgements. The work was supported by the German Aerospace Center (DLR, Grant 50WB9923), the PRODEX program of the European Space Agency, ESA, by the Italian Space Agency, ASI, and by the Swiss Federal Institute of Technology.

18.4 References

1   H. Machemer, R. Bräucker, Acta Protozool. **31**, 185 (1992).

2   R. Hemmersbach, D.-P. Häder, FASEB J. **13**, S69 (1999).

3   R. Hemmersbach, D. Volkmann, D.-P. Häder, J. Plant Physiol. **154**, 1 (1999).

4   S. Machemer-Röhnisch, R. Bräucker, H. Machemer, J. Comp Physiol A **171**, 779 (1993).

5   R. Bräucker, A. Murakami, K. Ikegaya, K. Takahashi, S. Machemer-Röhnisch, H. Machemer, J. Exp. Biol. **201**, 2103 (1998).

6   R. Hemmersbach-Krause, W. Briegleb, D.-P. Häder, K. Vogel, Acta Protozoologica **32**, 229 (1993).

7   R. Hemmersbach-Krause, W. Briegleb, D.-P. Häder, K. Vogel, D. Grothe, I. Meyer, J. Euk. Microbiol. **40 (4)**, 439 (1993).

8   H. Planel, G. Richoilley, R. Tixador, J. Templier, G. Gasset, Acta Astronautica **17**, 147 (1988.

9   R. Hemmersbach, R. Voormanns, W. Briegleb, N. Rieder, D.-P. Häder, J. Biotech. **47**, 271 (1996).

10 R. Bräucker, S. Machemer-Röhnisch, H. Machemer, J. Exp. Biol. **197**, 271 (1994).
11 R. Hemmersbach, R. Voormanns, D.-P. Häder, J. Exp. Biol. **199**, 2199 (1996).
12 M. Verworn, Exp. Unters. Jena **1**, 1889.
13 J. Dembowski, Arch. Protistenk. **66**, 104 (1929).
14 K. Fukui and H. Asai, Biophys. J. **47**, 479 (1985).
15 A.M. Roberts, J. Exp. Biol. **53**, 687 (1970).
16 H. Winet, T.L. Jahn, J. Theor. Biol. **46**, 449 (1974).
17 L. Kuznicki, Acta Protozool. **6**, 109 (1968).
18 D.-P. Häder, S.-M. Liu, Curr. Microbiol. **21**, 161 (1990).
19 E. Stallwitz, D.-P. Häder, Europ. J. Protistol. **30**, 18 (1994).
20 P. Jensen, Physiol. Inst. Jena, **428**, 1891.
21 C.B. Davenport, *Experimental Morphology*. Macmillan Publ. Co, New York, 1, 1897.
22 E.P. Lyon, Am. J. Physiol. **14**, 421 (1905).
23 O. Koehler, Arch. Protistenk. **45**, 1 (1922).
24 H. Machemer, S. Machemer-Röhnisch, R. Bräucker, K. Takahashi, J. Comp. Physiol. A **168**, 1 (1991).
25 S.A. Baba, Biol. Sci. Space **5**, 290 (1991) (in Japanese).
26 H. Machemer, A. Ogura, J. Physiol. **296**, 49 (1979).
27 H. Machemer, J.W. Deitmer, Progress in Sensory Physiol. **5**, 81 (1985).
28 M. Krause, Diplomarbeit Ruhr-Universität Bochum, 1999.
29 M. Gebauer, D. Watzke, H. Machemer, Naturwiss. **86**, 352 (1999).
30 M. Lebert, P. Richter, D.-P. Häder, J. Plant Physiol. **150**, 685 (1997).
31 M. Lebert, D.-P. Häder, Nature **379**, 590 (1996).
32 R. Hemmersbach, R. Voormanns, B. Bromeis, N. Schmidt, H. Rabien, K. Ivanova, Adv. Space Res. **21** (8/9), 1285 (1998).
33 D.C. Neugebauer, S. Machemer-Röhnisch, U. Nagel, R. Bräucker, H. Machemer, J. Comp. Physiol. A **183**, 303 (1998).
34 T. Fenchel, B.J. Finlay, J. Exp. Biol. **110**, 17 (1984).
35 S. Machemer-Röhnisch, U. Nagel, H. Machemer, J. Comp. Physiol. A **185**, 517 (1999).
36 U. Kowalewski, R. Bräucker, H. Machemer, Microgravity Sci. Technol. **XI/4**, 167 (1998).
37 D.-P. Häder, A. Rosum, J. Schäfer, R. Hemmersbach, J. Biotech. **47**, 261 (1996).
38 D.-P. Häder, M. Porst, H. Tahedl, P. Richter, M. Lebert, Micrograv. Sci. Techn. **10**, 53 (1997).
39 S. Machemer-Röhnisch, H. Machemer, R. Bräucker, J. Comp. Physiol. A **179**, 213 (1996).
40 M. Sokabe, F. Sachs, Advances Comp. Environm. Physiol. Springer, **10**, 55 (1992).
41 B. Millet, B.G. Pickard, Biophys. J. **53**, 155a (1988).
42 H. Tahedl, P. Richter, M. Lebert, D.-P. Häder, Micrograv. Sci. Techn. **XI/4**, 173 (1998).
43 U. Nagel, H. Machemer, Europ. J. Protistol. **36**, 161 (2000).
44 H. Sackin, Kidney International **48**, 1134 (1995).
45 J. Howard, W.M. Roberts, A.J. Hudspeth, Ann. Rev. Biophys., **17**, 99 (1988).
46 T. Hennessey, H. Machemer, D.L. Nelson, Europ. J. Cell Biol. **36**, 153 (1985).
47 A. Cogoli, A. Tschopp, P. Fuchs-Bislin, Science **225**, 228 (1984).
48 N. Longdon, V. David (Eds.) *Biorack on D1*, ESA SP 1091, 1988.

49   M.L. Lewis, J.L. Reynolds, L.A. Cubano, J.P. Hatton, B.D. Lawless, E.H. Piepmeier, FASEB J. **12** (11), 1007 (1998).

50   L.A. Cubano, M.L. Lewis, Exp. Geron. **35**, 389 (2000).

51   M. Cogoli-Greuter, M.A. Meloni, L. Sciola, A. Spano, P. Pippia, G. Monaco, A. Cogoli, J. Biotech. **47**, 279 (1996).

52   L. Sciola, M. Cogoli-Greuter, A. Cogoli, A. Spano, P. Pippia, Adv. Space Res. **24**, 801 (1999).

53   C. Papaseit, N Pochon, J. Tabony, Proc. Natl. Acad. Sci. U.S.A **97**, 8364 (2000).

54   M. Limouse, S. Manié, I. Konstantinova, B. Ferrua, L. Schaffar, Exp. Cell Res. **197**, 82 (1991).

55   R.P. de Groot, P.J. Rijken, J. Den Hertog, J. Boonstra, A.J. Verkleij, S.W. de Laat, W. Kruijer, J. Cell, Science **97**, 33 (1990).

56   D.A. Schmitt, J.P. Hatton, C. Emond, D. Chaput, H. Paris, T. Levade, J.-P. Cazenave, L. Schaffar, FASEB J. **10**, 1627 (1996).

57   I. Walther, P. Pippia , M.A. Meloni, F. Turrini, F. Mannu, A. Cogoli, FEBS Letters, **436**, 115 (1998).

58   T.G. Hammond, F.C. Lewis, T.J. Goodwin, R.M. Linnehan, D.A. Wolf, K.P. Hire, W.C. Campbell, E. Benes, K.C.O'Reilly, R.K. Globus, J.H. Kaysen, Nature Medicine **5**, 359 (1999).

19 Gravistimulated Effects in Plants

Heide Schnabl

Plant growth and development are affected by a lot of different environmental abiotic factors such as light, temperature and water supply. Immediately upon germination another physical stimulus, gravity, strongly influences the growth of plant organs, root and shoot, in order to ensure their correct orientation in space and the survival of the young seedling. Since plants have evolved under the constant stimulus of gravity, its presence is one of the most important prerequisites for their growth and spatial orientation. Because of the importance of gravity it is astonishing that sensitivity to the gravitational vector, the mechanisms of graviperception and gravisresponse which are still partially controversially discussed, are unclear up to now. Indeed, we know that the framework of the plant body is disturbed without the permanent g-vector, its physiological balance is disrupted. But we do not know the way the gravitational force as an abiotic signal is translated into gravity-dependent phenomena and definite plant structures maintaining adaptive strategies for survival. The biological effects of gravity (studies under microgravity) would be easier understood if this abiotic factor could be reduced or minimized as other common environmental parameters can be modified in order to study their effects on plant differentiation and growth.

Since we have no ways to reduce the gravity vector on our planet without inducing stress, we may only speculate on the strategies of plant adaptation mechanisms during the evolution. Leaving the original life space, the water environment, and occupying the land environment 400 million years ago, the plants have been forced to adapt to the new conditions characterized by missing convection. Thereby land-plants were affected by a force 1000 times bigger than in water. The local orientated plants had to optimize their life conditions with respect to water balance, to nutrition, to mechanical problems and to reproduction changing the typical horizontal orientated vegetative thallus into the upright orientated plant body, the cormus. This statically local orientated life way resulted in a high adaptation strategy to environmental factors (Fig. 19.1). In order to overcome mechanical problems induced by the deficiency of the convection, the collapse of the plant cormus had to be avoided by the formation of tissues, delivering special structural rigidity, such as the deposition of lignin in the secondary wall. The supply of water, minerals and nutrition materials had to be organized by the development of a system for the conduction of water and minerals from the soil via the roots in the rest of the plants (xylem) and for the transport of the products of photosynthesis from the leaves throughout the plant (phloem). The total vessels are known as the vascular bundles. Moreover, the plants had to develop suitable nutrition sources, resulting in the formation of leaves which are able to fix CO_2 and to store carbohydrates. The shoot functions as the mediator between root and leaves. Other cell types were necessary to be built up, such as tissues covering and protecting plants from wounding, pathogen infections, water loss and dryness, such as epidermal cells covered by waxy layers, the cuticle, in order to restrict the loss of moisture from the plant to the environment. Within the epidermis guard cells occur, which surround stomata, giving access to the tissues beneath the epidermis. Variation in turgor of the guard cells causes variation in the size of the stomatal aperture and therefore, allows the plant to regulate the rate of water loss. On the basis of the well adapted morphological cormus of the plant the ability was developed to restrict water loss or to tolerate total dryness. Moreover, the newly-adapted plants had to guarantee the conditions of reproduction by pollen and the expansion by seed and fruits.

The highly structured, immotile autotrophic organisms are characterized by well organized plant bodies on the basis of sensitively regulated orientation mechanism in space which were developed under the common supervisor "gravity". For perceiving the signal of the abiotic factor "gravity", a suitable stimulus-response system in a sequence chain – graviperception, signal gravity transduction and graviresponse - was developed in the process of evolution, which made the plants capable for distinguishing between up and down using gravity for orientation.

In the following sections the three processes of graviperception, gravi-signal transduction and graviresponse in uni- and multicellular plants are described on the basis of the actual literature.

Fig. 19.1 The plant bodies with differently structured organization. Left: The freshwater alga *Chara*, anchored with specific cells to the underground of water, living in ponds. Right: The upright standing tree (Pinus) as an example for a plant structured with gravity and antigravity sensing cells, the growth is stabilized by specialized cell tissue (see text).

19.1 Graviperception
as the Primary Gravity-Sensing Mechanism

It is a central question how plants sense gravity. The classical statolith hypothesis focuses on dense organelles such as amyloplasts (statoliths, [4]) which are sedimented on specific endoplasmatic reticulum layers, effecting a pressure and thus, triggering gravity sensing [5] (Fig. 19.2).

A revision of this hypothesis is, however, required, because the kind of mass, which will be effective as the gravity-stimulus, has to be investigated in more detail. Besides the statoliths, (and other sedimenting particles) the whole protoplast is discussed as a dense gravisensor. Furthermore, non-sedimenting particles are focused with pressing or pulling effects such as the nucleus, and components of the extracellular matrix. For perceiving gravity at cell surface and transducing signals in the cell interior they are coupled by receptors of plasma membrane. Plants should respond to gravity in different ways. It sounds understandable that the shoots and roots of multicellular plants show another way of graviperception as unicellular rhizoids or protonemata [6]. There may exist a spectrum of receptors in different systems, since the diversity of graviperceptions is suggested to be evolved in several systems in parallel.

19.1.1 Graviperception in Unicellular Plants

In order to test whether the statolith mechanism is involved in graviperception of the flagellate *Euglena gracilis*, the cells were subjected to increasing density of the outer medium (Ficoll). The gravitaxis of the organism was disturbed at 5%, which corresponds to a cell density of 1.04 g ml^{-1}, higher concentrations of Ficoll inverted the direction of movement [7-9]. These data indicate that intracellular sedimenting

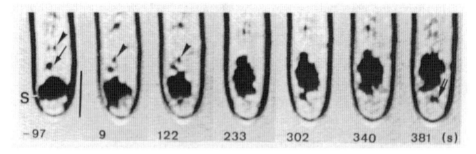

Fig. 19.2 A series of graphs of the freshwater alge *Chara* during a shuttle rocket flight (6 min under microgravity).The pictures were directly transferred via telecommunication. From left to right: 97sec just before starting (-97) up to 381 sec after starting. Conditions under microgravity are created after 122 s, 233 s, 302 s, 340 s, 381 s. The complex of statoliths is changing to a longer shape and is transferred into the opposite direction of the gravity vector. S=statoliths (taken from [86]).

organelles are not involved in graviperception of this unicellular plant and strengthens the idea that the whole cell acts as the sedimenting mass. Recent results show that the pressure of the sedimenting cytoplasm may open and close stretch-activated, mechano-sensitive ion channels in the plasma membrane, which probably pass calcium ions [10]. Using a calcium imaging system the involvement of this second messenger in sensory transduction of graviperception should be studied in rocket missions in future.

Also in the internodal cells of the alga *Chara* (see also 19.3.1), the gravity sensing is described by the gravitational pressure model, which hypothesizes that the whole protoplast acts as the gravity sensor. The cell perceives the g-vector by sensing the differential tension/compression between plasma membrane and the extracellular matrix at the top and the bottom of the cell via peptides spanning plasma membrane and extracellular matrix and activating thereby Ca-channels [11-13].

19.1.2   Graviperception in Multicellular Plants

The time course of the gravitropic stimulus-response chain in multicellular systems such as in seedling roots of cress (*Lepidium sativum*) is composed of different phases [14, 15].

19.1.2.1   *Statocyte Polarity and Gravisensing*

During the initial stage (signal perception) the physical stimuli - ×g and μg vector, respectively - are perceived via statoliths in statocytes, the signal is transformed into an intracellular signal without being transduced. In the subsequent period, the signal is transduced producing a message outside the statocytes (signal transduction) thereby inducing an effect. The biochemical signal is transmitted to the response zone, where the root curvature is initiated as a result of differential growth of upper and lower side of the root axis [14, 15]. Thus, the graviresponse in roots is preceded by the signal perception and transduction.

In shoots and grass pulvini, amyloplast-containing statocytes are associated with starch parenchyma cells surrounding vascular tissues, in roots they are localized in the columella of the cap [4]. The gravisensing mechanism of seedling roots, for example cress roots, is discussed to be localized in statocytes [14]. They are characterized by the structural polarity (Fig. 19.3) differentiated by the position of the nucleus in the proximal side of the cell, the endoplasmatic reticulum (ER) at distal side and dense amyloplasts sedimented at the physical bottom [16]. The amyloplasts performing a statolithic function are sedimented on ER-membranes when root is placed in vertically orientation under g-conditions.

Amyloplast sedimentation is hypothesized to activate receptors triggering signal transduction chain, which leads to a physiological signal and induces the bending of the organ as the consequence of the graviresponse. The polar arrangement of organelles does not depend on gravity [16] and is maintained by means of the cytoskeleton (Fig. 19.4) [17-19, 21]. Under microgravity this strict structural polarity of statocytes was displaced [15, 20], the amyloplasts move towards the proximal side of the cell, the nucleus was observed to migrate further away from the proximal plasma membrane (Fig. 19.5). Using cytochalasin the displacement of statoliths does not occur [21]. Thus, it could be shown that the organelles are attached to the actin filaments. A tension produced by dense amyloplasts is hypothesized on the basis of a connection between the statolith membranes and the actin filaments [15], which can be released under μg-conditions, a relaxation of actin network in the absence of gravity is

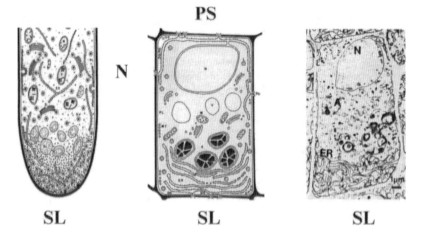

Fig. 19.3 Polarity of the gravity-perceiving cell. Left: Polar organization of the growing tip of a gravity-perceiving cell of the alge *Chara*. The statoliths (SL) sediment with high density above the vesicles and membranes. Perception and response are associated with in the tip. Middle: Gravity-perceiving cell from cress root cap. The statoliths filled with starch are sedimented above the membrane system, the nucleus (N) is located in the upper side of the cell. Plasmatic strings (PS) are participated in the signal transport. Right: Polar organisation of a gravity perceiving cell under microgravitation, showing the nucleus and the membrane system (ER). The statoliths are almost starch-free induced by microgravity (taken from [86]).

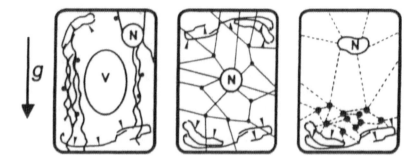

Fig. 19.4 The actin-cytoskeleton of a root in three cell types with different function, left: cell from the area of cell elongation, middle: cell from the area of cell division, right: gravity-perceiving cell; the sedimented statoliths (black points) are located within a network of actin-strings, N=nucleus, membranes=arrows, V=vacuole (taken from [86]).

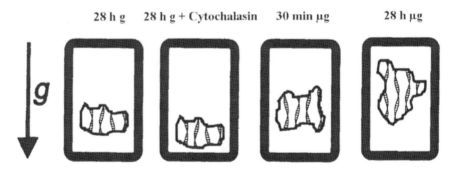

Fig. 19.5 A complex of statoliths in a gravity-sensing root cell under different conditions, from left to right: 28 h under gravity (control), 28 h under gravity with fungal toxin cytochalasin damaging the actin filaments; 30 min under microgravity; 28 h under microgravity. (taken from [86]).

hypothesized [22]. So, space research experiments have brought a lot of new insights resulting in a new graviperception hypothesis based on tension of actin filaments which are built up from interaction with statocytes and transduced to stretch-activated ion channels in the plasma embrane [22]. This proposal was confirmed by a Spacelab experiment flown on the Second International Microgravity Laboratory Mission (IML-2) showing that the sensitivity of roots treated with 1 g-conditions in orbit is less than in those roots grown under μg -conditions. Different tensions in the actin network are discussed [15].

19.1.2.2   The Plastid-Based Gravisensing

The identity of the mass, which specifically functions in gravisensing mechanism should be different in unicellular and multicellular plants as well as in shoots and roots on the one hand, and in tip-growing systems such as rhizoids on the other. Sedimentation of endogenous organelles, for example amyloplasts, was studied to take place in cells with specific locations and developmental stages, as in the endodermis of stems and in the root caps [4]. When the starch content in the amyloplasts was reduced by certain manipulations, the potential of gravitropic response was also varied. A reduced gravitropic sensitivity was shown in starchless and starch-deficient mutants [24, 25], leading to the conclusion that starch plays an important role in gravitropic sensitivity. These investigations demonstrated that kinetics of the bending response in roots is strictly correlated with the starch content. However, a residual gravitropism could be shown to occur without starch, thus, starch is proposed not to be absolutely necessary for gravisensing. Laser-ablation experiments have shown that *Arabidopsis* roots lose a part of graviresponse after removing of amyloplast-containing root tips [23].

Thus, it has to be concluded that amyloplast sedimentation is not the sole mechanism in graviperception. At least, there exist two hypotheses about the mass responsible for gravisensing: the model of Wayne et al. [26] postulating that the entire mass of a specialized cell is able to provide a signal for compression or tension (Fig. 19.6, left). In the other model an intracellular mass, presented as plastids [4, 27], exert tension on a receptor in the plasma membrane, for example the ER (Fig. 19.6, right). The ER is discussed as a high accumulation of receptors pressed by sedimented plastids, which contribute as a amplification signal mechanism [27]. Staves et al. [12] and Wayne and Staves [11] propose a perception of gravisensing by gravireceptors of the plasma membrane, the integrins, coupling the extracellular matrix and intracellular compounds. Meanwhile, integrin-similar proteins were identified in *Arabidopsis* and *Chara* [30].

19.2 The Gravity Signal Transduction Pathway

The question, which still remains open is how can the amyloplast sedimentation be transduced into a physiological signal leading to the bending of organs. Although the evidence for the interaction between amyloplasts and micro-filaments in statocytes could not be detected up to now, the association of statoliths with micro-filaments by myosin-like proteins was postulated [28]. Also, the molecular pathway responsible for the transformation of gravity signals in statocytes is not known. Therefore, the involvement of mechano-sensitive channels for opening and closing of cytoplasmic Ca^{2+}-concentrations functioning as a second messenger is hypothesized. Recent experiments with Ca^{2+} could not deliver evidence supporting gravity-induced transient Ca^{2+}-changes [29]. However, modulations in Ca^{2+}-levels are known to be required for producing auxin gradients across gravistimulated organs. Since Trewawas and his coworkers found strong similarities between the growing *Chara* rhizoid and growing pollen tubes, they describe a possible model for the events, which could initiate gravity-induced curvature of *Chara* rhizoids [31-33]. They measured a permanent

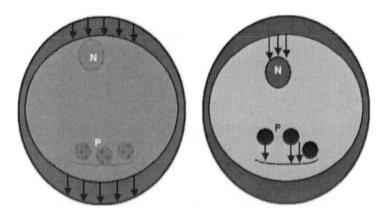

Fig. 19.6 Two models presenting mass symbols, which are responsible for gravisensing: at left, the whole cell provides mass for producing tension/compression to induce signal transduction; at right, intracellular organelles such as plastids (P) function as mass factors; they are attached to plasma membrane via cytoskeleton, thus creating tension on receptors (ER).

characteristic Ca^{2+}-gradient in growing pollen tips resulting from an accumulation of open Ca^{2+}-channels. The pollen tube reoriented its growth towards the side of the apical dome with higher Ca^{2+}-levels. A Ca^{2+}-oszillation was generated during the releasing process affecting gravitropic signal transduction by interacting with calcium binding proteins [34] and a Ca^{2+}-dependent protein kinase [35]. The authors suggest that this kinase may be associated with the establishment of an auxin gradient by the activation of a proton-pump leading to a redistribution of auxin.

The phosphoinositide pathway generating another second messenger, inositol 1,4,5 trisphosphate (Ins(1,4,5)P_3), may be also involved in gravitropic responses of plants [36, 37]. The second messenger Ins(1,4,5)P_3 is produced by phospholipase C, which cleaves phosphatidyl inositol 4,5 bisphosphate into 1,2 diacylglycerol and Ins(1,4,5)P_3. The latter one releases Ca^{2+} from Ins(1,4,5)P_3-sensitive Ca^{2+}-stores.

The first evidence for an auxin-dependent Ins(1,4,5)P_3 synthesis was given in 1988 [38], showing a 3-fold increase within 60 sec after auxin treatment. Recently, the activity of the biosynthetic precursor of Ins(1,4,5)P_3, the phosphatidylinositol-4-phosphate-5-kinase was found to be increased in the lower side of maize pulvini [39]. The increase of Ins(1,4,5)P_3 was detected in the lower halves of grass pulvini within 10 sec after gravistimulation. The authors interpret fluctuations in Ins(1,4,5)P_3 level as a part of the process that enables the cells to elongate [40]. The initial signal for intracellular processes is released within short time intervals, in contrast the sustained signal co-ordinates intercellular growth responses in maize pulvini. In order to provide data for an Ins(1,4,5)P_3-dependent signal transduction pathway in sunflower, protoplasts have to be exposed to an altered gravity vector [41].

Using isolated protoplasts as a suitable model system, the phosphoinositid-system was shown to be an element of studying signal transduction chain [41, 42, 82]. Sunflower hypocotyl protoplasts treated under real microgravity (TEXUS-sounding rocket flight, 6 min weightlessness; 1 ×g, 8 ×g, µg, 13 ×g; Fig. 19.7) showed an increase of Ins(1,4,5)P_3. The data deliver strong evidence that the increase of the second messenger, Ins(1,4,5)P_3, during a short period of weightlessness (6 min) was due to the microgravity period. These results provide hints for a direct involvement of signal transduction chain linking graviperception to gravisresponse.

Thus, we could show that protoplasts as the smallest functioning unit of plant are able to sense a modulation in g-vector without participating of starch. According to the gravitational pressure model the protoplast was investigated to perceive change in orientation of the gravity vector.

19.3 The Graviresponse: The Curvature of Organs

The perception of environmental stimuli via specific receptors is followed by regulation of physiological processes leading to adaptation to modified environmental conditions (response). Signals are transported from the locus of perception to the locus of response as first messengers. The translation of signals into a physiological response is induced via second messengers. In contrast to multicellular cells, in unicellular plants the processes of graviperception and graviresponse take place in identical cells. In this section, the g-induced responses in single cell systems as well as in multicellular plant roots and shoots are discussed.

19.3.1   Graviresponses in Unicellular Plants or Plant Systems

Flagellates and ciliates have been shown to orient to light (phototaxis) and gravity (gravitaxis) moving towards the center of the Earth (positive gravitaxis) or negative gravitaxis. Under terrestrial conditions the light-oriented responses of green flagellates are modulated by gravity, whereas under microgravity the responses show a more precise orientation with respect to light than at 1 g.

Experiments using sounding rockets as well as the fast rotating clinostat indicated that the g-vector is used for orientation in dark [43], the threshold of this response was found between 0.08 and 0.16 g. That means, the indication of an orientation was documented above 0.16 $\times g$ [7].

Gravitropic tip-growth is limited to a small number of tip-growing cell types, i.e. rhizoids of the characean alga [44]. Although these cell-types are attached to a multicellular organism, tip-growing cell type is discussed in this section as a term of "single-cell-system" because graviperception and graviresponse occur in the same cell type. The positively gravitropic *Chara* rhizoids are tube-like cells, in which cytoskeletal elements are involved for maintaining structural polarity, positioning of statoliths and

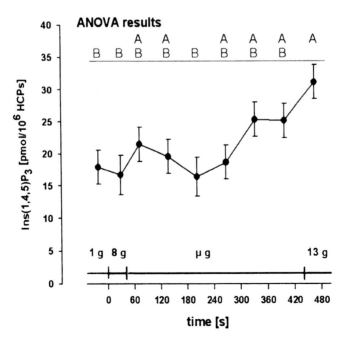

Fig. 19.7 Contents of Ins(1,4,5)P3 in sunflower hypocotyl protoplasts under real microgravity (TEXUS-flight; 6 min weightlessness conditions).

tip-growing response [16, 45, 46]. The negatively gravitropic bending of *Chara* protonemata is hypothesized to be based on different properties of actin cytoskeleton in a
similar way. Differences in the positioning and transport of statoliths strengthen the
role of actin cytoskeleton in the process of gravitropic tip-growth and reflect the diversity of gravitropic mechanisms.

Gravitropic bending of fruiting-body stems of *Flammulina* (Fig. 19.8) is based on
the differential growth of the transition zone between stem and cap [47, 48]. In this
system, graviperception and graviresponse are restricted to a small part of the apical
zone, the reorientation becomes visible already after 2 hours of gravistimulation. A
differential volume increase at one flank is able to induce a bending in the opposite
direction. Each hypha is hypothesized to be equipped with individual gravisensors as
well as a signaling system secreting a growth factor into the corresponding flank. A
resulting gradient is translated into differential growth leading to a gravitropic bending
response.

Apical cells of moss protonemata (*Ceratodon purpureus*) are unique among single-
celled systems in that way they perceive and react to light (positive and negative pho-
totropism) and to gravity (negative gravitropism). Gravitropism occurred at irradiances lower than 140 nmol m^{-2} s^{-1}, at irradiances higher than this threshold phototro-
pism predominates gravitropism despite the presence of amyloplast sedimentation

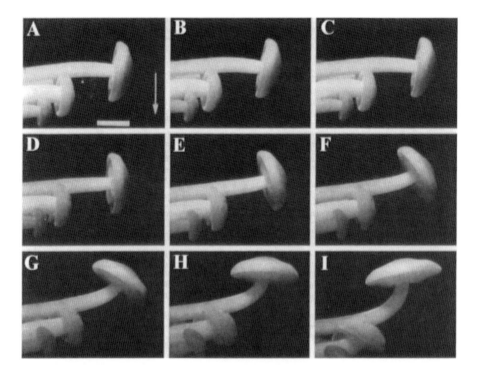

Fig. 19.8 Graviresponse of the fruiting bodies of the fungi basidiomycete *Flammulina* after
reorientation to the horizontal position at the time intervals: 0 h (A), 0,5 h (B), 1 h (C), 2 h
(D), 3 h (E), 4 h (F), 7 h (G), 10 h (H), 16 h (I), (taken from [48]).

[49]. In the dark, the protonemata grew in clockwise spirals under microgravity, which are suggested to be coupled with distinct zonation of starch-filled plastid orientation effective as early gravisensors.

19.3.2 Graviresponses in Multicellular Plants

19.3.2.1 Modulation of Auxin-Induced Gravitropic Responses

One of the most important mechanisms for the survival of the germinating seedling is the growth of the root in the direction of the gravity vector, which guarantees the young plant a rapid supply with water and minerals. The over-ground part of the plant is growing to the opposite direction reaching a position, in which CO_2 fixing processes of photosynthesis are able. The basis for this gravity orientated growth regulation is a complicated stimulus-response-chain, the beginning of which is represented by the transformation of the physical stimulus gravity into a biochemical signal. The places of transformation are highly specialized cells characterized by differences in densities of their organelles. During the process of graviresponse, the biochemical signal is transformed in a symptom of curvature as an indicator of a controlled growth process of the alga *Chara* (Fig. 19.9, left). In multicellular plants of cress the signal is transported to a target tissue via cell to cell communication (Fig. 19.9 right).

A general model of root/shoot gravitropism is based on the model of the asymmetric redistribution of the phytohormone, auxin, inducing a differential cellular elongation on opposite flanks of the central elongation zone (CEZ). The downward bending of roots and the upward bending of shoots of multicellular plants in response to

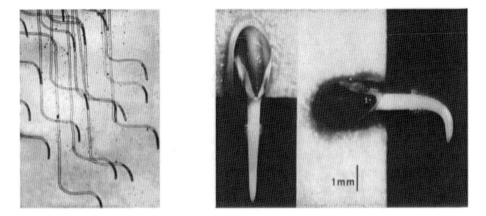

Fig. 19.9 Graviresponse in unicellular and multicellular plants. Left: Gravity-responding cells of the alga *Chara*. Right: Seedlings of cress: gravity-controlled growth of a root in vertical position and after 2 h in horizontal position (taken from [86]).

gravity result from induction of asymmetric redistribution of auxin within the elongation zone which is correlated with an accumulation of elevated auxin levels along the lower side. In roots, auxin functions as an inhibitor of cell growth and elongation, thus a downward curvature is the consequence. In contrast, in shoots the upward bending is induced due to the growing stimulation effect of auxin, resulting in an upward curvature [50]. However, this model of root/shoot bending is too simple to account for the complex spectrum of gravisresponse. Especially, some data showed that the gravitropic curvature functions without establishing the auxin gradient. Therefore, it seems to be necessary to find out the specialized mechanism of graviresponse in different organisms.

The reaction chain leading to curvature of cress roots is composed of four different phases, the perception phase (gravisensing of physical signal in statocytes, duration 1 s), the transduction phase (transport of the message out of the statocytes, lasts 20-30 s), the transmission phase (the asymmetrical transmission to the response zone, duration about 10-20 min) and the gravisresponse (the bending as the result of differential growth in the upper and lower flanks of the roots, after 1.5-3 h [15]. The gravitropic response, which can be used as a parameter for measuring the effect of gravistimulation depends on the duration of the stimulus, on mass acceleration and on the angle of stimulation [51]. Root elongations are similar both under microgravity and on the 1-g centrifuge [52], the sensitivity of roots is greater when the organs develop in simulated or in real microgravity than when they grow in 1 $\times g$ [53, 54]. The difference in sensitivity could be due to the polarity of the statocyte: in real and in simulated microgravity the amyloplasts are located in the proximal half of the statocyte, under 1 $\times g$ they are located in the distal flank, in which the statoliths are sedimented onto the ER.

Although the bending of an organ is the final event during gravitropism, the analysis of the visible part of the bending reaction delivers more information about the time-scale of the biochemical reactions involved. On the basis of sensitive video analyses the growth rate at the distal elongation zone (DEZ) was documented to be accelerated along the upper side of the root, affecting a bending of the root tip. The inhibition of cell elongation at the lower side of CEZ was followed by a reduced growth rate along the lower side of a gravistimulated maize root [50, 55, 56]. The root approached the vertical, followed by an overshoot reaction and the final organ elongation. In order to obtain a clearer understanding of the mechanism of gravitropic curvature, the physiological properties of the root sub-zones were characterized with regard to their hormonal sensitivities, to calcium, to pH and electrical changes. The accumulation of growth-inhibiting levels of auxin along the lower side of gravistimulated roots account for some, but not for all features of the differential growth pattern of organs: Maize roots treated with growth-inhibiting levels of auxin still responded to gravistimulation [56]. These observations indicate that the gravi-induced differential growth pattern can be provoked also by induction of growth asymmetry via auxin-independent mechanisms.

Alternative candidates are discussed as polar shifts of calcium toward the lower side of the apical region of maize roots, resulting in the built-up of an apoplasmatic Ca^{2+}-gradient in root tips [57, 58]. Thus, calcium is proposed to mediate the establishment of differential growth patterns in the DEZ and the apical part of the elongation zone during gravitropism. It can not be excluded that the built-up of a Ca^{2+}-

gradient is dependent on an active auxin transport [57, 58]. A further alternative hypothesis for participating in differential growth in the DEZ is acid efflux or influx leading to pH asymmetry across the root. Auxin was shown to affect a cell stretching by activating the plasma membrane associated H+-ATPase via a phospholipase A2 leading to a proton transport from the cytosol into the apoplast [59, 60]. *In vitro* experiments demonstrated a cell wall extension in growing tissue in a pH-dependent way [61] which, however, is not correlated directly with pH value. Cosgrove and his co-workers [62] suggest that a group of proteins, the expansins, are involved inducing extension and enhancing stress relaxation of isolated walls in a pH-dependent manner. *In vivo* expansin activity is associated with growing tissues of stems, leaves and roots. They were found in both grasses and dicotyledonous species [63-65], thus, these proteins were concluded to transduce the auxin-induced pH asymmetry into a growth asymmetry during gravitropism [66].

19.3.2.2 *Modulation of the Polar Auxin Transport*

The modulation of a polar auxin transport during gravitropic responses is mostly unknown. Up to now, a chemi-osmotic gradient is postulated, along which auxin is distributed asymmetrically via specific influx-and efflux carriers of plasma membrane [67]. The uptake and accumulation of protonated auxin-form from the apoplast is proposed to be stimulated via an influx-carrier. In contrast, the dissociated form is transported out of the cells by auxin-efflux carriers [68]. The chemiosmotic hypothesis was recently confirmed by Yamamoto and Yamamoto [69] and Bennett et al. [70] who identified an AUX1 gene in *Arabidopsis*, possibly encoding an auxin-influx-carrier in roots. Gene mutations showed a reduced auxin-sensitivity, the roots being agravitropic. Other experiments show the distribution of an efflux-carrier responsible for the establishment of an auxin gradient. The *Arabidopsis* AGR1/EIR1/PIN2 locus was cloned and shown to be essential for root gravitropism [67, 71-73]. It encodes as a component of the auxin-efflux carrier expressed mostly in roots [67]. The protein is localized in the basal membranes of DEZ and CEZ epidermal and cortical cells [71]. The patterns of AUX1 and AGR1/EIR1/PIN2 expression are correlated with the involvement in polar auxin transport and in root signal transduction. Both genes are expressed in the DEZ and CEZ of *Arabidopsis* roots [67]. However, neither of these genes is expressed in the root cap, so one can speculate that other gene products are necessary in participating on establishment of the auxin-gradient.

19.3.2.3 *The Regulation of Cellular Elongation in Graviresponding Flanks*

As mentioned above, the effects of auxin are demonstrated as cell growth inhibition in roots and cell elongation in shoots. The auxin accumulation at the lower flank of a gravistimulated organ induces an upward bending in shoots (Fig. 19.10 and 19.11) and a downward bending in roots (Fig. 19.9). The auxin-induced cellular elongation is regulated by modulation of the plasma membrane-associated ATPase, by cell wall extensibility and by the expression of auxin-responsive genes, i.e. ABP1, functioning as an auxin receptor during the cellular expansion [68].

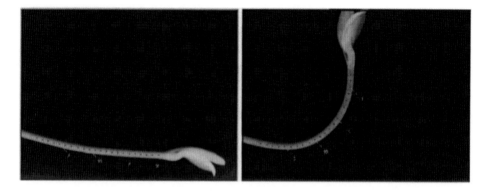

Fig. 19.10 Graviresponse of a sunflower shoot, left: in a horizontal position, 1 min; right: a gravistimulated shoot after 2 h in a horizontal position.

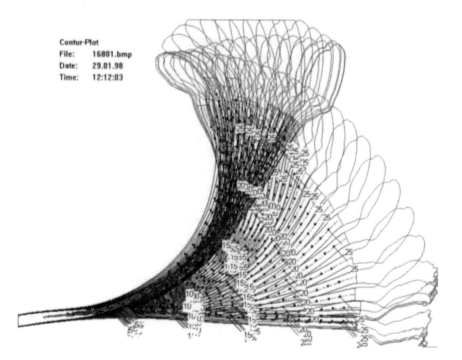

Fig. 19.11 Kinetics of a gravireponse (2,5 h): a bending sunflower seedling.

Although numerous auxin-binding proteins were identified and characterized [74] up to now, it is still unknown whether these proteins are functioning as auxin-receptors within a signal transduction chain regulating an auxin-dependent gene expression [75]. However, several studies focus the regulation of specific gene expression by auxin in different organs and tissues, the molecular mechanisms of auxin are still unknown [75]. The involvement of auxin-dependent genes, AXR1 and AXR3, in gravitropic responses could be identified [67, 76-78]. Moreover, for shoot gravitropism six independent loci were identified [79-81]. A specifically expressed gene, which was up/down-regulated during graviresponse of sunflower hypocotyls was found using differential display reverse transcriptase polymerase chain reaction (RT-PCR) [41, 82]. The identification of auxin-induced genes as opposed to genes, which are expressed in response to an altered gravity vector might supply additional information about mechanism involved. But, we still know very little about the molecular nature and function of the gravity receptors and the way of acting in the signal formation.

A comprehensive study of the role of gravity could not be made before it was not possible to vary the constancy of gravity. In order to understand the gravity-induced perception, transduction and response mechanisms and its physiological-biochemical reactions in roots and shoots of plants, it is a prerequisite to treat plants or their organelles under definite $\times g$ and μg-conditions. Since the variety of abiotic/biotic stimuli are normally elucidated by comparing its responses with and without the corresponding signals, it is also necessary to undertake experiments under microgravity [82-84] such as spaceflights, rocket programs, shuttle and spaceship missions.

Acknowledgement. The financial support and flight opportunities in D2 mission and a series of rocket programs (TEXUS) is gratefully acknowledged to DLR (Deutsches Zentrum für Luft- und Raumfahrt), Köln (Germany) and to MWF (Ministerium für Wissenschaft und Forschung) Düsseldorf (Germany). I thank Dr. G. Müller for critical reading and discussions.

19.4 References

1   E.L. Kordyum, Int. Rev. Cytol. **171**, 1 (1997).

2   W. Briegleb, Adv. Space Res. **4**, 5 (1984).

3   W. Briegleb, ASGSB Bull **5**, 23 (1992).

4   F.D. Sack; Int. Rev. Cytol. **127**, 193 (1991).

5   G. Haberlandt, Ber. Deutsch. Bot. Ges. **18**, 26 (1900).

6   A. Sievers, B. Buchen, D. Hodick, Trends Plant. Sci. **1**, 273 (1996).

7   D.P. Häder, A. Rosum, J. Schäfer, R. Hemmersbach, Jour. Plant. Physiol. **146**, 474 (1995).

8   M. Lebert, D.P. Häder, Nature **379**, 590 (1996).

9   D.P. Häder, R. Hemmersbach, Planta **203**, 7 (1997).

10   D.P. Häder, 14th ESA Symp. Europ. Rocket Ballon Progr. Rel. Res. **SP 437**, 131 (1999).

11   R. Wayne, M.P. Staves, Phys. Plant. **98**, 917 (1996).

12   M.P. Staves, R. Wayne, A.C. Leopold, Amer. J. Bot. **84**, 1522 (1997).

13   M.P. Staves, Planta **203**, 79 (1997).

14   D. Volkmann, H.M. Behrens, A. Sievers, Naturwissenschaften **73**, 438 (1986).

15   G. Perbal, D. Driss-Ecole, M. Tewinkel, D. Volkmann, Planta **203**, 57 (1997).

16   A. Sievers, M. Braun, in: Y. Waisel, A. Eshel, V. Kafkafi (Eds.) *Plant. roots. The hidden half*, Dekker, NY, 1996, pp. 31.

17   W. Hensel, Planta **162**, 404 (1984).

18   W. Hensel, Planta **173**, 142 (1988).

19   G. Lorenzi, G. Perbal, Bot. Cell **68**, 259 (1990).

20   D. Volkmann, B. Buchen, Z. Hejnowicz, M. Tewinkel, A. Sievers, Planta **185**, 153 (1991).

21   B. Buchen, M. Braun, Z. Heijnowicz, A. Sievers, Protoplasma **172**, 38 (1993).

22   A. Sievers, B. Buchen, D. Volkmann, Z. Hejnowicz, in: C.W. Lloyd (Ed.) *The cytoskeletal basis of plant growth and form*, Academic Press, London, N.Y., 1991, pp. 169.

23   E.B. Blancaflor, J.M. Fasano, S. Gilroy, Plant. Physiol. **116**, 213 (1998).

24   J.Z. Kiss, F.D. Sack, Planta **180**, 123 (1989).

25   J.Z. Kiss, J.B. Wright, T. Caspar, Physiol. Plant. **97**, 237 (1996).

26   R. Wayne, M.P. Staves, A.C. Leopold, Protoplasma **155**, 43 (1990).

27   F.D. Sack, Planta **203**, 63 (1997).

28   C. Wunsch, D. Volkmann, Eur. J. Cell Biol. Suppl. **61**, 46, (1993).

29   V. Legue, E. Blancaflor, C. Wymer, G. Perbal, D. Fantin, S. Gilroy, Plant. Physiol. **114**, 789, (1997).

30   W.J. Katambe, L.J. Swatzell, C.A. Makaroff, J.Z. Kiss, Phys. Plant. **99**, 7 (1997).

31   R. Malho, N.D. Read, A.J. Trewawas, S. Pais, Plant Cell **7**, 1173 (1995).

32   R. Malho, A.J. Trewawas, Plant Cell **8**, 1935 (1996).

33   V.E. Franklin-Tong, B.K. Drobak, A.C. Allan, P.A.C. Watkins, A.J. Trewawas, Plant **79** (1996).

34   W. Sinclair, I. Oliver, P. Maher, A.J. Trewawas, Planta **199**, 343 (1996).

35   Y.T. Lu, L.J. Feldman, H. Hidaka, Planta **199**, 18 (1996).

36   L.J. Swatzell, R.E. Edelmann, C.A. Makaroff, J.Z. Kiss, Plant Cell Physiol. **40**, 173 (1999).

37   R. Ranjeva, A. Graziana, C. Mazars, FASEB J. 13, 135 Cell **8**, 1305 (1999).

38   C. Ettlinger, L. Lehle, Nature **331**, 176 (1988).

39   I.Y. Perera, I. Heimann, W.F. Boss, Proc. Nat. Acad. Sci. USA **96**, 5838 (1999).

40   J.M. Stevenson, I.Y. Perera, I. Heilmann, S. Persson, W.F. Boss, Trends in Plant Sci. **5**, 252 (2000).

41   G. Müller, F. Hübel, H. Schnabl, 14th ESA Symp. Europ. Rockets and Baloon Progr. and Related Research, Potsdam, Germany, 1998, pp. 487.

42   H. Schnabl, in: M.H. Keller (Ed.) *Symp. Research under Space Conditions*, DLR Köln 67, 1998, pp. 43.

43   D.P. Häder, K. Vogel, K. Kreuzberg, Micrograv. Sci. Technol. **III**, 110 (1990).

44   A. Sievers, B. Buchen, D. Hodick, Trends Plant. Sci **1**, 273 (1996).

45   M. Braun, A. Sievers, Eur. J. Cell Biol. **63**, 289 (1994).

46   M. Braun, Planta **203**, S11 (1997).

47   V. Kern, A. Rehm, B. Hock, Adv. Space Res. **21**, 1173 (1998).

48   V. Kern, K. Mendgen, B. Hock, Planta **203**, 23 (1997).

49   V. Kern, F.D Sack, Planta **209**, 299 (1999).

50   M.L. Evans, H. Ishikava, Planta **203**, S115 (1997).

51 D. Volkmann, A. Sievers, in: W. Haupt, M.E. Feinleib (Eds.) *Encyclopedia of plant physiology*, Vol7, Springer Berlin, 1979, pp. 573.

52 V. Legue, F. Yu, D. Driss-Ecole, G. Perbal, J. Biotechnol. **47**, 129 (1996).

53 D. Volkmann, M. Tewinkel, J. Biotechnol. **47**, 253 (1996a).

54 D. Volkmann, M Tewinkel, Plant Cell Environ. **19**, 1195 (1996b).

55 H. Ishikava, K.H. Hasenstein, M.L. Evans, Planta **183**, 381 (1991).

56 H. Ishikava, M.L. Evans, Plant. Physiol. **109**, 725 (1993).

57 J.S. Lee, T.J. Mulkey, M.L. Evans, Planta **160**, 536 (1984).

58 T. Björkman, R. Cleland, Planta **185**, 379 (1991).

59 G.F.E. Scherer, J. Plant. Physiol. **145**, 483 (1995).

60 H. Yi, D. Park, Y. Lee, Physiol. Plant. **96**, 356 (1996).

61 D.J. Cosgrove, Planta **177**, 121 (1989).

62 S. McQueen-Mason, D.J. Cosgrove, Plant. Physiol. **107**, 87.(1995).

63 D.J. Cosgrove, Z.C. Li, Plant. Physiol. **103**, 1321 (1993).

64 Z.C. Li, D.M. Durachko, D.J. Cosgrove: Planta **191**, 349 (1993).

65 E. Keller, D.J. Cosgrove, Plant Journal **8**, 795 (1995).

66 D.J. Cosgrove, Planta **203**, 130 (1997).

67 R. Chen, E. Rosen, P.H. Masson, Plant. Physiol. **120**, 343 (1999).

68 A.M. Jones, K.H. Im, M.A. Savka, M.J. Wu, N.G. DeWitt, R. Shilito, A.N. Binns, Science **282**, 1114 (1998).

69 M. Yamamoto, K.T. Yamamoto, Plant Cell Physiol. **39**, 660 (1998).

70 M.J. Bennett, A. Marchant, H.G. Green, S.T. May, S.P. Ward, P.A. Millner, A.R. Walker, B. Schulz, K.A. Feldman, Science **273**, 948 (1996).

71 A. Müller, C. Guan, L. Gälweiler, P. Tanzler, P. Huise, A. Marchant, G. Parry, M. Bennett, E. Wisman, K. Palme, EMBO Journal **17**, 6903 (1998).

72 J. Sedbrook, K. Boosirichai, R. Chen, P. Hilson, R. Pearlman, E. Rosen, R. Rutherford, A. Batiza, K. Carrol, T. Schulz, Gravitational Space Biol. Bull. **11**, 71 (1998).

73 K. Utsuno, T. Shikanai, Y. Yamada, T. Hashimodo, Plant Cell Physiol. **39**, 1111 (1998).

74 R.M. Napier, M.A. Venis, New Phytol. **129**, 167 (1995).

75 T.J. Guilfoyle, G. Hagen, T. Ulmasov, J. Murfett, Plant. Physiol. **118**, 341 (1998).

76 J.C. del-Pozo, C. Timpte, S. Tan, J. Callis, M. Estelle, Science **280**, 1760 (1998).

77 D. Rowse, P. Mackay, P. Stirnberg, M. Estelle, O. Leyser, Science **279**, 1371 (1998).

78 H.M.O. Leyser, F.B. Pickett, S. Dharmasiri, M. Estelle, Plant Journal **403**, 413 (1996).

79 H. Fukaki, H. Fujisawa, M. Tasaka, Plant. Physiol. **110**, 945 (1996).

80 H. Fukaki, H. Fujisawa, M. Tasaka, Plant Cell Physiol. **38**, 804 (1997).

81 Y. Yamauchi, H. Fukaki, H. Fujisawa, M. Tasaka:, Plant Cell Physiol. **38**, 530 (1997).

82 H. Schnabl, in: H.J. Rehm, G. Reed, A. Pühler, P. Stadler (Eds.) *Biotechnology, Special Processes* Vol.10, 2nd Edition, VCH, Weinheim Germany, in press.

83 A.D. Krikorian, Physiol. Plant. **98**, 901 (1996).

84 E.K. Tuominen, Eur. J. Biochem. **263**, 85 (1999).

85 D. Volkmann, in: H. Rahman (Ed.) *Mensch, Leben, Schwerkraft, Kosmos*, G. Heimbach, Stuttgart, 2001, pp.162.

86 D. Volkmann, F. Baluska, in: C.J. Staiger et al. (Eds.) *Actin: A Dynamic Framework for Multiple Plant Cell Functions*, Kluwer Acad. Publ., Dordrecht, 2000, pp. 557.

20   Gravitational Zoology: How Animals Use and Cope with Gravity

Ralf H. Anken and Hinrich Rahmann

Since the dawn of life on Earth some four billions of years ago, gravity has been a more or less stable environmental factor thus influencing the phylogenetic development of all living organisms. On the one side, gravity represents a factor of physical restriction, which compelled the ancestors of all extant living beings to develop basic achievements to counter the gravitational force (e.g., elements of statics like any kind of skeleton - from actin to bone - to overcome gravity enforced size limits or to keep form). On the other side, already early forms of life possibly used gravity as an appropriate cue for orientation and postural control, since it is continuously present and has a fixed direction.

Due to such a thorough adaptation to the Earthly gravity vector, both orientation behavior as well as the ontogenetic development of animals is impaired, when they have to experience altered gravity (Δg; i.e. hyper- or microgravity). Nevertheless, animals still can cope with Δg in a certain range based on their physiological plasticity, which varies among the different animal phyla.

20.1 Gravity as a Factor of Physical Restriction: A Brief History of Evolutionary Challenges To Surmount It

As a matter of fact, the non-linear self-organizing dynamics of biological systems are inherent in any physical theory that satisfies the requirements of both quantum mechanics and general relativity [1]. Gravity therefore has always been a challenge for biological systems to adapt or/and to cope with it. Concerning single cells, the earliest life forms, it has been stated that average cell size results, in part, from the physical equilibrium between the destructive influence of the force of gravity and the protective role of diffusion and the cytoskeleton [2]. At increased forces of gravity the cell size would thus be decreased, whereas at lower gravitational forces and weightlessness cell size would be expected to increase. Mechanisms of protection of giant cells against internal sedimentation are based on protoplasmic motion, thin and elongated shape of the cell body, increased cytoplasmic viscosity, and a reduced range of specific gravity of cell components, relative to the ground-plasma. The nucleolus, due to its higher density, is considered as a possible trigger of mitosis. Although gravity limits the size even of single cells, its impact became especially apparent with the evolution of multi-cellular animals. There is not much known about the first multicellular animals inhabiting our planet in the late Precambrium (earlier than 570 millions of years ago) prior to the so-called "Cambrian explosion", which showed an extremely rapid evolutionary radiation with the development of almost all nowadays phyla of invertebrate animals. With only a few exceptions, the Precambrian animals (e.g., the forms of the Ediacara-fauna, named after the Ediacara Hills north of Adelaide, Australia) did not have any sort of inner or outer skeleton. All of them were small and many species had a worm- or jellyfish-like appearance (Fig. 20.1). In most cases, their relationships to the nowadays present invertebrate groups is unclear.

Obviously, elaborate anti-gravity systems had not yet been fully evolved, which would have allowed these animals to grow larger, to develop a directed locomotion and even to cope with the terrestrial impact of gravity at the stage of their exit from water to land (Fig. 20.2). Development of a directed locomotion might have been one of the most important evolutionary inventions. When heterotrophic animal life decreased the nutrients (e.g., autotrophic plants) in the oceans, animals were forced to cope with this evolutionary pressure; directed locomotion was therefore developed to predate other animals or, *vice versa*, to escape from predators. For exercising directed locomotion especially gravity was - besides other environmental factors such as radiation (especially light), atmospheric conditions/composition, sound and electromagnetic as well as mechanical impacts - one of the most important morphogenetic factors of animal evolution which pushed the gene to elaborate adequate mechanisms for surmounting it [3]. Directed locomotion generally requires any sort of skeleton to allow the insertion of muscles; such a skeleton then could (pre-adaptively) act as a prerequisite for animals to turn from their aquatic habitat to a terrestrial life some 440 millions of years ago, then following the green plants as a further source for their heterotrophic lifestyle (the first terrestrial animals were early ancestors of our nowadays spiders, belonging to the arthropods).

The first vertebrate animals which were able to cope with the terrestrial impact of Earth's gravity were early ancestors of fish some 350-400 millions of years ago. Concerning modern bluefish, it has been found that these animals can accelerate at 3 ×g

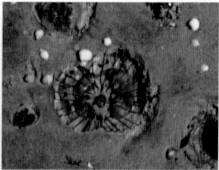

Fig. 20.1 This Ediacara-fossil may possibly represent an early jellyfish.

Fig. 20.2 Jellyfish certainly cannot cope with the terrestrial impact of gravity.

during swimming and that the vertebral column is strong enough to withstand this force [4]. This finding strongly indicates that the muscles and body structure of a bluefish would be able to withstand the force of gravity if the fish were otherwise equipped for terrestrial life (hypergravity experiments revealed in this context that development of larval fish is not impaired by 3 *g*; see 20.3.2.2). One can therefore speculate that early fish may also have evolved some degree of strength to overcome gravity-based inertia and drag during aquatic locomotion, and this evolution may have been a prelude to vertebrate terrestrial locomotion.

Terrestrial locomotion, of course, requires further special adaptations due to an animal's particular lifestyle. Gravitational force influences musculoskeletal systems, fluid distribution, and hydrodynamics of the circulation, especially in larger terrestrial vertebrates. The disturbance of hydrodynamics and distribution of body fluids relates largely to the effects of hydrostatic pressure gradients acting in vertical blood columns. These, in turn, are linked to the evolution of adaptive countermeasures involving modifications of structure and function. Comparative studies of, e.g., snakes [5] suggest that there are four generalizations concerning adaptive countermeasures to gravity stress that seem relevant not only to lower vertebrates: first, increasing levels of regulated arterial blood pressure are expected to evolve with some relation to gravitational stresses incurred by the effects of height and posture on vertical blood columns above the heart; second, aspects of gross anatomical organization are expected to evolve in relation to gravitational influence incurred by habitat and behavior; third, natural selection coupled to gravitational stresses has favored morphological features that reduce the compliance of perivascular tissues and provide an anatomical "antigravity suit"; fourth, natural selection has produced gradients or regional differences of vascular characteristics in tall or elongated vertebrates that are active in high gravity stress environments.

These generalizations can explain, why the position of the heart in relation to the head and the tail in different types of snakes vary: an aquatic snake will not be that affected by gravity. Consequently, the heart can be positioned relatively far away from the head and thus the brain. The heart of a terrestrial snake is positioned closer to the head. This is especially obvious in snakes living on trees, where the heart is situated almost directly behind the head in order to allow blood supply to the brain even during climbing upwards, i.e. in a direction opposite to the gravity vector [6]. Similar problems arise concerning extremely large or high-growing animals [7, 8]. The distance from the heart to the brain in the giraffe is ca. 2.8 m, whereas it reaches some 7.9 m in some herbivorous dinosaurs (human: 0.3 m), requiring an enormous blood-pressure to make the blood reach the brain, especially when the head eventually is being raised.

Bipedal walking was another challenge to cope with the force of gravity. There is not yet complete agreement, what the evolutionary pressure for bipedal walking actually had been, since bipedal walking is rather costly in terms of energetics: the short, flexed hindlimbs of chimpanzees younger than 5 years are not able to lift the body center of gravity high enough, so that these infants have a considerable energy output during bipedal walking [9]. Extension of the hindlimb is one of the bases for energy economy in human bipedalism (although lifting the legs in humans still consumes some 90% of the energy needed for locomotion [6]) and thus an important component of the evolution of human bipedalism.

An even higher ability to surmount gravity was necessary to come into being when animals made their way into the air. For certain, the potential diversity for evolution in large species (irrespective of them being invertebrates or vertebrates) is less than for medium-sized or small ones, and dwindles to zero above a body mass of about 14 kg [10]. The "World's largest flying bird", the Miocene (8-15 Ma ago) fossil *Argentavis* would require improbably high values for stress and strain in level flight, unless the air density were much higher in the Miocene times than at present, and/or the strength of gravity were much less. *Argentavis* therefore will not have been able to actively fly by flapping its wings, but will have mostly soared as do present-day condors.

Summarized, it was most probably the evolutionary pressure to develop directed locomotion which then, as a sort of prelude, allowed animals to cope with terrestrial gravity. Active, directed locomotion, and especially active maintenance of equilibrium during bodily movement (e.g., in locomotion) requires, however, appropriate sensory functions. Although many animals usually maintain their bodies with the long axis horizontal (backside up), humans being a notable exception, there are frequent departures from the usual position. A fish may dive steeply downward and a man may alter his normal orientation by lying down at full length. In no case, however, need there be any loss of equilibrium. Every deviation means an equilibrium disturbance and evokes compensatory reflex movements.

Gravity, therefore, did not only act as a factor of physical restriction, but was an environmental factor readily being available as an appropriate cue for orientation and postural control.

20.2 Gravity as a Cue for Orientation and Postural Control

Maintenance of equilibrium is based upon contact of the animal with the external world; several sensory systems may play a role in this context. When an animal moves over a solid surface, tactile stimuli usually predominate as cues. It has to be noted that also proprioceptors (i.e. sense organs allowing the perception of stimuli relating to the animal's own position, posture, equilibrium, or internal condition) in vertebrates and arthropods can also contribute to spatial orientation; bodily tissues like the club-shaped sensilla of arthropods (Figs. 20.3, 20.7) under gravity weigh vertically down and stimulate internal mechano-receptors in a way that depends on, and varies with, the animal's spatial position. When they are out of contact with the ground, many animals orient themselves in space by keeping their back (dorsal) side turned up toward the light, e.g., in the course of the dorsal light response (DLR) of fish (see 20.3.2.2). Visual cues also can serve equilibration; for example, through compensatory body movements (optomotor reflexes like the vestibulo-ocular reflex in fish and amphibians [11]) brought about by the shifts of the image of the environment over the retina of the eye. For the receptors mentioned thus far, however, equilibration is not the unique function. There are other sensory structures that are genuine organs of equilibrium in that they primarily and exclusively serve orientation of posture and movement in space: gravity receptors.

In general, gravity - or any sort of acceleration - can be transformed to a biological signal in several different ways (Fig. 20.3): already synthetic membrane bilayers con-

taining incorporated ion-channels respond to gravity [12, 13]. The exact signal transduction chain remains, however, hitherto unresolved. Unicellular protozoans (e.g.,
Paramecium) perceive gravity obviously by their membrane or have already developed minute, intracellular crystals that function as "heavy bodies" (*Loxodes*) (compare
20.2.1 as well as Chap. 18, Bräucker et al.).

20.2.1 Graviperception in Unicellular Animals

Already archaic extantly living eukaryotic organisms (particularly phytoflagellates and
ciliates) were shown to be gravitactic (cell orientation) or even gravikinetic (adjustment of swimming speed) (for review [14]; see also Chap. 18, Bräucker et al.). In the
ciliate *Paramecium*, gravikinesis is obviously regulated by a pressure-gradient between the membrane and the surrounding aqueous medium. This gradient is only at
equilibribium when the animal swims horizontally. Vertical tilts result in the open-
ing/closing of mechano-sensitive ion-channels, which effect a local de- or hyperpolari-
sation, which in turn selectively activates cilia for propulsion. In contrast to *Paramecium*, another ciliate - *Loxodes* – seems to perceive gravity via particular intracellular
organelles, called "Müller-bodies", which consist of a membraneous pouch, containing a "heavy body" or "statolith" of $BaSO_4$ (Fig. 20.4) [15] (see also Chap. 18,
Bräucker et al.).

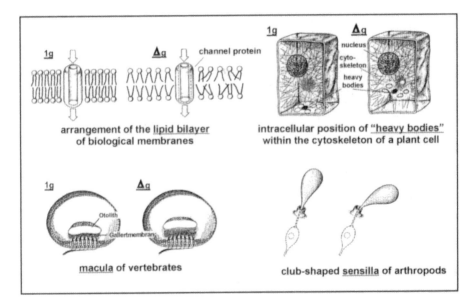

Fig. 20.3 Possibilities of graviperception in biological systems. Already (simulated) biological
membranes can respond to the gravity vector. Plants and most animals use "heavy bodies"
(called statoliths in plants, invertebrates and most vertebrates; the compact statoliths of fish are
called otoliths), whereas arthropods use body extensions to perceive gravity.

20.2.2   Graviperception in Multicellular Animals

All multicellular animals hitherto analyzed, who exhibit active (not necessarily also directed) locomotion, use so-called stato- or otoliths, being localized in differently specialized organs (e.g., the vestibular organs within the inner ear of vertebrates) or body extensions (e.g., club-shaped halters and sensilla of insects; i.e. proprioceptors, see above) for the transformation of an acceleration into a body-own signal.

Specific sensory abilities do not show a clear evolutionary progression, most likely because the development of any type of sense depends on many other factors in the total ecology of a given organism. Vision, for instance, is sometimes poor or absent in a species of a class in which other members have a highly developed visual system: examples include cave-dwelling species, who are relatives of sighted emergent species. Since mechanical stimuli rather than optical ones are effective in all forms of life, specialized organs already appear very early in animal evolution. In accomplishment of the "heavy body"-strategy of the unicellular protozoans, some fungi, some lower but most of the higher plants, almost all invertebrate animals and virtually all vertebrates use different types of heavy bodies to orient themselves towards the direction of earth gravity.

Depending on the function-morphologic characteristics of the body, the respective sense organs (sensilla, inner ear maculae etc.) have been positioned in different ways during the phylogenetic development (Figs. 20.5-20.7). Gravity receptors already appear in jellyfish. Ctenophoran jellyfish (the so-called Comb Jellies), e.g., are biradial-symmetric (two axes of symmetry) and possess one statocyst at the top of the body (Fig. 20.5). The calcareous statolith is supported by four, long, spring-like tufts of cilia called balancers. The whole structure is enclosed in a transparent dome that is apparently derived from fused cilia. From each balancer arises a pair of ciliated furrows, each of which connects with one of the so-called comb rows. Thus, each balancer innervates the two comb rows of its particular quadrant. Tilting the animal causes the statolith to press more heavily on one of the balancers, and the resulting stimulus elicits a vigorous beating of the appropriate comb rows to right the body.

Radial-symmetric cnidarian jellyfish with multiple axes of symmetry have many statocysts located around the mantle (Fig. 20.6). In contrast to the circumstances as observed in Comb Jellies, the statoliths in cnidarians have connections to epidermal neurons which transmit the sensory information to a nerve ring, which connects to muscles stimulating rhythmic pulsations of the bell and thus locomotion and postural control.

A bilateral-symmetric organization (one axis of symmetry) is found in most of the other animals. In the more advanced members of the phyla Mollusca and Arthropoda, greatly developed sense organs occur; whereas arthropods use mostly proprioceptors (Figs. 20.3, 20.7), the gravity-sensing organ of, e.g., many snails such as *Aplysia* consists of bilaterally paired statocysts. They are composed of supporting cells and receptor cells, forming a sac which contains calcium carbonate inclusions (called statoconia). The receptor cells are hair cell-like neurons (an analogue to the hair cells in the inner ear of vertebrates) whose cilia are motile and mechanosensory [16]. In the statocyst, the continuos beating of the mechanosensory cilia keeps the statoconia in constant motion. Gravity pulls the statoconia down, obstructing the beating of the cilia on

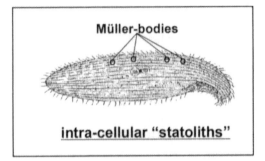

Fig. 20.4 The ciliate *Loxodes*, who possibly perceives gravity via intracellular statoliths.

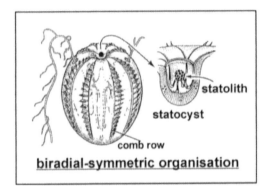

Fig. 20.5 Comb Jellies (ctenophorans) reveal two axes of symmetry, exhibiting one statocyst on the top of their body which directly regulates the beating of the comb row cilia.

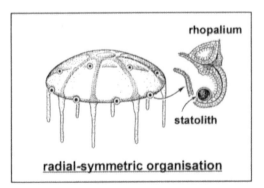

Fig. 20.6 Jellyfish (medusae of cnidarians) are organized radial-symmetrically (many axes of symmetry), exhibiting numerous statocysts (i.e. rhopalia). The sensory cells are hair cell-like neurons transmitting the gravity information to a nerve ring, which in its turn contacts to muscles.

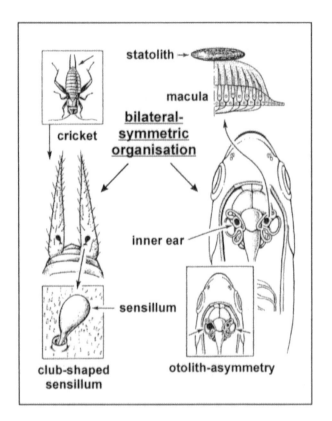

Fig. 20.7 Bilaterally organized animals (one axis of symmetry) usually exhibit gravireceptors on both sides of the body such as proprioreceptors (e.g., club-shaped sensilla in arthropods) or macula organs (vertebrates). An otolith asymmetry in the inner ear of vertebrates is believed to be the basic cause of motion sickness (e.g., sea-sickness, space sickness; see 20.3.2.2).

the bottom of the statocyst, which causes an increase in membrane conductance to Na^+ and the formation of an action potential [17]. The gravity receptors of some other mollusks show an amazingly close resemblance to vertebrate organs, e.g., the semicircular canals for equilibrium in *Octopus*.

In both inner ears of vertebrates (Figs. 20.7, 20.8), three semicircular canals (or cupular organs), located perpendicularly to each other for the detection of angular acceleration, are completed by three pouches containing either three oto-/statolith organs in lower vertebrates (utricle, saccule and lagena for the perception of linear acceleration/gravity as well as for sound) or two statolith organs and a lagena / cochlea without a stato-/otolith for sound perception in higher vertebrate animals. Sound or the movement of a statolith results in the bending of hair cell-cilia, which mechanically opens or closes ion channels [18] altering the electrical current of the respective sensory cell in the inner ear. Here, the transformation to computable action potentials takes place. Finally, a signal transduction on the level of the brain causes a motor response (Fig. 20.9). Generally, the brain integrates informations from the inner ear vestibular organs together with tactile, proprioceptive and visual cues for spatial orientation and postural control in the environment [19] (comp. Fig. 20.10).

Not only in vertebrates but in all animals exhibiting a central nervous system, there is always a close relationship between the presence of highly developed sense organs and a specialized region of the brain, since the latter is needed to process the incoming information in order to abstract the cues of importance to a given animal, which lastly results in behavior (comp. Fig. 20.9). An example in vertebrates is their cerebellum, which is responsible for the regulation of postural control; it is especially large and efficient in such animals who need to orient themselves in all three directions of space, like birds. (The fact that such elaborate systems exist does not exclude the possibility of much shorter and simpler pathways, which provide for more localized and quicker reactions; for instance, gravireceptor-mediation of postural control in Comb Jellies seems not even to require neurons.)

Summarized, animals did not only acquire systems to cope with gravity during evolution (comp. 20.1), but they also evolved receptors (and respective brain parts) in order to use gravity as the only virtually constant environmental factor for orientation, maintenance of equilibrium and postural control. On the one hand, gravitational environments which are not experienced by animals, disturb the neuronal processing of incoming information. This is especially the case, when at altered gravitational sensations (such as hypergravity or microgravity/weightlessness) the statolithic or otolithic organs transmit information to the brain, which do not necessarily match with others, e.g., the visual cues needed for a correct postural control. Thus "normal" behavior is inevitably affected. In human subjects, such an "intersensory conflict" can result in orientation problems, often accompanied by motion sickness and vomiting (kinetosis). On the other hand, altered gravity may have a strong influence on some aspects of development, since all systems (from sub-cellular organelles to complete organs) should be fully adapted to normal earth gravity.

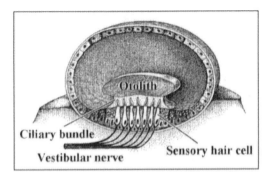

Fig. 20.8 Macula of a bony fish. Note the compact otolith. Amphibians, reptiles, birds and mammals reveal so-called statoliths, which are composed of hundreds of thousands of minute calcium carbonate cristals. A dislocation of the oto- or statolith causes shearing forces on the ciliary bundles of the sensory hair cells. These shearing forces alter the cellular ion current, which results in an increase or decrease (depending on the direction of the shearing) of the spike rate being transmitted to the brain by nerve fibers.

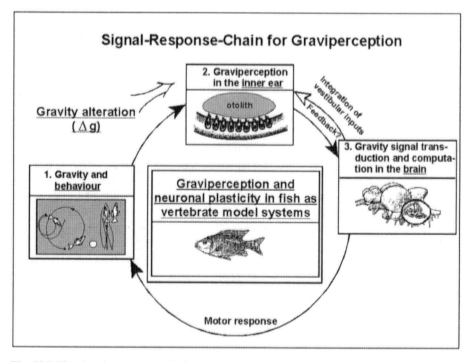

Fig. 20.9 The signal-response-chain for graviperception in vertebrate animals (fish). The basic stept of this response-chain (i.e. graviperception followed by a gravity signal transduction and computation, resulting in a motor response/behavior) are found in all animals who exhibit a central nervous system.

In order to understand the role of gravity on the (ontogenetic) development of animals as well as on their behavior, experiments at altered gravity (Δg) yielded valuable clues and insights.

20.3 Behavior and Differentiation of Animals at Altered Gravity

Whilst studies investigating the influence of Δg (hypergravity – hg – in centrifuges and microgravity – μg – during drop tower, parabolic aircraft flight, sounding rocket and spaceflight experiments) on unicellular systems (protozoa, cell cultures) and on plants are focused mainly on basic gravireceptive mechanisms on subcellular and cellular level (see Chaps. 18, Bräucker et al., and 19, Schnabl), the aim of a large number of experiments using animals, particularly vertebrates, is concerned with the analysis of the physiology of cardiovascular, respiratory, intestinal, endocrine, immune and muscular systems as well as with the calcification of the skeleton (for review, see [20]) in order either to elucidate aspects of the role of gravity during evolution or/and to understand the effects of weightlessness concerning medical problems of human space exploration.

The so-called microgravity syndrome [21], a prognosticated complex arising in reduced gravity environments such as the surfaces of the Moon and Mars and principally encompassing muscle atrophy, cardiovascular deconditioning and bone demineralization ("space osteoporosis"), stands to replace physics and rocketry as the fundamental challenge of interplanetary astronautics. Mirroring our past few million years of changing climate and resources, the mobility of humans between diverse gravitational environments on the high frontier will critically depend on our ability to adapt.

One major goal of gravitational zoology therefore is to elucidate the effect of long-term Δg on the behavior and development of animals as model systems and by this to clarify particularly the basic mechanisms of perception, transformation and central computation of a gravitational stimulus within the organism, including the clarification of the animals' possibilities to adapt and compensate. In this regard, results obtained using vertebrate animals – due to the homology of the morphological and physiological systems – can widely be transferred upon the conditions given in humans (the present review will, however, not deal with mammals, since respective biomedically relevant reviews abound).

20.3.1   Invertebrates

The morphogenetic development of various invertebrates in general is not heavily impaired by Δg: the shell of fresh-water snails (*Biomphalaria*), e.g., obviously develops normally at microgravity [22] indicating that the basic mineralization processes in mollusks are not affected by 16 days microgravity (this result is strongly contrasting to findings concerning human space osteoporosis and bone growth in rats!), which in its turn leads to the assumption that gravity did not play an important role in this context during evolution. Concerning squids (*Loligo* and *Octopus*), it has been argued that altering gravitational forces might not necessarily lead to morphogenetic aberrance but just to gradual deviations from the normal developmental speed of the embryos [23]. Microgravity, on the other hand, was shown to negatively affect the efficacy of fertilization processes in sea urchins [24].

In addition to this, major focus has been attributed to the physiology of the gravity sense system, to the corresponding neuronal computation and the resulting behavior: altered gravity influences statolith growth in the marine snail *Aplysia*, accompanied by changes in urease activity [25] indicating that the latter might regulate $CaCO_3$ deposition in statoliths. In contrast, Δg seems to have no effect on the development of gravireceptors in crickets, in which gravity is internalized by club-shaped sensilla; microgravity, however, mediates the sensilla-induced compensatory head-rolls [26, 27].

In conclusion, the results concerning the investigation of the effect of altered gravity – particularly on insects – clearly demonstrate the usefulness of these animals since they are by far less complex than, e.g., vertebrates. Due to their different lifestyles, the various invertebrate groups reveal a varying resistance/adaptability towards gravity. However, invertebrates, as a matter of fact, allow a comparison with vertebrate (including human) gravity related systems only in the broadest sense.

20.3.2   Vertebrates

Deeper clues and insights into the basic causes of gravity-stimulated effects in human beings can only be expected from studies performed using vertebrate animals, since the peripheral and central vestibular systems as well as the basic mechanisms of development are homologous among all animals with backbones and by this with humans.

With regard to this, most studies on the effect of altered gravity particularly on mammals had been focused on biomedical aspects comprising the physiology of cardiovascular, respiratory, intestinal, endocrine, immune and muscular systems as well as on calcification of the skeleton and related topics on which comprehensive reviews abound [7, 20, 28, 29]. Results using mammals as test subjects will therefore not be reviewed here.

Rather, in the following, we will briefly review some major results hitherto obtained using amphibians, reptiles and birds. Thereafter, focus will be laid on research results using fish as vertebrate model systems with regard to the components of their signal response chain for graviperception.

20.3.2.1   Amphibians, Reptiles, Birds

For over a century, embryologists using amphibians have debated as of whether gravity is required for normal embryonic development and, in particular, for the establishment of embryonic polarities such as pattern formation, morphogenesis and organogenesis (for review, see [30, 31]). Since their normal development at 1 ×g earth gravity is comparatively well understood, amphibians such as the clawed toad *Xenopus* and the salamander *Pleurodeles* are well suited animals to answer the aforementioned questions. The first fertilization of a vertebrate (*Xenopus*) in microgravity was successfully carried out in 1988 [32] showing that fertilization was monospermic as it is on Earth, and that development proceeded up to gastrulation (= end of the respective sounding rocket experiment). 1 ×g earth gravity thus seems not to be required for the early ontogenetic axis formation [33-35]. In elder developmental stages of amphibians, microgravity, however, normally results in non-inflated lung buds and tracheae, possibly being the result from the failure of the animals to inflate their lungs in a timely adequate fashion. Furthermore, microgravity induced some kind of malformed (typically) lordotic tails [31] with consequences for behavior (e.g., optomotoric responses [33]) and retarded larval growth. The effect on optomotor response suggests that tadpoles raised at microgravity may receive less vestibular information to control their position [11, 30].

Summarized, experiments both with salamanders and frogs indicate that amphibian egg maturation, fertilization, and embryonic development is not significantly influenced by microgravity, but larval growth might be retarded or even abnormal in microgravity. So far, however, no spaceflight mission had been long enough to investigate as of whether these animals can complete a full life cycle under space conditions. Comprehensive clues on the complete developmental life cycle are thus still lacking.

Few research at altered gravity has hitherto been conducted with reptiles and birds. There have been reviewed disorientation responses of various vertebrate animals exposed to microgravity produced by parabolic aircraft flights and in space experiments [36]. Like in mammalian species and frogs, coordinated performance can be easily compensated by visual function also in turtles. Like in fish (see below), in birds (pigeons, Japanese quail), who can move three-dimensionally in their environments, exposure to parabolic flight microgravity induces irregular tumbling with the eyes open and regular looping with the eyes closed, although the loop direction is the opposite in these two animals. A centrifuge experiment – conducted on Japanese quail – showed that hatchability is negatively affected by hyper-g [37]. Further, it has been reported based on a spaceflight experiment that the difference in specific gravity between the yolk and the albumen appeared to play a critical role in early chick embryogenesis [38]. In elder embryos, all the tissues, including cartilage and bone, however, were formed normally. Microgravity also seems to have no effect on normal eye development [39].

20.3.2.2 Fish

Fish have been proved to be the most suited vertebrate animals for basic gravity-research [40]: they can be characterized by an absence of body weight related proprio-perception (in comparison with surface-bound terrestrial vertebrates), a reduced influence of gravity on supporting tissue, muscles, vascular tonus system etc., a relative higher sensitivity for gravity due to larger otoliths etc., a high reproduction rate (combined with higher genetic homogeneity of individuals), and mostly an external development, thus enabling better access to defined developmental stages. Moreover, there is rich information about genetics and developmental mutations available (Zebrafish *Danio*, Medaka *Oryzias*, Swordtail Fish *Xiphophorus*), the developmental pattern of large numbers of genes are known and there exists an extensive homology to mammals not only at the molecular level but also concerning the central and peripheral vestibular system. Last not least, fertilization and development is not significantly impaired by altered gravity [41].

Basically, fish use visual and vestibular cues for postural equilibrium maintenance and orientation as do all other vertebrates and many invertebrates. Already in 1935, the so-called DLR (Fig. 20.10) was described [42]: illuminated from the side at normal 1 ×g earth gravity, a fish tilts its back towards the light source. In general, the DLR expresses a balance between the tilting force induced by visual information and the so-called vestibular righting response (VRR) [43] induced by gravitational information. Interestingly, the performance of the DLR depends on the visual performance (visual acuity) of an individual, suggesting that there are more "vestibular" and more "visual" individuals [44].

During microgravity, fish are often seen performing an abnormal swimming behavior, such as down- or upward pitching, inward loopings, spinning movements etc., especially following the transition from 1 ×g to microgravity [36, 45, 46]. This behavior has most likely the same source like motion sickness (a kinetosis) in humans. Subsequent experiments at microgravity using fish provided clues and insights into the understanding of the neurobiological basis of vertebrate gravity sensation and kinetosis.

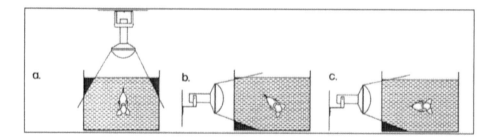

Fig. 20.10 The dorsal light response of fish. Illuminated from the side (light from above is the normal case, a) at normal 1 ×g earth gravity (b), a fish tilts its back towards the light source. The angle of the tilt increases when the light intensity is being raised. Increasing the force of gravity in a centrifuge at constant illumination decreases the tilt angle. After bilateral labyrinthectomy or under microgravity conditions, the tilt is consequently guided by light alone (c). It is a general feature of animals that the brain integrates informations from the inner ear vestibular organs together with further sensory cues for spatial equilibrium.

If, for instance, the sensitivity of the otolith system of humans in space is increased as compared to the ground, any brisk translation of the head would be interpreted erroneously. Such illusionary tilts in turn would conflict with visual and other information and possibly generate motion sickness [47]. Especially the spinning movements, which are induced by microgravity in some individual fish are assumed to indicate a possible source of illusionary tilts: individually asymmetric vestibular maculae (e.g., based on side to side differences in the weight of otoliths) might, at rest and under 1 ×g earth gravity, possibly cause asymmetric shearing forces on the sensory epithelia. A normal posture then would require neurovestibular compensation for the asymmetric discharge rates. At microgravity, however, there would be no weight differences in the otoliths from one side of the body to the other, and the (primary) discharge rates should no longer be asymmetrical. A continuing, but now unnecessary compensation, however, would then cause erroneously (secondary) asymmetric discharge rates.

It is the particular strength of this "asymmetry-hypothesis" [48] based on [49], that it could explain the great interindividual differences in kinetotic behavior of fish or even in motion sickness susceptibility of humans.

Several findings speak in favor of the asymmetry-hypothesis (Fig. 20.11): indeed, asymmetrically weighed otoliths are a common feature in fish (for review [43, 50]). Larval cichlid fish, who were hatched at 3 ×g hypergravity, morphogenetically developed normally and showed a normal swimming behavior. However, as soon as the hyper-g-centrifuge was stopped, many individuals revealed looping responses and spinning movements, as had also been frequently observed in the course of space-flight- and parabolic aircraft flight-experiments after the transfer from 1 ×g earth gravity to microgravity conditions [45, 46, 51, 52]. This kinetotic behavior normally disappears within several hours or days like the so-called space adaptation syndrome of humans [19, 51, 52]. As predicted by the asymmetry-hypothesis, kinetotically behaving individual fish larvae after hyper-g and at microgravity in the course of para-

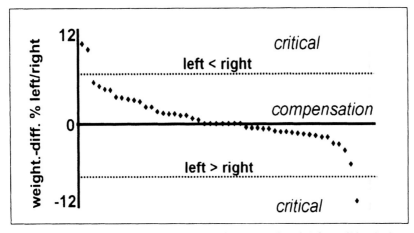

Fig. 20.11 Otolith asymmetry (weight differences between left and right otoliths) in Swordtail Fish, which were maintained in a normal aquarium. According to the "asymmetry hypothesis", a moderate range of asymmetry can be compensated for by the brain (compensation). Asymmetries on a level beyond the respective compensation capabilities (critical), however, might induce kinetotic behavior at altered gravity.

bolic aircraft flights indeed revealed a larger otolith asymmetry than normally behaving ones [48, 52].

Recently, evidence has been provided for the existence of a feedback mechanism adjusting size, asymmetry and Ca-content of fish inner ear otoliths towards altered gravity [48, 53]: both in the hypergravity- (hg-) animals and in the 1 ×g controls, the morphogenetic development was identical. Evaluating the otoliths´ growth during hg in comparison to the 1 ×g controls, however, it was found that the growth both of utricular and saccular stones (lapilli and sagittae, respectively) was slowed down by hg and that the development of bilateral asymmetry was considerably decreased.

This can be explained from a functional point of view as follows: under increased gravity, given otoliths will be heavier and thus cause increased shearing forces during tilts. However, any tilts need to be computed by the brain and asymmetric shearing forces need to be compensated (neuro-vestibular compensation). A feedback mechanism thus might slow down otolith growth at hg, so that an otolith formed at hg will possibly cause about the same shearing forces during tilts as a respective normally sized otolith at 1 g. Without a feedback mechanism, a given bilateral asymmetry between two otoliths would be increased at hg, but for a correct interpretation of the afferent inputs to the brain for postural control and spatial orientation, any asymmetry ought not to be too pronounced in order to stay in the range of the compensation.

Therefore, the feedback mechanism mentioned above might reduce a hg-based increased functional asymmetry in order to fulfill the requirements for a sufficient compensation. Moreover, the feedback ought to be activated immediately after the onset of altered gravity. Indeed, recent investigations showed that otolith growth is obviously regulated by the brain: Ca-incorporation (and thus growth) stops after transection of the vestibular nerve [54].

Summarized, the basis of a disorientated (kinetotic) behavior in individual fish at a sudden reduction of the g-force is relatively well understood concerning the level of swimming behavior and concerning morphometrical studies on otoliths. The neuronal mechanisms, however, that possibly underly the adaptation to altered gravity by means of a feedback mechanism, remain so far poorly understood.

General neurochemical effects due to altered gravity abound (for review [51]). Respective investigations yielded the general finding that, e.g., enzyme activities are affected by altered gravity (from approx. 0.01 $\times g$ in the fast rotating clinostat to 3 $\times g$ in the centrifuge) in a dose dependent manner and that the effects are especially pronounced in larval fish as compared to adult ones [55]. This latter finding, according to which larval fish may react gross-biochemically to a larger extent to altered gravitational forces than adult animals may be due to their wider range of neuronal plasticity. Histochemical and electronmicroscopical investigations focused on single brain centers showed that special inner ear related nuclei in the brain react specifically towards altered gravity [56, 57], even on morphological level (form and number of synaptic contact zones [58, 59]).

Taken together, the results of these investigations suggest that the excitatory inputs from the inner ear are resembled in the vestibular system of the brain by its neuronal activity. Effects of altered gravity on the plasticity of the inner ear sensory epithelium were by far less pronounced [60].

Taken together, investigations of fish, who had developed at altered gravity, reveal a variety of short lasting, adaptive effects following Δg. The behavioral adaptation to Δg can be correlated with a compensatory otolith growth, based on a (negative) feedback mechanism between the inner ear and the brain. The neuronal control of this feedback mechanism seems to be effected by stimulation dependent enzyme activities (i.e. neuronal activity) as well as synaptic plasticity, especially in vestibular brain centers.

20.4 Conclusion

During evolution of animals, they have managed both to cope with and use the environmental gravity vector for orientation. Mechanisms concerning the latter (function of gravireceptors and the central nervous computation of respective inputs) is comparatively well investigated. Yet, however, it is not completely clear, to which extent gravity is necessary for a normal development and in which range adaptive mechanisms are efficient enough at altered gravity to guarantee a normal lifestyle.

The various data being obtained using especially vertebrates as model systems concerning gravistimulated effects reveal first of all that the normal development seems not to be significantly influenced by altered gravity. The studies undertaken so far, however, do not yet answer the question whether a complete life cycle can be completed for instance under weightlessness. Secondly, the results regarding gravity-effects, particularly on fish, speak in favor of the following concept of possible interactions: short-term altered gravity (up to around 1 day) can induce transitional aberrant behavior due to malfunctions of the inner ear, originating from asymmetric otoliths or, generally, from a mismatch between canal and otolith afferents. The vanishing aberrant behavior is due to a reweighing of sensory inputs and neurovestibular compensation, probably on bioelectrical basis. During long-term altered gravity (several days and more), step by step neuroplastic reactivities on molecular basis (i.e. molecular facilitation) in the brain and inner ears possibly activate feedback mechanisms between the CNS and the vestibular organs for the regain of normal behavior. It was shown that such a mechanism is involved in adjusting (assimilating) the otolith weights in fish.

The experimental data on the effect of altered gravity on developing and adult animals help understanding the role of gravity during evolution. The consequences for animal life at altered gravity are well in concordance with a concept about an evolution of the gravi-resistance and gravi-reactivity of animals of the known Ukrainian space physiologist N. Sirotonin (1896-1977): *There is a high tolerance to altered gravity in arthropods (especially insects), the presence of responses in fish and amphibians mainly in most early developmental stages and a wide range of the responses in birds and mammals* [61]. The reasons for this varying range of gravity-tolerance have hitherto not been disclosed.

Acknowledgements. The present study was kindly supported by a grant from the German Aerospace Center (DLR) e.V. (FKZ: 50 WB 9997).

20.5 References

1   M. Conrad, Biosystems **42**, 177 (1997).
2   P. Moroz, J. Theor. Biol. **107**, 303 (1984).
3   I. Vinnikov, Aviakosm. Ekolog. Med. **29**, 4 (1995).
4   A. DuBois, G. Cavagna, R. Fox, J. Exp. Zool. **195**, 223 (1976).
5   H. Lillywhite, J. Exp. Zool. **275**, 217 (1996).
6   K. Kirsch, H.-C. Gunga, in: H. Rahmann, K. Kirsch (Eds.) *Mensch-Leben-Schwerkraft-Kosmos: Perspektiven biowissenschaftlicher Weltraumforschung in Deutschland,* Günter Heimbach Verlag, Stuttgart, 2001, pp.210
7   K. Kirsch, in: P. Sahm, M. Keller, B. Schiewe (Eds.) *Research in Space: The German Spacelab Missions,* Wissenschaftliche Projektführung D-2, Aachen, 1993, pp. 187.
8   H.-C. Gunga, K. Kirsch, F. Baartz, L. Röcker, Naturwissenschaften **82**, 190 (1995).
9   T. Kimura, Folia Primatol. (Basel) **66**, 126 (1996).
10 C. Pennycuick, J. Biomech. **29**, 577 (1996).
11 C. Sebastian, K. Eßeling, E. Horn, Exp. Brain Res. **112**, 213 (1996).
12 A. Schatz, R. Reitstetter, A. Linke-Hommes, W. Briegleb, K. Slenzka, H. Rahmann, Adv. Space Res. **14**, 35 (1994).
13 W. Hanke, Adv. Space Res. **17**, 143 (1996).
14 H. Machemer, Adv. Space Res. **17**, 11 (1996).
15 D. Neugebauer, H. Machemer, Cell Tissue Res. **287**, 577 (1997).
16 M. Wiederhold, Brain Res. **78**, 490 (1974).
17 M. Wiederhold, Annu. Rev. Biophys. Bioeng. **5**, 39 (1976).
18 A. Hudspeth, P. Gillespie, Neuron **12**, 1 (1994).
19 G. Clément, A. Berthoz, Proc. Eur. Symp. Life Sci. Res. Space **5**, 333 (1994).

20 D. Moore, P. Bie, H. Oser (Eds.) *Biological and Medical Research in Space,* Springer, Berlin, Heidelberg, 1996.

21 D. Sieving, Aviat. Space Environ. Med. **68**, 346 (1997).

22 W. Becker, J. Marxen, M. Epple, O. Reelsen, J. appl. Physiol. **89**, 1601 (2000).

23 H.-J. Marthy, Proc. Eur. Symp. Life Sci. Res. Space **5**, 177 (1994).

24 H. Schatten, A. Chakrabarti, M. Taylor, L. Sommer, H. Levine, K. Anderson, M. Runco, R. Kemp, Cell. Biol. Int. **23**, 407 (1999).

25 H. Pedrozo, Z. Schwartz, M. Luther, D. Dean, B. Boyan, M. Wiederhold, Hearing Res. **102**, 51 (1996).

26 E. Horn, H. Agricola, S. Böser, S. Förster, G. Kämper, P. Riewe, C. Sebastian, Adv. Space Res., in press.

27 E. Horn, C. Sebastian, J. Neubert, G. Kämper, Proc. Eur. Symp. Life Sci. Res. Space **6**, 267 (1996).

28 R. Gerzer, P. Hansson (Eds.) *The Future of Humans in Space*, Proc. 12th IAA Man in Space Symposium in Washington, D.C., U.S.A., June 1997, special issue of Acta Astronautica, Elsevier, Oxford, 1998.

29 A. Berthoz, A. Güell (Eds.) *Space Neuroscience Research*, Proc. of a workshop held in Paris, France, April 1997, special issue of Brain Res. Rev., Elsevier, Amsterdam, 1998.

30 A.-M. Duprat, D. Huson, L. Gualandris-Parisot, Brain Res. Rev. **28**, 19 (1998).

31 R. Wassersug, M. Yamashita, Adv. Space Res. **25**, 2007 (2000).

32 G. Ubbels, ESA Microgravity News **2**, 19 (1988).

33 K. Souza, S. Black, R. Wassersug, Proc. Natl. Acad. Sci. U.S.A. **92**, 1975 (1995).

34 L. Gualandris-Parisot, S. Grinfield, F. Foulquier, P. Kan, A.-M. Duprat, Adv. Space Res. **17**, 265 (1996).

35 C. Aimar, A. Bautz, D. Durand, H. Membre, D. Chardard, L. Gualandris-Parisot, D. Husson, C. Dournon, Biol. Reprod. **63**, 551 (2000).

36 S. Mori Nagoya J. Med. Sci. **58**, 71 (1995).

37 V. Sabo, K. Bod'a, V. Chrappa, Vet. Med. (Praha) **40**, 147 (1995).

38 T. Suda, Bone 22 (Suppl.), 73S (2000).

39 J. Barrett, D. Wells, A. Paulsen, W. Conrad, J. Appl. Physiol. **88**, 1614 (2000).

40 H. Rahmann, K. Slenzka, Proc. Eur. Symp. Life Sci. Res. Space **5**, 147 (1994).

41 K. Ijiri, The First Vertebrate Mating in Space, RICUT, Tokyo, 1995.

42 E. v. Holst, Pubbl. Staz. Zool. Napoli **15**, 143 (1935).

43 S. Watanabe, A. Takabayashi, M. Tanaka, D. Yanagihara, in: S. Bonting (Ed.) *Advances in Space Biology and Medicine*,Vol. 1, JAI Press, Tokyo, 1991, pp. 99.

44 R. Hilbig, T. Schüle, M. Ibsch, R. Anken, H. Rahmann, Proc. C.E.B.A.S. Workshop Conf. **12**, 51 (1996).

45 R. v. Baumgarten, G. Baldrighi, G. Shillinger, Aerospace Med. **43**, 626 (1972).

46 H. DeJong, E. Sondag, A. Kuipers, W. Oosterveld, Aviat. Space Environ. Med. **67**, 463 (1996).

47 R. v. Baumgarten, Exp. Brain Res. **64**, 239 (1986).

48 R. Anken, T. Kappel, H. Rahmann, Acta Otolaryngol. (Stockh.) **118**, 534 (1998).

49 R. v. Baumgarten, R. Thümler, Life Sci. Space Res. **17**, 161 (1979).

50 R. Anken, M. Ibsch, D. Bremen, R. Hilbig, H. Rahmann, Proc. C.E.B.A.S. Workshops **11**, 73 (1995).

51 H. Rahmann, K. Slenzka, R. Anken, R. Appel, J. Flemming, R. Hilbig, T. Kappel, K. Körtje, G. Nindl, U. Paulus, in: P. Sahm (Ed.) *Scientific Results of the German*

Spacelab Mission D-2, Wissenschaftliche Projektführung D-2, Aachen, 1995, pp. 621.

52 R. Hilbig, R. Anken, G. Sonntag, S. Höhne, J. Henneberg, N. Kretschmer, H. Rahmann, Adv. Space Res., in press

53 R. Anken, K. Werner, J. Breuer, H. Rahmann, Adv. Space Res. **25**, 2025 (2000).

54 R. Anken, E. Edelmann, H. Rahmann, NeuroReport **11**, 2981 (2000).

55 K. Slenzka, R. Appel, V. Seibt, R. Anken, H. Rahmann, Proc. C.E.B.A.S. Workshop Conf. **10**, 42 (1994) .

56 R. Anken, K. Slenzka, J. Neubert, H. Rahmann, Adv. Space Res. **17**, 281 (1996).

57 R. Anken, H. Rahmann, Adv. Space Res. **22**, 281 (1998).

58 H. Rahmann, K. Slenzka, R. Hilbig, J. Flemming, U. Paulus, K. Körtje, A. Bäuerle, R. Appel, J. Neubert, W. Briegleb, A. Schatz, B. Bromeis, Proc. Eur. Symp. Life Sci. Res. Space **5**, 165 (1994).

59 R. Anken, R. Hilbig, M. Ibsch, P. Vöhringer, H. Rahmann, Proc. Eur. Symp. Life Sci. Res. Space **7**, 124 (1999).

60 G. Nindl, K. Körtje, K. Slenzka, H. Rahmann, J. Brain Res. **37**, 291 (1996).

61 L. Serova, Aviakosm. Ekolog. Med. **27**, 15 (1993).

Part V

Complexity and Life

21 Scaling Phenomena and the Emergence of Complexity in Astrobiology

Juan Pérez-Mercader

The history of the Universe started about 14.5 Ga ago, when the period known as the Big Bang took place. Its evolution in space and time went on, and in the region where we live, about 3.5 Ga ago, the first manifestations of the phenomena we call Life were already taking place. Today living things (human beings) are capable of asking meaningful questions about the origin, evolution and even the future of the Universe and its organization [1], all the way from apparently less "complicated" structures to structures as complex as living organisms and ecologies (Figs. 21.1 and 21.2). The science that describes such interpolation is Astrobiology. Perhaps, Astrobiology will be for Biology in the future what Cosmology is for Astronomy today.

In the beginning the original components of the Universe interacted to give rise to the nuclei of the light primeval elements (H, He, Li and some of their isotopes) during the first three minutes [2]. It took until the Universe was 300,000 years old for its size to be large enough so that the free electrons were captured by these nuclei and their atoms formed. This is the era of de-coupling between matter and radiation which led from domination by radiation to a matter-dominated Universe. In some regions of the Universe the energy density was slightly superior to others and matter tended to accumulate there˙ (remember that in general mass is equivalent to energy), thus giving rise to huge structures that in a rather short period of time fragmented and became super-clusters of galaxies. In the galaxies the primeval gas atoms accumulated in some large enough regions, and stars began to form that due to the compression effects of self-gravity became gigantic nuclear furnaces where the heavier elements began to "cook". Sometimes, the materials were reprocessed inside other stars that eventually formed and even heavier elements, in turn, formed. Clouds of gas and dust collapsed and fragmented inside galaxies, giving rise to associations of gravitationally bound structures such as globular clusters or open star clusters. In other locations the gas and dust accreted into rotating blobs which flattened into discs out of which planetary systems arose: this must have had happened where we live about 4.6 Ga ago. For reasons that we do not understand in detail, the material in the cloud, out of which the Solar System formed, organized itself into a huge gas ball (Sun) and into rocky planets (such as

---

˙ The "top-down" scenario for structure formation in the Universe will be described: from super-clusters of galaxies to galaxies and stars. There is a reversed scenario with the smaller structures forming first and then accreting into large structures: the "bottom-up" scenario. It still is not known which of these two scenarios actually took place.

Fig. 21.1 Iconic representation of the evolution of the Universe: from the Big Bang to the emergence of DNA/RNA. Major events such as the formation of galaxies, the cooking of the heavy elements in supernovae explosions, the formation of planetary systems, separation of planets into gas giants and rocky planets, the presence of carbon molecules in the interstellar gas, or the emergence of DNA are all represented here. This picture does not show the evolution of life on Earth. Each of the major events represented in the picture is known to have been associated with its power law phenomenology; why this is so is known in some cases, in other cases it remains unknown. Time runs from the upper right corner (the Big-bang) to the origin of planetary systems and to the formation of DNA (lower right) on Earth. Taken from the NASA Jet Propulsion Laboratory website.

the Earth or Mars) and gas giants (for example Jupiter or Saturn) [3]. After its formation 4.56 Ga ago, the Earth must have been very hot, with additional heat being released due to the disintegration of radioisotopes in the original material. Probably large quantities of water vapor and other volatiles where captured around the planetary bodies [4]. As the Earth cooled and its surface was less subject to impacts from large-size bodies which were orbiting around the Sun, the conditions became such that chemical evolution of molecules could take place, membranes enclosing chemical reactors arose and by about 3.9 Ga to 3.5 Ga ago living objects in the form of pro-karyotic cells were already present.

The chemical and biological evolution of these cells took place as a chemical/biological response to the changing environmental conditions on the young planet Earth. More complex structures emerged as a consequence of the adaptive evolution of these living agents, so that new species populated the various available niches [5]. Eventually, man and many other complex systems arose on this planet [6].

Why all this happened, and more specifically the events leading to life, is the subject matter of Astrobiology, which tries to test wether "Life is a cosmic imperative" [7].

Fig. 21.2 Continuation of Fig. 21.1 into the evolution of life on Earth. Taken from the NASA Ames Research Center website.

In the history that we have just described we note several important regularities that have taken place. There is a pattern of evolution in time and in space; accompanying this pattern, there is a series of events that can be identified. These events mark changes in the local state of the Universe where a transition has clearly taken place from one state (for example, a gas and dust cloud) to another qualitatively different state (for example a planetary system). In addition, we observe that these events take place at similar scales: no planetary system has the size of a galaxy, just as there are no ants having the size of a dinosaur. Galaxies of each type are roughly equivalent in size and mass, ants of the various families are roughly equal in size and mass, and so on [8]. Finally, among the main patterns we can identify a systematic presence of systems within systems, within systems: planetary systems, within galaxies, within clusters of galaxies, or bases within DNA molecules, within chromosomes, within cell nuclei, within cells, etc .

These patterns have to do with phenomena known as "emergence", "scaling" and "hierarchies" [9]. All of them are part of what is now known as "complexity science" or "complexity theory". Complexity theory tries to identify non-obvious patterns of self-organization in nature that occur in complex systems, i.e. in systems where "the whole is more than the sum of its parts". Thus, trying to understand if life is a cosmic imperative, that is, that it emerges everywhere in the Universe where there is an opportunity for chemical evolution, requires that we ask these questions within the framework of "complexity theory", which becomes the basic analytical tool for describing, understanding and unifying astrobiological phenomena.

21.1 Astrobiology as the Realm of Emergence, Hierarchies and Scaling

When observing the evolution of the Universe we see that we are dealing with "complex systems". That is, systems made up by many parts (components) in interaction with each other (the parts) and between the systems themselves. As a result of the interaction among the components of the system, the behavior of the system as a whole is "more than the sum of its parts". More specifically, this means that given the properties of the individual parts, the laws of their interactions and the boundary conditions (including the nature of the environment) to which the system is subject, the behavior of the whole cannot be predicted by a mere simple superposition. The system can thus not be investigated by the usual reductionist approach so common in science. The properties of the whole "emerge" [9, 10] because the complex system has access to new states which a collection of simple systems could not reach.

For these systems, notions such as "emergence", "evolution", "transition", "organization" and "collective behavior" are commonplace and have a very real meaning. How does this happen? By describing with words how this takes place, we will uncover a very deep connection between the emergence of complexity, hierarchies and scaling behavior. Although a lot is already understood about this connection not all details are worked out, and they constitute a very active area of research [11-14].

Let us begin by considering a system made up of just a few parts (which we will consider to be simple, i.e. non-divisible) and which are weakly aware of each other: that is, they are diluted and weakly interacting. For such a system the whole may be thought of as a superposition of the parts: nothing but additivity is relevant in this uncomplicated system. However, if we now increase the number of parts to the point where we no longer can enumerate them separately, new situations come up as a consequence of the rise in relevance of the various interactions (largely due to the nonlinear character of the interactions) among the components associated with the increase in the number of parts. By doing this, it may come to a point where the number of components is so large that we cannot keep track of each individual component, and we need to "average" in some way; perhaps by introducing "effective components" as a means of keeping track of the increase in complications. An "effective component" would then be a selected association of the components. At some point in this process, we will also have to attempt to consider some effects that we observe in the system but whose exact cause we cannot accurately and uniquely pin-point because they are maybe, the consequence of processes taking place at scales smaller than our resolution scale. These latter effects we will classify as "noise", so that at this stage we will have to introduce a statistical description of the system which, in turn, is dependent upon our resolving power or observation scale [15].

Several features begin to become prominent in a complex system: the presence of many individual (or "simple" at some level of description) components, stochastic effects which we cannot describe precisely but which we can characterize as "noise" with some statistical properties and the "environment" within which the complex system is submersed. In addition, interactions absent when only two-bodies (or just a few bodies) were present may begin to develop activity as we increase the number of components and as non-linearity begins to dominate the system. These features: many components, non-linear interactions, stochastic effects, interaction with (and in response to) the environment are the hallmark of a complex system [16, 17]. Such systems can display qualitatively new behaviors depending on the circumstances and even their history. These behaviors are called "emergent behaviors", and although their precise description is not well understood, we are now in the process of arriving at such a description using "complexity" theory [9].

As is clear, emergent behavior is accompanied by a change of state in the system. Before the system emerged it was in a state different from the one it was before it actually emerged into the new state. Such transitions between states occur in a dynamical fashion, i.e. they take place in the course of time [18, 19]. Some times the system fragments, breaks-up, in the course of these transitions; other times the system remains whole but changes its properties, and in yet other systems, the system aggregates with other systems and gives rise to larger entities. It is an observational fact that in this process of emergence, fragmentation / coagulation many of the key properties of the system follow what is mathematically known as a "power-law" or "scaling behavior". Furthermore, many times the fragmentation process is such that within the apparent ensuing disorder there is a "hierarchical" order, according to which the system breaks-up into autonomous subsystems which, in turn, break-up into subsystems which are made-up of further sub-systems until some "elementary" system level is reached (see Fig. 21.3) [20].

There is then a relationship between Emergence, Hierarchies and Scaling behavior. This scaling behavior has been detected for all of the major transitions that have taken

Fig. 21.3 The popular Russian craft "matryushkas" provide a good example of a hierarchical system. In this example each hierarchy level corresponds to a particular doll. Note also that the various dolls can be very different in their individual details.

place in the evolution of the Universe such as was described in the introduction; in addition, in many transitions in the history of Life on Earth, where this class of analysis has only recently been introduced, we are also beginning to detect the presence of power law (or scaling) behavior associated with emerging behaviors. This should not cause any surprise for, as it has been known in physics for almost a century, transitions between states (phase transitions or critical phenomena) where emergence is at work, are characterized by power law (scaling) phenomenology. It took until the 1960's and 70's to explain how these took place in condensed matter physics, where a tool called "Renormalization Group" was essential to understand critical phenomena [21, 22].

But, "what is a power law?" A power law is a particular mathematical relation where an observable property of a system, say $N(m)$, written in terms of a variable m, is given by Eq. 21.1:

$$N(m) \propto m^{\alpha} \tag{21.1}$$

that is, property $N(m)$ is proportional to m raised to the power α. Here $N(m)$ could be the number of objects in the system with a property of value m, or any other relation between measurable properties; for example $N(m)$ could be the correlation between two objects separated by a distance m. The quantity α is called the exponent of the power law [18, 23].

A power law distribution of a quantity in terms of another is highly non-trivial and its presence indicates emergence, criticality and the presence of collective behavior, as has already been mentioned. Furthermore, as will be described in more detail below, the value and nature of the exponent in a power law is of fundamental importance for, at least, attempting to identify the nature of the phenomenon that generates such behavior.

Experimentally one knows of power laws with integer, real and complex exponents. Each has its own twist. For example, all known fractal behavior corresponds to α being a real number, but understanding lies in being able to calculate the value the exponent by applying reasoning based on first principles. It took 50 years to clarify this in the case of phase transitions and critical phenomena [24, 25]. In astrobiology, with the variety of phenomena and processes that underlie the known power laws and the many that are being discovered, it is a monumental challenge to make sense of them. But such progress will help greatly in improving our understanding of life and possibly its cosmic nature.

Phenomenologically, power laws are detected for example when studying the statistics of some property of a system for which a time series is available, or when one is studying some property of the system as a function of some control variable. Power laws are "seen" as a straight line in a log-log plot; the slope of the line is the exponent in the power law. However, much caution needs to be exercised: double logarithmic plots tend to smooth out variations in the data and one may be tempted to identify as power law behavior some kind of behavior which in reality is not. This calls for the use of multiple statistics and for extensive testing. Realistic power law behavior tends to hold over several orders of magnitude, and this is something that one can take as an indication that a true scaling phenomena is present in the phenomena being analyzed. This word of caution is particularly important when trying to discriminate power-law behavior in a probability distribution function (pdf) from log-normal behavior. Due to the properties of the exponential function, the log-normal distribution can mimic a power law distribution over a non-trivial range of the independent variable, and what in actuality it is a log-normal distribution could be mistaken by a power law with an exponent made-up of a constant part and a slowly varying function of the independent variable [26].

Note also that for power law distributions, large, or "catastrophic", events are much more frequent than for gaussian or poissonian distributions, where these events are exponentially suppressed.

21.2 Implications of Power Law Behavior

Many of the features of systems described by power law behavior follow from a simple (and important!) mathematical property of a power law: self-similarity or scale invariance. This is clearly seen and understood from its mathematical form. In fact, denoting the proportionality constant in a power law by A, we obtain Eq. 21.2

$$N(m) = A \times m^{\alpha} \tag{21.2}$$

The proportionality constant can be re-written as Eq. 21.3 and we obtain Eq. 21.4

$$A = m_0^{-\alpha} \tag{21.3}$$

$$N(m) = (m/m_0)^{\alpha} \tag{21.4}$$

and m_0 can be interpreted as a reference value for m, i.e. where we normalize $N(m)$ to be 1, that is $N(m_0) = 1$. Performing a change of scale in the system, i.e. choosing a different set of units for m (and m_0), by using λm and λm_o instead of m and m_0, we see that

$$N(\lambda m) = (\lambda m/\lambda m_0)^{\alpha} = (m/m_0)^{\alpha} = N(m) \tag{21.5}$$

That is $N(m)$ remains unaffected by the scale change on m; furthermore, since the above is true for any value of λ, we see that $N(m) = A\, m^{\alpha}$ is scale invariant (or self-similar).

The above is all the mathematics one needs in order to understand the gross features of systems with power law behavior.

A first observation one can make is that these phenomena are such that many different scales in the system must be involved in producing power law phenomenology. Hence, we are dealing with a system where its many components are in interaction and share in the dynamics; both, the short distance (short time) and the long distance (long time) behavior of the system are somehow involved when power laws are detected. The small scales (fast variables) and the large scales (slow variables) participate in the phenomenon which in this way becomes a collective or cooperative phenomenon. Hence we see right away the connection with the complex behavior of many-component systems. Furthermore, we also infer immediately that short distance inter-actions can effectively lead to a long range correlation, as happens in the case of phase transitions and critical phenomena.

The above properties are seen in a variety of behaviors, some of which are under-stood in great detail while others are still poorly understood. The first place where power law phenomenology was seen in the context of physical phenomena was in phase transitions, where the specific heat of the material undergoing the transition, taken as a function of the deviation of temperature from the critical temperature, goes like a power law. A thorough understanding of this class of phenomena took more than half a century, and the classes of critical phenomena have increased to dynamical critical phenomena and to phenomena beyond what was traditionally understood as condensed matter, such as biological or astrophysical phenomena.

Another realm of phenomena where power law behavior is known to occur is within "self-organized" systems. For reasons which one did not at all understand in general, these systems go into self-organized niches while displaying scale invariance [27, 28]. It is as if long-range correlation and short distance interaction conspired to give rise to a new collective system made-up of many "elementary" components. Sometimes, these transitions are equivalent to changes in complexity in the system, and which final state the system goes into is a function of the initial values in the pa-rameters that characterize the system: such changes can often be viewed as examples of emerging behavior or of "emergence". In yet other classes of systems what happens is that the "landscape" (i.e. the environment on which the system exists) matches bet-ter certain ranges of values of the parameters (such as viscosity) and there is an evolu-tion in their values which amounts, to driving the system into different emerging states, giving rise to the category of phenomena known as "Complex Adaptive Sys-tems" [10], so typical of living systems. For these systems feedback plays an essential role, but the generalities of how these mechanisms actually work are far from being understood.

During the process of self-organization and the concomitant long-range correlation, there is the possibility that the system dynamics is such that it breaks-up (fragments) into smaller subsystems or that it adheres (accretes) more components. This, again, gives power law phenomenology. It is equally possible that the initial conditions on the values of system parameters or that evolution by adaptation lead the system into power law behavior with a complex exponent; in this latter case the system develops a hierarchical character associated with the log-periodic behavior of correlation func-tions [29]. Furthermore, it can be shown that the phenomena of "punctuation" and "Devil's staircases" [27] so important in evolution and adaptive behavior also emerge in the presence of power laws with complex exponents. The observed "lumpiness" of systems within a hierarchical structure is also generated for power laws with this class of exponents [30].

21.3 Exponents and Their Meaning

The exponents in a power law have very deep significance. Not only because de-pending on their nature the system follows one behavior or another, but also because they are a direct reflection of the geometrical context (e.g., bulk density is proportional to spatial dimension raised to –3) in the simplest cases, and because in general they are indicators of a concurrence of the regular (deterministic) and the random (fluctuating) components present in all physical systems. Quite generally, an exponent is made up of the additive contributions of a geometrical part (or canonical dimension) and a dynamical part with a deterministic piece on which the effect of fluctuations is superimposed. Out of the competition between the various pieces of the exponents in a variety of circumstances there results an effective value for the exponent, which is why several emergent behaviors for a single system can happen; it is the task of the experimentalist to actually identify the dominant component, and the task of the theoretician to identify mechanisms based on first principles which can explain the relative contributions identified by the experimentalist.

Given a phenomenon, power law behavior of the pdf percolates into all parameters in the model. Different properties of the system may show apparently unrelated values for their exponents, but their common origin (from the pdf) manifests in the existence of relationships between the exponents and therefore their values. This observation plays an essential role in the classification of phenomena and the grouping together of seemingly unrelated phenomena. Through these relationships phase transitions have been classified, and out of it has come to light the notion of "universality" [21, 24]. Phenomena such as for example magnetization and the condensation of a fluid which physically are so different have been shown first experimentally and then theoretically, to have power law behavior with exponents obeying certain relationships, but also with identical values for the exponents. The property underlying such behavior is "universality" and it translates into the tendency for the local and short distance details of a system to become less and less relevant as the system approaches the transition to criticality or self-organized behavior.

In fact, it turns out that it is the dimensionality of space or the number of neighbors involved in the interaction that end up really playing a role in the critical behavior of the system. More specifically, if the transition between states is characterized by some "order parameter" (such as a difference in density) which goes to zero at the transition, the value of the exponents in these phenomena depends on the dimensionality of space and in the symmetry properties of the "order parameter". Since all that matters is the symmetry properties of the order parameter and not its details, one can see how very different phenomena end up being classified in a particular universality class, for the detailed properties of the order parameter are not relevant. The existence of universality and universality classes is a very powerful practical tool for the study of apparently unrelated phenomena: once the appropriate relations between exponents have been found, and the phenomenon classified as belonging to a certain universality class, and at least a member system in the universality class is well understood, then one can hope to describe and understand the properties of all systems in the universality class.

A typology of scaling phenomena and their general classification does not yet exist, but astrobiology is a prime candidate field to benefit from such a classification, since the many power laws observed in astrobiology would fall into place, and their relationships would be clarified.

21.4 Where and How Have Power Laws Been Used?

As we have mentioned above, power law behavior has many uses in the sciences (see below for specific examples). One of the first places where scaling behavior was discovered was in the physics of phase transitions, where the coexistence (of the various phases of a fluid near its critical point) curves (a plot of T/T_c vs. ρ/ρ_c) of many fluids were shown to collapse into a single curve under simple rescaling of the temperature and density into the dimensionless variables T/T_c and ρ/ρ_c, with T_c the critical temperature and ρ_c the critical density. Since all the data points for all the substances studied collapse into a single curve, this means that all the materials can be described by a single, universal exponent. The materials ranged from noble gases to carbon monoxide and methane, and it was a completely unexpected result that all curves collapsed under rescaling. This opened the way for a revolution of the detailed understanding of the physics of phase transitions [21, 24].

In the field of elementary particle physics the tool needed for understanding scaling was first developed. This tool is the "renormalization group" (actually a semi-group!) and it was in this field where the uses of scaling and scale-invariance in connection with a fundamental understanding of the underlying dynamics were first put to use [31]. The discovery of "colored" quarks and of the nature of the strong force was a direct consequence of understanding the connection between scaling, fluctuations and fundamental dynamics in particle physics.

Of essential importance in polymer chemistry, power laws have been exploited to increase our understanding of the fascinating world of polymers [32]. There, a generalization of a quintessential scaling phenomenon, Brownian motion, to the case of self-avoiding random walks has led to a huge understanding of the processes involved in polymer chemistry and physics.

More recently, power laws have been seen to play important roles in helping to understand the nature of the large scale structure of the Universe [33-35], allometries in biology [36] and even to understand the evolution of ecologies [37]. In some cases a measure of detailed understanding is available, but in most examples there is little or no fundamental understanding of the ubiquity of power law behavior. In the following we give a brief discussion of some selected examples of power law behavior that range from astrophysics to ecology.

21.5 Examples of Power Laws: From Cosmology to Biology

Power law behavior is phenomenologically well documented in many fields, but understanding its origins is, often, a formidable challenge, as in the case of Kolmogorov scaling in the inertial regime of turbulent flows [38]. Furthermore, the number of new phenomena displaying some form of power law behavior is simply staggering, which makes their understanding even more pressing. The choice of power law phenomena that will be described below is a very small sample of what is known, and the intention is only to give a brief introduction for astrobiologists; we will go from very large scales to smaller scales, from apparently less complex to more complex systems.

21.5.1 Examples from Cosmology and Astrophysics

In cosmology, the power spectrum of density fluctuations is known to obey a composite power law: one at small moment (corresponding to very large spatial coherence scales typical of the epoch of de-coupling) and another at larger moments typical of the size of galaxies. The power spectrum for small moments k, goes like $k^{+1.0}$ and at large moments (corresponding to the sizes of galaxies) goes like $k^{-1.65}$ (see Fig. 21.4). The $\alpha = 1.0$ exponent at small moments has been predicted by inflationary cosmological models and it is the best evidence in their favor; the $\alpha = 1.65 \pm 0.15$ exponent at the scales of galaxies was a mystery for more than a quarter century, but it is now understood in terms of a phase transition in the Universe as it expanded and cooled. Understanding the change in slope (change in α) from 1.0 to -1.65 ± 0.15 is a major open problem which affects our understanding of the transition from the largest scales to shorter scales in the Universe or, equivalently, the transition from a homogeneous to an inhomogeneous Universe [39, 40].

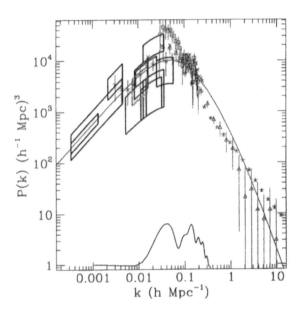

Fig. 21.4 The observed power spectrum for the largest scale structures in the Universe. The power spectrum is related to the Fourier transform of the 2-point correlation function, and a power law power spectrum implies a power law for the correlation function. In simpler language, this implies that at the scales where the power law holds, the Universe is self-similar. The transition from the largest structures to galaxies takes place where the slopes change: this is a major puzzle for cosmology.

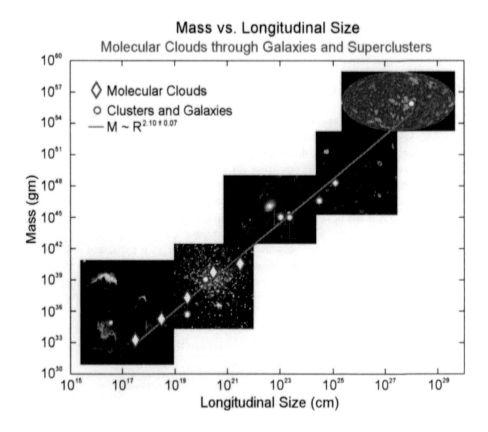

Fig. 21.5 The structures seen in the Universe arrange themselves as hierarchies. It is remarkable that their mass/size relationship can be fitted with a power law; furthermore, both their mass and size can be additionally classified according to a hierarchical law similar to the Titius-Bode law (taken from [41]).

Still in astrophysics, it is known that the mass of objects which range from the largest clusters of galaxies to the smallest open star clusters, is proportional to the longitudinal size of the object raised to $\alpha = 2.10 \pm 0.07$ as seen in Fig. 21.5. This is again a mystery, but there are well founded reasons to believe that this relationship together with the associated fragmentation phenomenology can be understood through the theory of dynamical critical phenomena [41].

At shorter distances and still within the realm of astrophysical phenomena, it is very well documented that the number of clouds of mass in the interstellar medium goes like the mass raised to $\alpha \approx -1.8$. This is another unexplained power law, and it becomes even more mysterious when one notices that assuming either gravitational collapse of the cloud or reorganization of the material due to turbulent flow give the same exponent. This is part of a very vigorous debate of the issues associated with the physics of the interstellar medium [42-45].

Perhaps the "queen" of power laws in astrophysics is the Titius-Bode law of planetary distances, shown in Fig. 21.6, which was discovered in the latter part of the 18th century and still is not understood from first principles [46]. The law states that the distance $d(n)$ to the n-th planet from the Sun is given by the modified power law (Eq. 21.6)

$$d(n) = A \times B^n \times f(n)$$ (21.6)

where $A = 44$ (if distances are expressed in terms of the radius of the Sun), $B = 1.73$ and the function $f(n)$ is an oscillatory function of n with an amplitude of less than 0.1. The $f(n)$ piece can be understood in terms of the chaotic evolution of the primeval solar system. This effect is very small. The large effect of the product $A \times B^n$ is not yet understood from first principles even though many explanations have been offered in terms of various phenomenology; but its origin has remained mysterious for a long time. Recently, however, an explanation has been offered [29] in terms of the hydro-dynamics of the primeval solar system, maximal probability and dynamical critical phenomena.

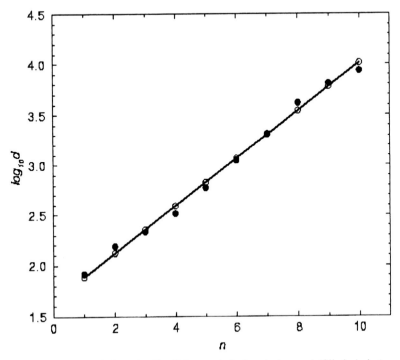

Fig. 21.6 The Titius-Bode law. Predicted (open circles) and observed (filled circles) positions of the planets in the Solar System. Predicted positions (and the straight line) are computed using the power law with $B = 1.73$, as derived without any object in location $n = 5$ (Asteroid Belt) and $n = 10$ (Pluto). The remaining points represent the other planets.

Still within astrophysics, but closer to the scales with which we are more familiar, it is also known that for objects within the solar system the number of impact craters versus their diameter follows a power law [47]. The exponent is not a constant for all the bodies in the Solar System, and has different values for the various bodies. Little, or almost nothing, is understood about the origin of this law, but it is believed that it could hold important clues to answer the question of the nature of the processes that were involved in the origin and early evolution of the Solar System.

21.5.2   Examples from Planetology

In the study of the physical behavior of our planet there are many power laws that have been detected in the phenomenology. We will briefly discuss only three power laws: one associated with rock mechanics, another one that emerges in earthquake phenomenology and a third one that holds in the realm of river geology [48].

When one considers the action of weathering on rocks and the ensuing fragmentation of the rock, one uses the cube root of the fragment volume to define and assign a longitudinal (or linear) size for each fragment. By a sieve analysis one can proceed and count how many fragments have a size larger than a certain value r. It has been known for many years that the number of fragments whose size is larger than r is given by

$$N(>r) \propto r^{-D} \tag{21.7}$$

where $D \approx 2.5$. This power law is known to hold with a similar exponent not only for rock fragments resulting from weathering, but for impact ejecta, rock fragmented in chemical or nuclear explosions and even in laboratory impacts of poly-carbonate projectiles impacting on basalt samples. The power law holds with a constant exponent over a few orders of magnitude. Similar power laws hold for a large variety of fragments produced by a variety of processes, and a certain dispersion in the value of the exponent has been detected which correlates exponent value with the origin and nature of the fragmentation process. No causal mechanism is nonetheless available to explain either the origin of the power law or the value of the exponent.

One of the best known power laws in geophysics is the famous Gutenberg-Richter frequency-magnitude law for earthquakes, Fig. 21.7. Its origin, however, is not understood in spite of decades of effort to unveil the nature of earthquakes and the geophysics underlying these important phenomena. The law is purely phenomenological, stating that the number or earthquakes whose magnitude (a logarithmic quantity) is greater than a certain value M, is proportional to a power of the magnitude. The law applies to individual fault systems and to earthquakes worldwide. The value of the exponent in the power law depends on the definition adopted for the magnitude of the quake, and this leads to difficulties in identifying a dynamical origin not only for the law, but also for the value of the exponent. Because of this there has been a tendency in the specialized literature to search for a more basic observable in earthquake dynamics and to refer to it the regularity expressed by the law. The basic observable that is used is the offset area of fault involved in the quake. More specifically, earthquakes occur, for example, when failure takes place in the stick-slip evolution of a fault or a

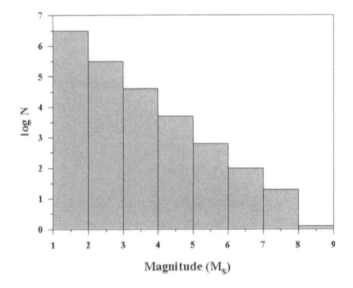

Fig. 21.7 The well known Gutenberg-Richter law relating number of earthquakes to magnitude.

fracture. The area of the slip between fault faces that leads to the quake is now used for the characterization of the Gutenberg-Richter law. A direct relationship with earthquake magnitudes can be established in this way.

The Gutenberg-Richter law has clearly an exponent for earthquakes occurring near the surface and a slightly different exponent for deep earthquakes. This difference in exponents, can be intuitively seen to have its origin in the following arguments: "small" earthquakes (up to magnitude 6) are contained within a volume of the schizosphere and, therefore, are "three-dimensional", on the other hand "large" earthquakes involve fractures with surface ruptures that result from their large size because they cannot be fully contained in the schizosphere, and are therefore "two-dimensional". Critical (or power law) exponents have different values in different number of dimensions. The values for the exponent in the area version of the Gutenberg-Richter law are $b(3) = -1 \pm 0.1$ and $b(2) = -1 \pm 0.2$, with $b(3)$ for "small" and $b(2)$ for "large" earthquakes respectively [49].

Finally we consider an example of a power law that applies to river basins and which, again, is lacking an explanation, this time from geologists. Let us consider the drainage basin of a river, i.e. the surface area containing creeks and rivulets transporting the water that eventually flows into larger rivers, and finally into an even larger river that flows into the ocean. It has been known for a long time that river networks follow a power law in their ordering. To make this explicit, we need to introduce an ordering system to describe the network: streams with no upstream tributaries are called first order streams; the combination of two first order streams defines a second order stream, whereas the combination of two second order streams gives rise to a third order stream and so on. When a lower-order stream joins a larger order stream we call the latter "side tributaries". If we define the bifurcation ratio as

$$R_b = \frac{N_i}{N_{i+1}} \tag{21.8}$$

where N_i is the number of streams of order i, and the length-order ratio R_r as

$$R_r = \frac{r_{i+1}}{r_i} \tag{21.9}$$

then we find that

$$R_b = R_r^D \tag{21.10}$$

where D = 1.8. Note that since river basins are almost two-dimensional systems, this value of the exponent is reasonable (it is close to 2) and the deviation from 2 can be attributed to the mechanisms of water flow in rivers: for short distances the water flows diffusely under the surface or on the surface without cutting streams and leaving big "scars" on the terrain. The precise origin of this exponent value and of the power law itself are far from being understood, however.

21.5.3   Examples from Biochemistry

In biochemistry we also find a huge number of power laws. The vast majority of these remain a mystery. They show-up everywhere, from the basic chemistry of bio-polymers to the properties of cellular membranes, including frequency of introns. Hierarchical behavior also proliferates in biochemistry but practically nothing is known about its origin; even less is known about the connection between morphology, functionality, hierarchy and scaling behavior, all of which are suspected to be linked but little more than that is available.

For reasons of space we will only consider here two power laws: the one associated with introns and one appearing in the case of cellular membranes.

In studying the sequence of bases in segments of chromosomal DNA it is a possible to introduce a version of a random walk which has been called a "DNA walk" [50]. It is a procedure to map the DNA sequence into a form of random walk by assigning a "step-up", in the walk at position i if a pyrimidine (C or T) occurs in position i of the DNA sequence, and assigning to position i a "step-down" if instead a purine (A or G) happens to be at location i. With this very simple definition of a random walk based on the chemical complementarity between pyrimidines and purines it is possible to introduce a means of analysis of correlation in DNA sequences. The resulting random walk when plotted gives rise to a "DNA landscape", Fig. 21.8. Because of the way in which we have established the correspondence (purine/pyrimidine) the random walk is a one dimensional random walk. Analysis of actual data show that the 2-point correlation function for points separated by a distance l in the "DNA walk" goes like l^α with $\alpha \neq \frac{1}{2}$. For uncorrelated random walks one has $\alpha = \frac{1}{2}$, whereas for "persistent" one-dimensional walks $\alpha > \frac{1}{2}$ and for "anti-persistent" walks $\alpha < \frac{1}{2}$. It turns out that for non-coding regions of the human genome $\alpha \sim 0.72$, while for coding regions it is found that $\alpha \sim 0.49$. Further study and analysis of the data shows indications of both some very special (and not understood!) polymer phenomena taking place as well as properties like entropy and redundancy typical of natural languages. Understanding all these phenomena provides a fascinating opportunity for astrobiologists.

Membrane phenomenology is another field where many power laws have been identified [51]. In this area there is a little more guidance than in some of the previously discussed phenomena because the connection between the phenomena and some underlying physical properties is easier to establish. Biological membranes are generally amphiphilic bilayers made up by phospholipids; in addition, they also contain a large variety of different components. In the case of both prokariotic and eukaryotic cells a structure made with protein complexes holds the membrane in place and forms the cytoskeleton. For Archaea the structure of the membrane is different both

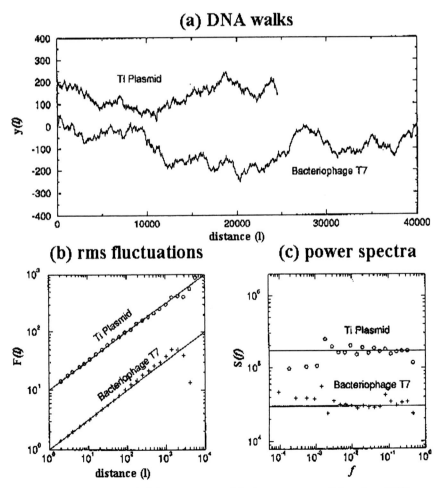

Fig. 21.8 Typical "DNA walks" for some parts of the human genome (taken from [50]).

chemically and structurally from that of Bacteria and Eukarya. As noted by several authors, understanding the origin of this difference is key to understanding the "early organization of Life" on Earth.

Of particular importance in understanding membrane phenomenology is the effect of fluctuations both on single membranes and in collective effects in cells immersed in a fluid. It is well known that fluctuations in the membranes of red cells play a significant role in creating an effective long-range steric repulsion between membranes. It is also well documented that fluctuations are essential in determining membrane stability. In fact the presence of fluctuations gives rise to modifications of the elastic constants of a membrane which manifest in crumpling, surface vibrations and affect all the properties of the membrane, including its "porosity". When the bilayer closes, it forms a vesicle. The membrane (and the vesicle) are held together by using free energy with an important contribution coming from the curvature of the membrane. The vesicle shapes follow from imposing a set of geometric constraints on the free energy of the membrane. The collective presence of these affects can be seen experimentally in the determination of the number of vesicles containing n amphiphilic molecules per unit volume. The size distribution is of the form

$$N(n) \propto n^{-\alpha} \tag{21.11}$$

where the exponent α is given in reference [52]. The origin of the exponent is closely linked to self-organization mechanisms for out-of-equilibrium systems, a formidable physics problem where again physics and complexity theory can be of help to astrobiology, especially in helping determine the differentiation between archea and prokariots-eukariots.

21.5.4   Examples from Biology

Literally hundreds of power laws have been identified in biology, including microbiology, developmental and evolutionary biology, ecology and paleontology (which we have chosen to include in Biology). They show up as neat power laws valid over many orders of magnitude, as concatenated sets of power laws that describe hierarchical behavior, or as trends in data for which there is little intuition [53]. How they work or why they exist is, to date, mostly a mystery, but progress is beginning to be made in this wonderful collection of problems. Here we can again touch upon a few selected examples; they are chosen from the biology of vertebrates, cell biology, ecology and paleontology. We will not be able to discuss the many fascinating examples linking morphology with environment and survival (e.g., in the so-called "Tiffany" wings) or aspects relating compartmentalization with network scaling.

In biology isometric (geometrically similar) bodies are not common, even through living things make use of many common patterns of organization. Because of this, Huxley coined the word "allometry" to describe the non-isometric nature of scaling in biological phenomena. ("allos" referring to different). Allometry in biology deals with an incredibly large number of phenomena where some physical property, such as aorta radius or population density, are related either to mass or size of the host system via a power law. Perhaps one of the most celebrated allometric relations is the one describing mammalian skeletons. As it is obvious, the bones of an elephant are proportionally heavier and thicker than those of a shrew. We wish to understand the "proportionality". This is an example already considered by Galileo and leads to interesting observations. If we consider leg bones, and since leg bones support the weight of the mammal, if we assume that we double the linear dimensions of the animal without distorting its shape, then its mass will increase by a factor of eight (mass is proportional to volume which goes like size to the cube). Under this scaling change the cross section of the bones also increases, and since we are dealing with an area (the cross-section of the bone), it must quadruple. But multiplying by 4 will not be able to compensate the factor of 8 needed for the increase in body mass! Thus the bones must be scaled out of the predicted "logical" scaling proportion! There must be a limit to how one can maintain such a proportion based on the compressive strength which bones can withstand. This led Galileo to say that the largest vertebrates would have "to float" in order to compensate for the compressive stresses in their skeleton if they stood on the surface: thus he understood that the largest mammals (whales) had to be marine animals. (A first prediction of Astrobiology?). Simple theoretical arguments based on the composition and geometry of skeletons lead one to postulate that the skeletal mass of a mammalian would have to increase like body mass raised to 1.33. This is not borne out in Nature, where for more than six orders of magnitude it is observed that the scaling exponent is 1.09 instead of the 1.33 required by the arguments just sketched. A similar scaling exponent (1.07) holds for the whole skeleton of the mammal, not only its legs. It is then clear that the skeletal mass of a mammal is not designed to only cope with gravitational loads, but other (not completely known) factors, like adaptability for locomotion, must be involved in the low value of the exponent. But there is no complete understanding of this remarkable regularity ... more than three centuries after it begun to be analyzed!

Another famous power law in biology is the one known as Kleiber's law [8]. It says that the metabolic rate (a power measured in watts) goes like the body mass raised to 0.75. This holds over many orders of magnitude and, again, it is not understood although some progress has been made recently. If the metabolic rate was in balance through heat exchange with the exterior, then it would go like the exposed surface of the animal which goes like the body mass raised to 2/3; since this is not the case, again other explanations have been sought but there is no general agreement, even though a model based on the "elastic similarity" of muscles has been proposed and has had for years a rather wide following. Recent progress attributes Kleiber's law to general features of biochemistry combined with a postulate on the fractal nature of transport systems in living things.

Our next example is drawn from the growth patterns of cell colonies [54]. The prototype experiment consists of inoculating a uniform agar substrate with some species. This has been done for example with *Escherichia coli* and with *Bacillus subtilis*. As the bacterial colony feeds on the agar it grows in size and a rough surface emerges at the edges of the bacterial colony, where it invades the agar. The edge of the rough surface serves as the phenomenological feature characterizing the growth process. Since this edge is non-uniform, one can determine the deviation $\sigma(l)$ of parts of the interfaces for various values of the separation l; then one can average over segments of the same length. Two features stand out: (1) first the deviation $\sigma(l)$ follows a power law, $\sigma(l) \propto l^{2H}$ and (2) that $H = 0.78 \pm 0.07$. The power law behavior for the correlation is surprising because one would expect the growth to be uniform or, at most, gaussian ($H = 1/2$); the value of the exponent is a complete surprise because it puts bacterial growth into a universality class of its own. Finally, we point out that the exponent has been measured for other bacterial species and the value is consistent (within errors) with the value quoted for H above. Why bacterial colonies follow this rule is a complete mystery. How can one tell from sedimentary processes? This perhaps could be used to develop criteria for markers of biological activity.

From bacterial colonies we now go to fully active ecologies [8]. Here again many power laws and hierarchies are known; also very little causal explanations are available. We will describe only one example: the population/body size law, also known as Damuth's law (Fig. 21.9). If we measure in an ecosystem the population density for the species in the ecosystem habitats and plot it versus typical longitudinal size for each species, one finds that population density goes like size raised to −2.25 all the way from bacteria to large mammals, a staggering 8 orders of magnitude! Since body mass goes like size to the cube, one finds that population density goes like mass raised to −2.25/3 = −0.75. This form of Damuth's law is the most widely used. Nobody has a good explanation for Damuth's law, but its interpretation is both elegant and useful:

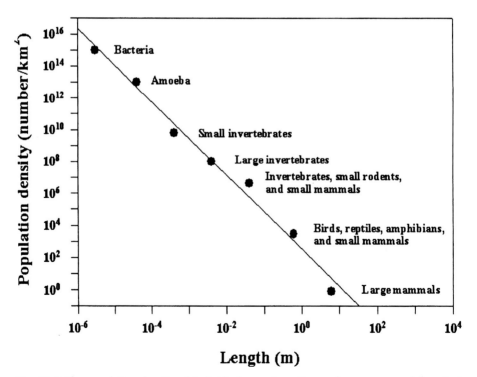

Fig. 21.9 The population density of individual organisms (in number per square kilometer) versus the typical length of the adult organism also follows a power law. Here the data from bacteria to the largest mammals are plotted on a log-log plot; the slope of the line indicating the power law is -2.25 (taken from [8]).

combined with Kleiber's law, Damuth's law implies that the amount of food generated per day and per square kilometer is consumed by the species in the ecosystem in such a way that resources are conserved. If this does not happen the ecosystem will evolve until it dies. It is one of the golden rules of ecology, even if nobody understands its origin at a basic level.

Even in paleontology we immediately discover the presence of power law phenomenology as soon as we try and submit it to mathematical analysis [55]. The number of species on planet Earth has had fluctuations in the course of time: there have been periods of extinction followed by periods of growth in the number of species and vice-versa. This we know by painstaking studies of the fossil record, including its diversity as a function of time. These data contain information on both the evolution of living beings and on their collective interactions, including the causes of their extinction. The two are very difficult to disentangle, but even so regularities emerge for example in the distribution of the geologic life-spans of fossil genera, or in the extinction intensity for genera. In both cases power law behavior is again, mysteriously, found. In the first case the number of fossil genera goes like their life span in million of years raised to –2.0 (Fig. 21.10). In the latter case, if we characterize extinction intensity by the number of geologic stages involved in the extinction and by the percentage of species wiped-out in the extinction, one finds again a power law. The number of geologic stages goes like the percent of the extinction raised to –2.0. The

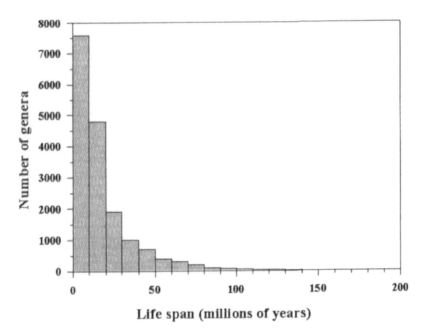

Fig. 21.10 Distribution of life spans for fossil genera [42]. As explained in the text, this distribution also follows a power law. This graph is not a log-log plot and therefore the power law does not show up as a straight line, however fitting of the data is best with a power law.

smooth grading of extinctions that results from this analysis suggests that there is no discontinuity between small and large extinctions, but nobody understands the reasons why this is so, nor why power laws appear, let alone the values of the exponents. What is known, however, is that the presence of power law behavior must be related to the existence of punctuated equilibrium in macro-evolution. The details of the connection are, once more, missing.

21.6 Conclusions

We have very briefly reviewed the presence of power law behavior in astrobiology. It is remarkable how such a simple phenomenology, so alien to chance (in the sense of gaussian disorder) is so ubiquitous. We have mentioned that power laws are related to changes of phases or to critical states, where a many body system evolves from one state of self-organization into a different state. Often these states are complex and they indicate patterns of self-organization in the evolution of complexity. Power laws show-up not only as a sign of correlation between microscopically unrelated parts, but also as indicators of ways into which a given system responds to the environment and evolves its organization; thus hierarchies, fragmentation, coagulation or network feedback give power law phenomenology: either with real or with complex exponents.

This phenomenology shows-up everywhere and we have illustrated it with just a small number of examples of the many that are known all the way from cosmology to astrophysics, from geology to biology. The specific causes for each of the phenomena known to display power law behavior must be different, since the forces and variety of components differs from phenomenon to phenomenon. However, the fact that such simple behavior stands out from such a huge variety of complex phenomena undoubtedly leads one to posit that there must be some general principles or set of rules at work that pervade the behavior of the Universe: from the non-living everywhere, to the living here on planet Earth and, maybe the living elsewhere. This indicates that understanding the evolution and emergence of complexity is essential in order to be able to determine wether "Life is a cosmic imperative".

Complexity theory is intimately related to scaling or power law behavior, and this, together with the ubiquity of power law phenomenology in Astrobiology, means that by scientifically applying complexity theory it may be possible to identify the principles of self-organization of extra-terrestrial life; it may be possible to submit such principles to mathematical/theoretical analysis, and, in the true tradition of the scientific method, it may be possible to use computers to apply the theoretical findings to study, codify or, eventually, even to predict the existence of extraterrestrial life. And understand *Life* on Earth ... as a byproduct.

Acknowledgement. The author thanks Murray Gell-Mann, Terry Goldmann and Geoffrey West for many discussions on power laws: their meaning, applications and interpretation, and also the late Jerry Soffen for his enthusiastic support of the application of complexity theory to Astrobiology. Finally, he thanks Gerda Horneck for inviting him to write this contribution and for her patience and interest in having scaling and complexity theory as part of this volume.

21.7 References

1   S. Weinberg, *Scientific American* **10**, 44 (1994).
2   J. Silk, *A Short History of the Universe*, Scientific American Library Paperback No. 53, W.H. Freeman and Co., New York, 1997.
3   J. Beatty, C. Petersen, A. Chaikin, *The new Solar System*, 4th Edition, Cambridge University Press, 1999.
4   J. Lunine, *Earth: Evolution of a Habitable World*, Cambridge University Press, 1999.
5   D. Deamer, G. Fleischaker, *Origins of Life. The Central Concepts*, Jones and Bartlett Publ., London, 1994.
6   W. Schopf, *Major Events in the History of Life*, Jones and Bartlett Publ., London, 1992.
7   C. De Duve, *Vital Dust. Life as a Cosmic Imperative*, Basic Books, New York, 1995.
8   T. MacMahon, J. Bonner, *Life and Size*, Scientific American Library, W.H. Freeman and Co., New York, 1983.
9   H. Simon, *The Sciences of the Artificial*, 2nd Edition, 8th Printing, MIT Press, 1994.
10 M. Gell-Mann, *The Quark and the Jaguar*, W.H. Freeman and Co., New York, 1994.
11 G.A. Cowan, D. Pines, D. Meltzer, Complexity: Metaphors, Models, and Reality, SFI Studies in the Sciences of Complexity, Addison-Wesley Publ., 1994.
12 R.F. Service, Science **284**, 80 (1999).
13 C. Zimmer, Science **284**, 83 (1999).
14 N. Goldenfeld, L.P. Kadanoff, Science **284**, 87 (1999).
15 E. Beltrami, *What is Random? Chance and order in mathematics and life*, Springer-Verlag, 1999.
16 D. Hochberg, C. Molina, J. Pérez-Mercader, M. Visser, Phys. Rev. **E60**, 6343 (2000).
17 D. Hochberg, C. Molina, J. Pérez-Mercader, M. Visser, J. Stat. Phys. **99**, 903 (2000).
18 M.C. Cross, P. Hohenberg, Rev. Mod. Phys. **65**, 851 (1993).
19 D. Sornette, *Critical Phenomena in Natural Sciences. Chaos, Fractals, Self-organization and Disorder: Concepts and Tools*, Springer-Verlag, 2000.
20 H. Pattee, *Hierarchy Theory. The challenge of complex systems*, George Braziller, Publ., New York, 1973.
21 J. Binney, N. Dowrick, A. Fisher, M. Newman, *The Theory of Critical Phenomena: an Introduction to the Renormalization Group*, Oxford University Press, 1992.
22 J. Yeomans, *Statistical Mechanics of Phase Transitions*, Oxford University Press, 1992.
23 B. Mandelbrot, *The Fractal Geometry of Nature*, W.H. Freeman and Co., New York, 1982.
24 K. Wilson, Scientific American **241**, 140 (1979).
25 K. Wilson, Rev. Mod. Phys. **47**, 773 (1975).
26 W. Feller, *An Introduction to Probability Theory and its Applications*, Vol. I and II, J. Wiley and Sons, New York, 1971.
27 P. Bak, *How Nature works*, Springer-Verlag, 1996.
28 H. Jensen, *Self-organized Criticality*, Cambridge University Press, 1998.
29 J. Pérez-Mercader, A. Giménez, *The Titius-Bode Law as a Dynamical Critical Phenomena*, talk given at the First ESA-CAB Eddington Workshop, Córdoba, June 2001, ESA Proc., submitted.
30 C. Holling, Ecol. Monographs **62**, 447 (1992).

31 M. Gell-Mann, F. Low, Phys. Rev. **95**, 1300 (1954).
32 P.G. de Gennes, *Scaling Concepts in Polymer Physics*, Cornell University Press, 1979.
33 A. Berera, F. Li-Zhi, Phys. Rev. Lett. **72**, 458 (1994).
34 D. Hochberg, J. Pérez-Mercader, Gen. Rel. Grav. **28**, 1427 (1996).
35 F. Barbero, A. Domínguez, T. Goldman, J. Pérez-Mercader, Europhys. Lett. **38**, 637 (1997).
36 G.B. West, J.H. Brown, B.J. Enquist, Science **276**, 122 (1997).
37 C. Allen, E. Forys, C. Holling, Ecosystems **2**, 114 (1999).
38 U. Frisch, *Turbulence. The Legacy of A. N. Kolmogorov*, Cambridge University Press, 1995.
39 F. Sylos-Labini, M. Montuori, L. Pietronero, Phys. Rep. **293**, 61 (1998)
40 J. Gaite, A. Domínguez, J. Pérez-Mercader, ApJ. Lett. **522**, L5-L8 (1999).
41 T. Goldman, J. Pérez-Mercader, in prep.
42 T. Gehrels, (Ed.), *Protostars and Planets,* University of Arizona Press, Tucson, AZ, USA, 1979, pp. 756.
43 D.C. Black, M.S. Matthews (Eds) *Protostars and Planets II,* University of Arizona Press, Tucson, AZ, USA, (1985).
44 E. Levy, J.I. Lunine (Eds) *Protostars and Planets III,* University of Arizona Press, Tucson, AZ, USA, (1993) pp. 1596.
45 V. Mannings, A. P. Boss, S.S. Russell (Eds) *Protostars and Planets IV*, University of Arizona Press, Tucson, AZ, USA, (2000) pp. 1422
46 M.M. Nieto, *The Titius-Bode Law of Planetary Distances: Its History and Theory*, Pergamon Press, 1972.
47 J.S. Lewis, *Physics and Chemistry of the Solar System*, Revised Edition, Academic Press, 1997.
48 D. Turcotte, *Fractals and Chaos in Geology and Geophysics*, Cambridge University Press, 2nd Edition, 1997.
49 C. Scholz, *The Mechanics of Earthquakes and Faulting*, Cambridge University Press, 1991.
50 H.E. Stanley, S.V. Buldyrev, A.L. Goldberger, S. Havlin, R.N. Mantegna, C.K. Peng, M. Simons, M.H.R. Stanley., *Long-range correlations and Generalized Lévy Walks in DNA Sequences*, in: M.F. Shlesinger, G.M. Zaslavsky, U. Frisch (Eds.) *Lévy Flights and Related Topics in Physics*, Lecture Notes in Physics, Vol. 450, Springer-Verlag, 1995.
51 L. Peliti, *Shapes and Fluctuations in Membranes*, in: H. Flyvbjerg, J. Hrzt, M.H. Jensen. (Eds.) *Physics of Biological Systems: from molecules to Species*, Lecture Notes in Physics, Vol. 480, " Springer-Verlag, 1997.
52 D. Morse, S. Milner, Europhys. Lett. **26**, 565 (1994).
53 K. Schmidt-Nielsen, *Scaling: Why is Animal Size so Important?*, Cambridge University Press, 1995.
54 T. Vicsek, M. Cserko, V. Horvath, Physica A **167**, 315 (1990).
55 P. Bak, M. Paczuski, *Mass Extinctions vs. Uniformitarianism in Biological Evolution*, in: H. Flyvbjerg, J. Hrzt, M.H. Jensen. (Eds.) *Physics of Biological Systems: from molecules to Species*, Lecture Notes in Physics, Vol. 480, Springer-Verlag, 1997.
56 D. Raup, *Extinctions: Bad Genes or Bad Luck?*, W.W. Norton and Co., New York, 1991.

22 Molecular Self-Assembly and the Origin of Life

Wolfgang M. Heckl

"All things began in order, so shall they end, and so shall they begin again; according to the ordainer of order and mystical mathematics of the city of heaven" (Sir Thomas Browne 1658)

Nanoarchitectonics is a new interdisciplinary field within the NanoSciences, which investigates the principles responsible for the formation of higher-ordered functional structures starting from their nanoscopic building blocks like atoms and molecules. Such a bottom up approach is new within the field of contemporary technology, which has used very successfully the top down strategy for the miniaturization of fabrication processes during the last hundred years. However, nature has always worked bottom up, where the principles of self-assembly lead to crystal growth in the inorganic world, and, via molecular self-assembly, to functional structures in biology. For instance, the three-dimensional architecture of a nanomachine called the ribosome comprises the natural molecular assembler, which organizes the transition from the DNA informational blueprint into polypeptides and other functional units. To understand and make technological use of the underlying mechanism of this process is one of the major goals of modern Proteomics, where the relation between the DNA base sequence and the respective protein must be mastered. One approach to this question is to simplify the process by transferring it into a two-dimensional scenario, thus reducing the complexity of the three-dimensional architecture to an in-plane problem. Such a reduced coordination space may also be adequate for a primordial soup scenario, where the spontaneous self-assembly of abiotically produced organic compounds may be facilitated. The formation of highly ordered monolayers of the purine and pyrimidine DNA bases through physisorption mediated molecular self-assembly at a solid-liquid mineral interface and the subsequent stereospecific adsorption of amino acids is an example for the spontaneous creation of nanoscale order. We have proposed a functional role of this process for the emergence of life that may also lead towards the construction of genetically based supramolecular architectures for modern technical applications [1-4].

22.1 Key Questions for the Emergence of Life

There are a number of key questions in the context with models for molecular self-assembly and the emergence of life, amongst them the following which have widely been discussed within the literature [5-14]:

-   Which key molecules are necessary for the first steps toward self-assembled molecular nanoscopic systems capable to reproduce and undergo some type of molecular evolution towards creation of higher-ordered structures? How are these elementary building blocks being synthesized in the absence of any biological machinery?
-   What type of primordial synthetic chemistry comprises a possible route to answer this question and is there evidence enough from primitive earth simulation experiments, from analysis of extraterrestrial debris or from spectroscopic analysis of interstellar gases?
-   Where are plausible locations, such as terrestrial and subterranean areas of geothermal activity or marine/non-marine (tidal) pool systems (to allow molecular sizzling) or submarine hydrothermal systems, for such a self-assembly?
-   What is the origin of the observed biological homochirality when simple chemical synthesis normally leads to racemic mixtures?
    Is the reason for the required symmetry break that leads the route to life in evolutionary forces (for example selective enzymes at some stage)?
    Might purely physical forces such as the known parity violation process in the radioactive ß-decay play a role?
    Can minute energy differences between two isomers (on the order of $\sim 10^{-17}$ kT for biomolecules) be responsible for a bias towards homochirality?
    Might the interaction with external chiral forces such as left or right handed polarized light drive the crystal growth towards one side?
-   How can the circular relationship between a coded enzyme and its code arise when each is needed for the synthesis of the other?

22.2 Directed Molecular Self-Assembly

Molecular self-assembly has been defined as the spontaneous emergence of highly organized functional supramolecular architectures from single components of a system under certain external condition [15]. Such conditions may include for example a suitable template, where molecules can adsorb to and the appropriate environmental conditions, such as the right mixture and concentration of molecules in a solvent like water, temperature etc. In contrast to molecular chemistry, that has established its power over the covalent bond, non-covalent intermolecular forces prevail in this field of supramolecular chemistry. The energetic and stereochemical properties of non-covalent intermolecular forces such as electrostatic interactions, van der Waals forces and, most prominent, hydrogen bridge bonding act like the joints of Lego game pieces because they direct the monomeric building blocks to spontaneously assemble into supramolecular structures with inherent higher order. The stereospecific hydrogen bridge bonds determine the structure that comprises the blueprint for this type of transition from monomers to two-dimensional polymeric flat crystalline organic layers. In nature, the main example for coded self-organization is the DNA-double helix. Molecules capable of making hydrogen bridge bonds, such as the DNA-bases may self-organize into two-dimensional crystals as shown in Fig. 22.1. Here the steric arrangement of functional groups such as the possible H-bridge donors N-H, O-H and the possible H-bridge acceptors such as N, O moieties are responsible for the specific periodic DNA-base surface. The electrostatic potential of adenine, where N-H and H

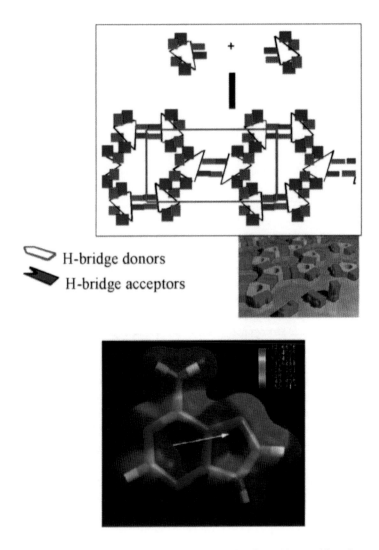

H-bridge donors

H-bridge acceptors

Fig. 22.1 Schematic of recognition directed molecular self-assembly of monomeric DNA-based molecules into a two-dimensional organic crystalline layer.

interactions are used to build a monolayer, is shown here as an example. It has been calculated in an *ab initio* molecular orbital calculation to exactly display H-bond acceptors (in blue) and H-bond donors (in red) [16]. Such a potential map can be considered as a representation of the chemical activity of the molecule. The arrangement of the molecules shown here is only one possible pattern, which fulfils the periodic space-filling requirement of an energetically minimized two-dimensional crystal structure. Subsequent experimental observations must verify the proposed models.

22.3 Direct Microscopic Verification of Self-Assembled DNA Bases on Mineral Template Surfaces

With the advent of scanning tunneling microscopy [17] it was possible to observe directly DNA base molecules with possible hydrogen bridge donors and acceptors to self-organize into such periodic organic molecular layers on mineral template surfaces [18,19]. Since then a large number of studies of these self-assembled DNA base systems have been published (for a comprehensive overview refer to [2]).

To mimic a primordial situation, where molecules (for example of the primordial soup) are dispersed in liquid solution and come into contact (for example by tidal movements) with the surface of a mineral template, also possibly present within a shallow sea scenario, the sizzling technique [19] has been applied to initiate the self-organization of the molecules. Here the molecules are dissolved in water and applied to a slightly heated (up to 130 °C) mineral template surface providing the energy to facilitate the two-dimensional crystal growth via molecular surface diffusion (Fig. 22.2). In order to be error tolerant, with respect to the right docking position of a new molecule adsorbing to a growing two-dimensional seed crystal, it is important that the molecular interactions are of weak second order chemical bond nature. Simultaneous occurring growth and dissolution processes are shifted towards the formation of the molecular crystal because the mineral template forces a seed crystal to grow due to the fixation of molecules upon physisorption.

The key to the spontaneous emergence of these supramolecular structures is the appropriate energetic interplay between intermolecular interactions in two dimensions. The in-plane molecular H-bonding is responsible for creating the repetitive motive of the base layer. This is comparable to the pattern, which arises through complementary H-bonding in the three-dimensional DNA-polymer. The physisorption energy of the mineral surface determines to what exact surface atomic positions adsorption occurs. Clearly, this relates to the question of whether clay minerals [6, 14], sulfur containing minerals with catalytic possibility [20, 21] or other mineral surfaces may have played the actual role as inorganic template. From our experimental point of view the mineral should have a surface energy low enough to facilitate physisorption in contrary to chemisorption where the growth of two-dimensional organic crystals is hindered by the restricted surface mobility of the molecules. Therefore, and in order to be able to do Scanning Tunneling Microscopy (STM), we have mainly used the conductive natural crystalline mineral surfaces graphite (001) and molybdenite (001) which can be found frequently on Earth.

In the following example the concept of organic monolayer structure determination as shown in Fig. 22.3 has been applied to adenine physisorbed to graphite [22]. Here the real space STM technique provides high resolution microscopic images of the molecules on a local scale and is combined with the low energy electron diffraction (LEED) technique or surface X-ray diffraction [23] providing a reciprocal space

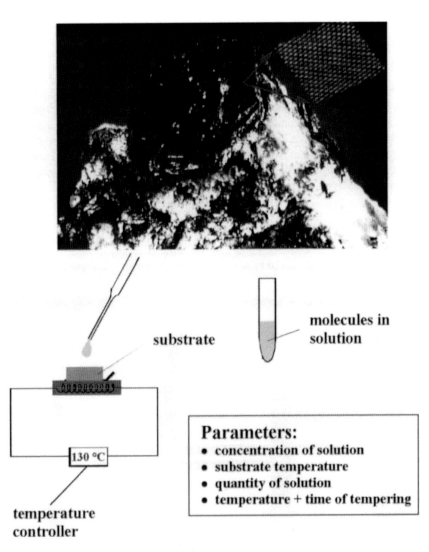

Fig. 22.2 Schematic of the sizzling technique applied to mimic the possible spontaneous self-assembly of two-dimensional molecular crystalline layers onto mineral surfaces within a primordial soup scenario where the elementary building blocks of life may be dissolved. Here a natural molybdenite crystal is shown together with the atomic scale resolution STM image of the surface.

image of the average two-dimensional molecular crystal structure. Together with molecular modeling force field calculations [24, 25] and the determination of the preparation parameters and thermodynamic variables, such as adsorption energy measured by thermal desorption spectroscopy (TDS), clear models can be derived as a result. Figure 22.4 shows the adenine on graphite layer in a high resolution STM image togetherwith the LEED pattern. Single molecules can be identified to form a periodic molecular arrangement with a rectangular unit cell with vectors a = 22.1 Å and b = 8.5 Å. The flat lying centro symmetric adenine dimers with p2gg symmetry heteroepitaxially grow on the mineral surface with 4 molecules per coincident unit cell with

$$\begin{pmatrix} a \\ b \end{pmatrix} = \begin{pmatrix} 9 & 0 \\ 2 & 4 \end{pmatrix} \bullet \begin{pmatrix} g_1 \\ g_2 \end{pmatrix} \qquad (22.1)$$

metric, where g_1 and g_2 are the hexagonal arranged graphite unit cell vectors of 2.46 Å in length. The coordination number for a molecular dimer is 6, representative for the dense packed energetically minimized stable structure where each dimer is surrounded by 6 next neighbor dimer molecules. Each molecule is sorrounded by 3 neighboring molecules with the maximum number of 6 hydrogen bonds per molecule. This type of arrangement also leads to cyclic hydrogen bonding, which stabilizes the monolayer via π-electron cooperativity.

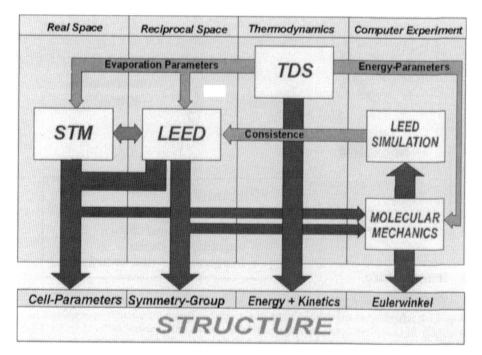

Fig. 22.3 Concept for the structure determination of self-assembled molecular layers by the combination of different complementary techniques.

Fig. 22.4 Scanning tunneling microscopy image of a spontaneously self organized adenine monolayer physisorbed to the mineral surface of graphite, together with the structure (below) determined from reciprocal space LEED-analysis (diffraction pattern show above right) and direct space STM image with rectangular unit cell with vectors a = 22.1 Å and b = 8.5 Å.

In some cases the spontaneous molecular self-assembly from liquid solution leads to localized chiral symmetry break, which may have some role in the origin of bio-molecular optical asymmetry. Sowerby et al. [1] have observed such a spontaneous symmetry break for example upon adsorption of adenine on molybdenite. Although the adenine molecule is achiral by the usual definition whether a stereochemical center is present or not, there are two mirror-symmetric possible ways when the molecules adsorb to the (symmetric) mineral template surface. Whenever the unit cell is oblique, it exists in possible left handed and right handed molecular crystals (Fig. 22.5) within the definition of two-dimensional chirality of not being superimposable by any trans-lations or rotation in two dimensions.

Whether such a symmetry break upon adsorption may gain a bias towards one handedness over the other due to a force from an external source, such as for example polarized light, is not known to date. STM however is the only technique to observe such differences on the local scale and the possibility that purine-pyrimidine arrays assembled in such a chiral fashion on naturally occurring mineral surfaces might act as chiral organic templates for subsequent biomolecular assembly of higher-ordered compounds is intriguing.

22.4 Genetically Based Supramolecular Architectures from Self-Assembled DNA Bases Coding for Amino Acids

We have presented an organic template model [2] where the key features are summa-rized in Fig. 22.6. As shown above for adenine, the spontaneous molecular self-assembly of DNA-base molecules from liquid solution on a mineral surface may lead to an (localized chiral?) organic template surface, capable of stereospecific interaction with subsequently adsorbed amino acids which may self-organize on top of the tem-plating nucleic acid layer. Thus the construction of supramolecular complexity through the control of the intermolecular bond, preferentially the H-bond is achieved. The proposed architecture may facilitate the polycondensation of the amino acid monomers and lead to polypeptides, leaving the surface after loss of the ammonium proton upon formation of the peptide bond, which can then no longer form a stabiliz-ing H-bond to the base template. As a consequence, this process is capable of cata-lyzing the formation of polypeptides in liquid.

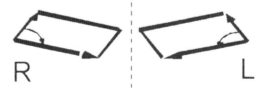

Fig. 22.5. Two-dimensional chirality of right handed and left handed molecular crystals.

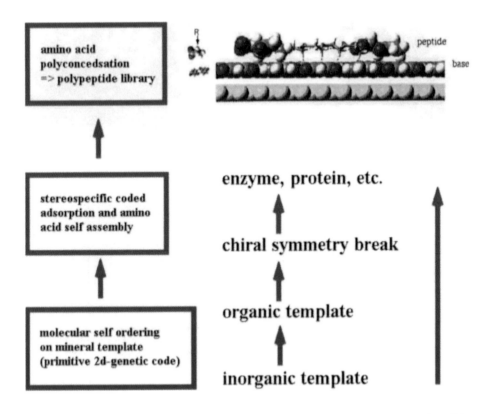

Fig. 22.6 Model for the assembly of polypeptides based on a 2d-DNA-base library at the solid-liquid interface.

Such a scenario can be regarded as primitive self-programmable, self- assembling two-dimensional genetic matter. An aperiodic mixture of different nucleic acids has been observed and may be used to construct a variable peptide library [1,4], thus reducing the complexity of the ribosomal RNA-mediated process of polypeptide synthesis in nature.

Whether the two-dimensional DNA-base layer can act as a primitive coding mechanism depends on the exact adsorption process of amino acids on top of the nucleic acid layer. This has recently been tested by predicting the adsorption of lysine on a flat adenine template surface via molecular mechanics simulation [26]. Here the lysine molecules order as shown in Fig. 22.7 with calculated adsorption energy of 13 kcal/Mol. The energetic landscape for this process exhibits a very quick and steep descent to a stable energetic minimum in total energy, which demonstrates the robustness of the proposed process. It has recently been shown experimentally, that the adsorption of different nucleic acids to graphite surfaces is specific, demonstrating the influence of the energy of adsorption, which is the prerequisite for a rational coding mechanism [27]. A coding-like discrimination of amino acids by purine bases adsorbed on an inorganic surface has just been shown [28].

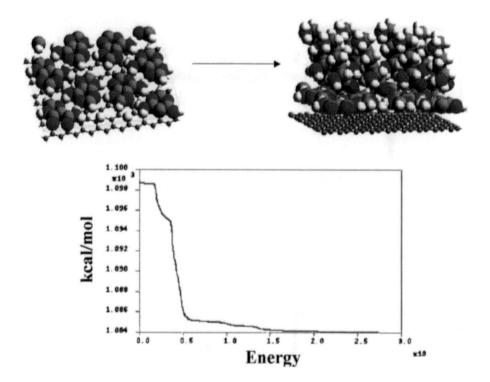

Fig. 22.7 Molecular mechanics calculated model of the observed self-assembly of nucleic acid adenine (left) and subsequent self-assembly of amino acid lysine on top of the templating nucleic acid layer (right). Bottom shows the development of the energetic hypersurface over time.

22.5 Conclusion

The application of near field microscopy techniques, namely scanning tunneling microscopy (STM), to self-assembled two dimensional nucleic acid crystals has allowed for the first time real space analysis of these systems with molecular scale resolution. This has stimulated the development of new concepts regarding the possible role of molecular self-assembly in the de novo emergence of higher-ordered supramolecular architectures, comprised of today's DNA and protein building blocks and eventually guiding a route to life under prebiotic conditions. We have suggested that purine and pyrimidine monolayers could be candidates for a stationary phase in organic molecule separation systems, and as templates for the assembly of higher-ordered polymers at the prebiotic solid-liquid interface. In some cases, such as adenine on molybdenite, a symmetry break can be observed which may have some role in the origin of biomolecular structural asymmetry. In the future it should be possible to test experimentally whether the proposed scenario actually may lead to the necessary compounds.

Acknowledgments: The theory described in this paper has been developed over a number of years as the result of a collaboration with Dr. Stephen Sowerby and Professor George Petersen of the Department of Biochemistry, University of Otago, New Zealand. I am grateful to my collaborators for reading the manuscript and for their suggestions. My students, M. Edelwirth, F. Jamitzky, F. Trixler and F. Griessl, have also contributed with various images.

22.6 References

1   S.J. Sowerby, W.M. Heckl, G.B.Petersen, J. Mol. Evol. **43**. 419 (1996).

2   S.J.Sowerby, W.M. Heckl, Origin of Life and Evolution of the Biosphere **28**, 283(1998).

3   S.J. Sowerby, M. Edelwirth, W.M. Heckl, J.Phys.Chem., **102**(30), 5914 (1998).

4   S.J. Sowerby, P.A. Stockwell, W.M.Heckl, G.B.Petersen, Origin of Life and Evolution of the Biosphere **30**(1), 81 (2000).

5   E. Schrödinger, *What is life?* Cambridge University Press, Cambridge, 1944.

6   J.D. Bernal, *The Physical Basis of Life*. Routlege and Keegan Paul, London, 1951.

7   C. DeDuve, *Bluprint for a Cell: The Nature and Origin of Life*, Neil Pattern Publ., 1991.

8   A. Brack (Ed.) *The molecular origins of life: assembling pieces of the puzzel*, 35, Cambridge University Press, 1998.

9   J.P. Ferris, A.R. Hill, R. Liu, L.E. Orgel, Nature **381**, 59 (1996).

10 S.L. Miller, Cold Spring Harbor Symposium LII, 17 (1987).

11 E.L. Shock, Orig. Lefe Evol. Biosph. **20**, 331 (1990).

12 R.J.C. Hennet, N.G. Holm M.H. Engel, Naturwissenschaften 79, 361-365(1992)

13 A. Brack, Pur Appl. Chem. **65**, 1143 (1993).

14 A.G. Cairns-Smith, *Genetic Takeover and the Mineral Origins of Life*. Cambridge Univ. Press, 1982.

15 J.-M. Lehn, in: *Supramolecular Chemistry, Concepts and Perspectives*, Wiley VCH, 1995.

16 M. Edelwirth, J. Freund, S.J. Sowerby, W.M. Heckl, Surface Science **417**, 201 (1998).

17 G. Binnig, H. Rohrer, C. Gerber, E. Weibel, *Phys. Rev. Lett.* **49**, 57 (1982).

18 M.J. Allen, M. Balooch, S. Subbiah, R.J. Tench, W. Siekhaus, R. Balhorn, *Scanning Microscopy* **5**, 625 (1991)

19 W.M. Heckl, D.P.E. Smith, G. Binnig, H. Klagges, T.W. Hänsch, J. Maddocks, *Proc. Natl. Acad. Sci. USA* **88**, 8003 (1991).

20 E. Blöchl, M. Keller, G. Wächtershäuser, K.O. Stetter, *Proc. Natl. Acad.Sci. USA* **89**, 8117 (1992).

21 M.J. Russell, R.M. Daniel, A.J. Hall, *Terra Nova* **5**, 343 (1993).

22 J. Freund, M. Edelwirth, W.M. Heckl, Physical Rev. B **55**, 3, 5394 (1997).

23 H.L. Meyerheim, F. Trixler, W. Stracke, W.M. Heckl, Z. Kristallogr. **214** (1999) 771 (1999).

24 M. Edelwirth, J. Freund, S.J. Sowerby, W.M. Heckl, Surface Science **417**, 201 (1998).

25 S.J. Sowerby, M. Edelwirth, W.M. Heckl, Appl. Phys. Lett. **A66**, 649 (1998).

26 F. Jamitzky, W.M. Heckl, to be published.

27   S.J. Sowerby, C.A. Cohn, W.M. Heckl, N.G. Holm, Proc. National Academy of Sciences, USA, **98.3** 820 (2001).

28   S.J. Sowerby, G.B. Petersen, N.G. Holm, pers. communication

23 Search for Morphological and Biogeochemical Vestiges of Fossil Life in Extraterrestrial Settings: Utility of Terrestrial Evidence

Manfred Schidlowski

Any veneer of life covering the surface of a planet will necessarily interact with the solid and fluid phases, with which it is in contact. Specifically, it is bound to impose a thermodynamic gradient on all planetary near-surface environments (inclusive of the atmosphere and hydrosphere), which ultimatively stems from the accumulation of negative entropy by living systems. Consequently, life acts as a driving force for a number of globally relevant chemical transformations. On Earth, typical examples of such life-induced chemical inequilibria are the glaring redox imbalance at the terrestrial surface caused by photosynthetic oxygen, or the release of large quantities of hydrogen sulfide by sulfate-reducing bacteria in the marine realm. Also, the dynamic persistence of metastable atmospheric gas mixtures (such as O_2, N_2 and CH_4 in the terrestrial atmosphere), and of isotopic disequilibria (e.g., between water-bound oxygen of the hydrosphere and atmospheric O_2), is ultimately sustained by the thermodynamic imbalance imposed by the biosphere on its environment. Conspicuous thermodynamic inequilibria within the gaseous and liquid envelopes of a planet may, therefore, be taken as *a priori* evidence of the presence of life [1, 2]. Applying this criterion to the present composition of the Martian atmosphere [3], the latter gives little, if any, indication of contemporary biological activity on that planet.

Moreover, a planetary biosphere is apt to leave discrete vestiges in the surrounding inorganic habitat. Relying on terrestrial analogues, it is safe to say that organisms commonly generate a morphological and biochemical record of their former existence in sedimentary rocks. Though in part highly selective, this record may survive, under favorable circumstances, over billions of years before being annealed in the wake of a metamorphic and anatectic reconstitution of the host rock. This is particularly true for relics of multicellular life (Metaphyta and Metazoa) but also - albeit with restrictions - for microorganisms, which had held dominion over the Earth during the first 3 billion years of recorded geological history.

In the following, an overview will be presented of the principal morphological and biogeochemical evidence indicating the presence of life on a planetary surface. The discourse both summarizes, and elaborates on, previous more exhaustive presentations of the subject by the author [4-7].

23.1 Paleontological Evidence

Given the validity of terrestrial analogues, the stratigraphic (sedimentary) record of any planet should serve as a potential store for both bodily remnants of former organisms and other possible traces of their life activities. This should also apply to Mars during the early stages of its history, when the planet was probably bathed in abundant water, and environmental conditions on the surface probably did not differ much from those on the early Earth [8, 9]. If all terrestrial planets – and notably Mars and Earth – had occupied comparable starting positions in terms of solar distance, condensation history and primary endowment with matter from the parent solar nebula [10], then the surface conditions on both planets should have been very similar in their juvenile states.

Specifically, with evidence at hand for a denser atmosphere and extensive aqueous activity on the Martian surface during the early history of the planet, a convincing case can be made that the primitive Martian environment was no less conducive to the initiation of life processes and the subsequent emplacement of prolific microbial ecosystems than the surface of the ancient Earth [11, 12]. Even if the evolutionary pathways of both planets had diverged during their later histories to the effect that life became extinct on Mars due to a gradual deterioration of surface conditions that consequently rendered the planet inhospitable to protein chemistry in the widest sense, the planet could still have started off with a veneer of microbial (prokaryotic) life comparable to that existing on the Archaean Earth [4, 5, 13, 14]. Given the apparent failure of the Viking life detection experiment for the present Martian regolith [15], the prime objective of a search for life on Mars should be to seek evidence of extinct (fossil) life, and the oldest Martian sediments would constitute appropriate targets for such efforts.

In any search for extinct life on Mars, the basic problem would not rest with the cognitive aspects of the identification of the fossil evidence, but rather with the serendipity inevitably involved in the selection and recovery of suitable sampling material from the vast stretches of potential host rocks exposed on the planetary surface. Among the two-thirds of the Martian surface deemed to be covered by rocks older than 3.8 Ga [16], there are several occurrences of well-bedded sediments (notably in the Tharsis region and the associated Valles Marineris canyon system) that are interpreted as lake deposits and believed to include thick sequences of carbonates [17]. If present at all in these early Martian formations, morphological relics of fossil life should lend themselves to as ready a detection as do their terrestrial counterparts in Archaean sediments, provided the host rock is accessible to either robotic sensing or direct investigation following a sample return mission.

In that regard, it seems to be a reasonable conjecture to resort to the paleontological inventory of the oldest terrestrial sediments (between 3.30 and 3.85 Ga ago) for guidance with regard to potential fossil evidence from coeval Martian rocks. On Earth, the earliest prokaryotic (bacterial and archaebacterial) microbial ecosystems have basically left two categories of morphological evidence, namely (i) stromatolite-type biosedimentary structures ("microbialites") and (ii) cellular relics of individual microorganisms ("microfossils").

23.1.1 Microbialites

Microbialites or "stromatolites" [18] are laminated biosedimentary structures that preserve the matting behavior of bacterial and algal (primarily prokaryotic) micro-benthos. Microbial build-ups of this type represent stacks of finely laminated lithified microbial communities that had originally thrived as organic films at the sediment-water interface, with younger mat generations successively superimposed on the older ones (Figs. 23.1 and 23.2). The structures derive from the interaction of the primary biologically active microbial layer with the ambient sedimentary environment, the fossilization of the laminae resulting from either trapping, binding or biologically mediated precipitation of selected mineral constituents.

In terrestrial sediments, microbialites represent the most glaring (macroscopic) expression of fossil microbial life, with a record extending back to the Early Archaean (~3.5 Ga ago). This constitutes *prima facie* evidence that benthic prokaryotes were widespread already in suitable aquatic habitats of the Archaean Earth. Both the morphological inventory of the oldest stromatolites and the observed microfossil content of the ambient rock or coeval sequences (see below) allow a fairly elaborate reconstruction of the Earth's earliest microbial ecosystems, indicating that Archaean stromatolite builders were not markedly different from their geologically younger counterparts (inclusive of contemporary species). It appears well established that the principal microbial mat builders were filamentous and unicellular prokaryotes capable of phototactic responses and photoautotrophic carbon fixation [19]. The unbroken stromatolite record from Archaean to present times attests, furthermore, to an astounding degree of conservatism and uniformity in the physiological performance and communal organization of prokaryotic microbenthos over 3.5 Ga of geological history. In spite of recently voiced reservations elaborating on morphological convergences between biologically induced and purely inorganic (evaporitic and microclastic) laminations [20], the balance of the currently available evidence suggests that the oldest stromatolites constitute a crucial (if not dominant) part of the early record of life [21].

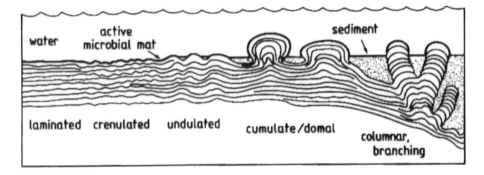

Fig. 23.1 Scheme of principal morphologies of laminated microbial ecosystems that thrive at the sediment-water interface. These structures subsequently lend themselves to lithification in the form of "stromatolites" (see Fig. 23.2). The mat-forming microbenthos is mostly made up of cyanobacteria.

Fig. 23.2 Typical microbialite (or "stromatolite") from the Precambrian Transvaal Dolomite (~2.3 Ga) that is made up of a succession of superimposed fossil microbial mats. The bunshaped and partially interfering laminae represent consecutive growth stages of the primary microbial community. Breadth of rock specimen is ~14 cm.

23.1.2 Cellular Microfossils

Apart from the macroscopic vestiges of past microbial activity in the form of stromatolites and related biosedimentary structures, there is a second (microscopic) category of paleontological evidence stored in the sedimentary record. This microscopic or cellular evidence is currently supposed to go back to at least ~3.5 Ga (as in the case of microbialites), with the biogenicity of microfossil-like morphologies reported from the ~3.8 Ga-old Isua metasedimentary suite from West Greenland [22] still being under debate.

While a wealth of authentic microbial communities has been reported from Early and Middle Precambrian (Proterozoic) formations, the unequivocal identification of cellular microfossils becomes notoriously difficult with increasing age of the host rock. In Early Precambrian (Archaean) sediments, both the progressive diagenetic alteration and the metamorphic reconstitution of the enclosing mineral matrix tend to blur the primary morphologies of delicate organic microstructures. This results in a large-scale loss of contours and other critical morphological detail. At the extreme of such alteration series lie the so-called "dubiofossils" of variable and sometimes questionable confidence levels. To ascertain the biogenicity of possible cellular morphotypes in Archaean rocks, a hierarchical set of selection criteria has been proposed, postulating that genuine microfossils (i) be authentic constituents of the rock as testified by their exposure in petrographic thin sections, (ii) occur in vast multiples, (iii) be associated with residual carbonaceous matter, (iv) equal or exceed the minimum size of viable cells and display a central cavity plus structural detail in excess of that resulting from inorganic processes. Moreover, it has been proposed that, as a matter of principle, putative evidence from metamorphosed sediments should not be considered.

In spite of the blatant impoverishment of the Archaean record there are, however, single reports of well-preserved microfossils that, for the most part, comply with the above criteria. Most prominent among these assemblages are chert-embedded microfloras from the Warrawoona Group of Western Australia, which closely approach the 3.5 Ga age mark. After an initial controversy about the authenticity of these fossil communities on grounds of imprecisely constrained petrographic background parameters for the host lithologies [23, 24], Schopf and Packer [25] and notably Schopf [26] have forwarded evidence prompting an acceptance of the observed morphotypes as *bona fide* microfossils. Conspicuous within the Warrawoona microbial community are both the coccoidal and filamentous (trichomic) micromorphologies that have been found to abound in the cyanobacterial precursor floras of Proterozoic formations. While the septate filaments stand for fossil trichomes that could be attributed to either filamentous cyanobacteria or more primitive prokaryotes (such as flexibacteria), the coccoidal aggregates described by Schopf and Packer [25] have been claimed to strictly exclude other than cyanobacterial affinities.

Given the remarkable degree of diversification of the Warrawoona microflora we must, of necessity, infer that the genetic lineages of the principal microbial species had emerged well before Warrawoona times. We may, therefore, reasonably assume that precursor floras had been extant prior to ~3.5 Ga when the preserved rock record becomes scant and increasingly metamorphosed. In this context, the observation of cell-like carbonaceous structures in the 3.8 Ga-old metasediments from Isua, West-Greenland, has attracted considerable attention. Described as *Isuasphaera isua* [22], the biogenicity of this morphotype (Fig. 23.3) has been violently disputed, specifically on grounds of the improbability of survival of delicate cell structure during the amphibolite-grade metamorphism of the host rock.

Meanwhile, however, there is ample evidence that fossils in general and microfossils in particular may - in variable degree - withstand obliteration in rocks subjected to medium-grade metamorphism. Therefore, we cannot *a priori* exclude microbial affinities for selected cell-like microstructures from the Isua metasediments, and notably for *Isuasphaera*-type micromorphs that show a striking resemblance to a possible counterpart of recognized biogenicity in the younger (Proterozoic) record described as *Huroniospora* sp. In spite of the uncertainty surrounding a large number of morphotypes described from Isua, and of occasional convergences with purely mineralogical features, there is a reasonable chance that the microstructure inventory as a whole

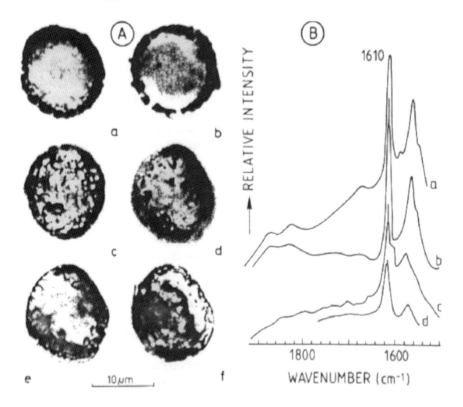

Fig. 23.3 A: Comparison of *Huroniospora* sp. from the ~2.0 Ga-old Gunflint iron formation, Ontario (a-c) with *Isuasphaera* sp. from the ~3.8 Ga-old Isua metasedimentary suite, West Greenland (d-f). The optically distinctive marginal rim may be explained as a relic of the original cell wall. B: Laser Raman spectra obtained from *Huroniospora* sp. as an isolated particle (a) and in thin sections (b) compared to those from *Isuasphaera* sp. (c, d) obtained under the same conditions. The close resemblance of the spectra suggests similarities in the composition of the residual organic component of the two types of microstructures. The prominent peak close to 1610 cm^{-1} is indicative of aromatic double bonds among the carbon atoms of the molecular structure (adapted from [44]).

includes at least some elements of a structurally degenerated microfossil assemblage such as might result from metamorphic impairment of a Warrawoona-type microflora. As fully detailed below, the existence in Isua times of microbial ecosystems would not only be consistent with, but also conditional for, the actually observed carbon content and carbon isotope geochemistry of the Isua suite.

23.2 Biogeochemical Evidence

Apart from morphological relics, organisms also leave a chemical record of their former existence. While the bulk of the body material degrades after the death of the organism (with the dominant carbon component remineralized as CO_2), a minuscule fraction of the organic substance (between 10^{-2} and 10^{-3}) usually escapes decomposition by burial in newly-formed sediments. Here, the primary biopolymers undergo a large-scale reconstitution ("humification" with subsequent transformation into inorganic carbon polymers), resulting finally in the formation of kerogen, a chemically inert (acid-insoluble) polycondensed aggregate of aliphatic and aromatic hydrocarbons that figures as end-product of the diagenetic alteration of primary biogenic matter in sediments [27]. Representing the residuum of living substances, kerogenous materials and their graphitic derivatives constitute *per se* first-order proxies of past biological activity.

A more specific category of biogeochemical evidence is represented by quasi-pristine organic molecules (mostly pigments or single discrete hydrocarbon chains) that have preserved their identity over the entire pathway of kerogen evolution from dead organic matter to the polymerized aggregates of solid hydrocarbons that mark the end of the maturation series of primary biogenic substances. The record of such "biomarker" molecules or "chemofossils", respectively [28, 29], goes far back into Precambrian times, gaining special significance in connection with the geologically oldest (Proterozoic) petroleum occurrences [30]. Current work on molecular fossils from the kerogen fraction of Precambrian rocks [31] holds the promise of considerable potential for the further elucidation of the early history of life.

Most importantly, the isotopic composition of the kerogenous carbon constituents gives additional testimony to their biogenicity as it reflects $^{13}C/^{12}C$ fractionations of a magnitude and direction that are typically obtained in the common assimilatory (photosynthetic) pathways, notably the ribulose bisphosphate (RuBP) carboxylase reaction of the Calvin cycle that operates in C3 photosynthesis. Particularly if seen in conjunction with the isotope record of coexisting carbonate, the $^{13}C/^{12}C$ compositions of sedimentary organic carbon convey a remarkably consistent signal of biological carbon fixation that derives, for the most part, from a kinetic isotope effect imposed on the first CO_2-fixing enzymatic carboxylation reaction of the photosynthetic pathway.

In the following, selected aspects of the biogeochemical record of life will be dealt with in more detail.

23.2.1   Sedimentary Organic Carbon as a Recorder of Former Life Processes

As set out above, reduced or organic carbon (C_{org}) stored in sedimentary rocks constitutes the fossil residue of biological matter, the bulk of which is made up of highly polymerised aliphatic and aromatic hydrocarbons ("kerogen") that figure as end-products of the diagenetic alteration of primary biogenic substances in the sediment. The C_{org}-content of the average sedimentary rock normally lies between 0.5% and 0.6% [32]. With a total mass of the Earth's sedimentary envelope of about 2.4×10^{24} g, this would imply that, over geologic time, between $1.2-1.4 \times 10^{22}$ g C_{org} had leaked out from the surficial carbon cycle and come to be stored in the crust, thus making kerogen the most abundant form of organic matter on this planet. The fossil relics of biogenic matter thus accumulated can be traced back to the very beginning of the sedimentary record some 3.8 Ga ago [33], with shales representing the C_{org}-richest (up to 12-16%) and sandstones the C_{org}-poorest lithologies. The mobile (volatile) fraction of the sedimentary C_{org}-burden has commonly oozed out from the source sediment during the maturation process of the kerogenous substances, assembling as oil and gaseous hydrocarbons in suitable host rocks.

Systematic assays for C_{org} carried out on Phanerozoic (<0.54 Ga) rocks indicate that the deposition rates of organic matter have, for the most part, moderately oscillated around a mean of about 0.5% C_{org} for the average sediment over this time span. Although both the preserved rock record and the available data base are progressively attenuated as we proceed into the older geological past, the evidence suggests that the scatter of C_{org} in Precambrian sediments is basically the same as in the Phanerozoic record [34, 35]. It is worth noting that the (largely graphitized) organic carbon load of the 3.8 Ga-old Isua metasediments of West Greenland that mark the very beginning of the record may considerably surpass the local average of 0.67% in the case of carbon-rich members of the suite.

Apart from constituting materialized residues of primary living substances in the widest sense, the elemental composition (H, C, O, N, S) of kerogens may offer additional information on the source organisms and their habitat. Plots of H/C vs. O/C ratios have allowed to differentiate between sapropelic kerogens of microbial and algal pedigree (characterized by high H/C and low O/C) and a humic moiety derived from higher plants (with low H/C and high O/C). When plotted against each other in the so-called "Van Krevelen diagram", both ratios decrease with increasing maturity of the respective kerogen species, their evolution paths finally converging in the overlap of H/C = 0.5-0.8 and O/C = 0.0-0.1. The degree of dehydrogenation as quantified by the H/C ratio serves, furthermore, as an index of aromaticity, reflecting the progressive replacement of aliphatic hydrocarbon chains by cyclic (aromatic) hydrocarbons during the polymerization process. At the very end of the kerogen evolution series stands graphite as the H- and O-free variant of reduced sedimentary carbon.

In sum, it can be stated that organic carbon in the form of kerogen and its graphitic derivatives is a common constituent of sedimentary rocks over the whole of the presently known record. Based on hitherto available data, a good case can be built that the C_{org}-content of the average sediment has stayed remarkably uniform from Archaean to present times, ranging broadly between 0.4% and 0.7%. The abundance of highly graphitized kerogenous materials in the 3.8 Ga-old Isua metasediments is likely to give eloquent testimony to the operation of life processes already during the time of formation of the oldest terrestrial sediments.

23.2.2 $^{13}C/^{12}C$ in Sedimentary Organic Matter: Index of Autotrophic Carbon Fixation

Carbon basically consists of a mixture of two stable isotopes, ^{12}C and ^{13}C; a third, short-lived radioactive nuclide, ^{14}C, occurs only in trace amounts. All transformations of the element in the geochemical cycle have been shown to entail thermodynamic and/or kinetic isotope effects that cause newly-formed phases to be isotopically distinctive from their precursor substances. The largest isotope effects are commonly brought about during the conversion of inorganic to organic carbon in autotrophic carbon fixation. This is primarily the assimilation of CO_2 and bicarbonate ion (HCO_3^-)

by plants and microorganisms that proceeds by a limited number of pathways, which invariably discriminate against the heavy carbon isotope (^{13}C).

With the uptake and intracellular diffusion of external CO_2 and the subsequent first CO_2-fixing carboxylation reaction, constituting the principal isotope-selecting steps in the primary metabolism of autotrophs, the essentials of biological carbon isotope fractionation may be summarized, with adequate approximation, by a two-step model,

$$CO_2(e) \underset{k_2}{\overset{k_1}{\rightleftarrows}} CO_2(i) \underset{k_4}{\overset{k_3}{\rightleftarrows}} COOH \tag{23.1}$$

in which $CO_2(e)$ and $CO_2(i)$ represent external (environmental) and internal (cell-hosted) CO_2, respectively, and k_1 - k_4 stand for the rate constants of both assimilatory and reverse reactions. In sum, the individual fractionations tend to bring about a sizeable increase of "light" carbon (^{12}C) in all forms of biosynthesized matter relative to the carbon dioxide of the environmental feeder pool. The largest single fractionation effect commonly derives from the isotope-discriminating properties of the carboxylating enzymes operative in the second step responsible for the first irreversible CO_2-fixing carboxylation reaction [cf. 36-38]. This latter reaction promotes the incorporation of CO_2 into the carboxyl group (COOH) of an organic acid, which, in turn, lends itself to further processing in subsequent metabolic pathways. As biochemical carbon-fixing reactions are largely enzyme-controlled, and living systems constitute dynamic states undergoing rapid cycles of anabolism and catabolism, it is generally accepted that most biological isotope fractionations are due to kinetic rather than equilibrium effects.

Quantitatively, the differences in the isotopic composition of carbon are expressed in terms of the conventional δ-notation that gives the permil deviation in the $^{13}C/^{12}C$ ratio of a sample (sa) relative to that of a standard (st), i.e.

$$\delta^{13}C = \left[\frac{\left(^{13}C/^{12}C \right)sa}{\left(^{13}C/^{12}C \right)st} - 1 \right] \times 1000 \qquad (‰, PDB) \tag{23.2}$$

The standard defining zero permil on the δ-scale is Peedee belemnite (PDB) with $^{12}C/^{13}C = 88.99$. Positive values of $\delta^{13}C$ indicate an enrichment of heavy carbon (^{13}C) in the sample, while negative values stand for a depletion in ^{13}C. Because of the discrimination against "heavy" carbon in the common carbon-fixing pathways, we observe a marked decrease of the $\delta^{13}C$ values in all forms of biogenic (reduced) carbon compared to the feeder pool of inorganic (oxidized) carbon made up mainly of atmospheric CO_2 and dissolved marine bicarbonate ion (HCO_3^-). Generally, the $\delta^{13}C$ values of average biomass usually turn out to be 20-30‰ more negative than those of marine bicarbonate, the most abundant inorganic carbon species in the environment.

The isotopic difference thus established between organic (biogenic) carbon and the surficial bicarbonate-carbonate pool is largely retained when organic and carbonate carbon enter newly-formed sediments. As is obvious from Fig. 23.4, both the carbon isotope spreads of extant primary producers and those of marine carbonate and bicarbonate are basically transcribed into the sedimentary record back to 3.5 or even 3.8 Ga ago. This would imply that organic carbon and carbonate carbon had always been transferred from the surficial environment to the crust with relatively little change in their isotopic compositions. For example, the $\delta^{13}C_{org}$ spread in recent marine sediments (Fig. 23.4) faithfully integrates over the spread of the contemporary living biomass with just the extremes eliminated, indicating that the effect of a later diagenetic overprint on the primary isotope values is rather limited (usually below 3‰) and, for the most part, gets lost within the broad scatter of the original values. Consequently, the kinetic isotope effect inherent in photosynthetic carbon fixation is propagated from the biosphere into the rock section of the carbon cycle almost unaltered, which opens up the possibility of tracing the isotopic signature of this process back into the geologic past.

With these relationships established, decoding of the vast body of isotopic information stored in the sedimentary record is fairly straightforward. There is little doubt that the conspicuous ^{12}C-enrichment displayed by the data envelope for fossil organic carbon (Fig. 23.4) constitutes a coherent signal of autotrophic carbon fixation over almost 4 Ga of recorded Earth history as it ultimatively rests with the process that gave rise to the biological precursor materials. Moreover, the long-term uniformity of the signal attests to an extreme degree of conservatism of the basic biochemical mechanisms of carbon fixation. In fact, the mainstream of the envelope for $\delta^{13}C_{org}$ depicted in Fig. 23.4 can be most readily explained as the geochemical manifestation of the isotope-discriminating properties of one single enzyme, namely, ribulose-1,5-bisphosphate (RuBP) carboxylase, the key enzyme of the Calvin cycle.

It is well known today that the carbon transfer from the inorganic to the organic world largely proceeds via the RuBP carboxylase reaction that feeds CO_2 directly into the Calvin cycle as a 3-carbon compound (phosphoglycerate). Most autotrophic microorganisms and all green plants operate along this pathway of carbon assimilation; higher plants relying on it entirely are termed C3 plants. As a result, the bulk of the Earth's biomass (both extant and fossil) bears the isotopic signature of C3 (or Calvin cycle) photosynthesis characterized by the sizeable fractionations of the RuBP carboxylase reaction that assigns a mean $\delta^{13}C_{org}$ range of -26 ± 7‰ to most biogenic matter.

Occasional negative offshoots from this long-term average (Fig. 23.4) are commonly restricted to the Precambrian [35, 38] and suggest the involvement of methanotrophic pathways in the formation of the respective kerogen precursors. Though, at first sight, these excursions might appear as oddities confined to side stages of the carbon cycle, a closer scrutiny of Fig. 23.4 reveals that the respective minima are superimposed on a markedly lowered margin of the $\delta^{13}C_{org}$ envelope that characterizes the early Paleoproterozoic and Archaean record, attesting to an important role of methane in biogeochemical carbon transformations on the early Earth.

A major apparent discontinuity in the $\delta^{13}C_{org}$ record depicted in Fig. 23.4 is the break between the ~3.8 Ga-old Isua metasediments from West Greenland and the whole of the post-Isua record. The observed isotope shift is, however, fully consistent with the predictable effects of an isotopic re-equilibration between coexisting organic carbon and carbonate in response to the amphibolite-grade metamorphism experienced by the Isua suite. Both currently available thermodynamic data on $^{13}C/^{12}C$ exchange between C_{org} and carbonate carbon (C_{carb}) as a function of increasing metamorphic temperatures and observational evidence from a host of geologically younger metamorphic terranes make it virtually certain that the "normal" sedimentary $\delta^{13}C_{org}$ and $\delta^{13}C_{carb}$ records had originally extended back to 3.8 Ga ago, and that the Isua anomaly is clearly due to a metamorphic overprint [33, 38]. Apparently pristine $\delta^{13}C_{org}$ values in the range –21‰ to –49‰ (average: –37 ± 3‰) have been recently reported for minor carbon inclusions in apatite grains from a possibly 3.85 Ga-old

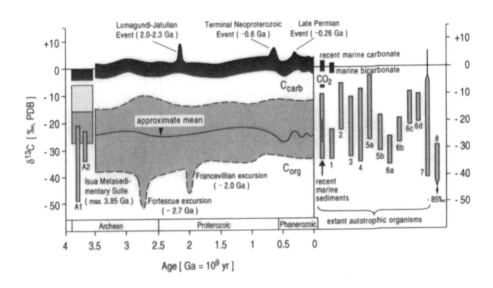

Fig. 23.4 Isotope age functions of organic carbon (C_{org}) and carbonate carbon (C_{carb}) as compared with the isotopic compositions of their progenitor substances in the present environment (marine bicarbonate and biogenic matter of various parentage, cf. Right box). Note that the $\delta^{13}C_{org}$ spread of the extant biomass is basically transcribed into recent marine sediments and the subsequent record back to 3.8 Ga, with the Isua values reset by amphibolite-grade metamorphism. In the displaced Isua segment, the lower shaded portion represents the range of values with *bona fide* biological signatures ($\delta^{13}C_{org}$ <-16‰) that have ostensibly survived the metamorphic reconstitution of their host rock; the upper light field covers those values, which, for the most part, have suffered secondary alteration as a result of high-T $^{13}C/^{12}C$ exchange. Superimposed on the Isua envelope for whole rock analyses are the spreads obtained for carbon inclusions in apatite grains from a 3.85 Ga-old Isua banded iron-formation (A1) and from other Isua iron-formations (A2) as reported by Mojzsis et al. [39]. The envelope for fossil organic carbon as a whole is an update of the data base presented by Schidlowski et al. [38], comprising the means of some 150 Precambrian kerogen provinces as well as the currently available information on the Phanerozoic record. The negative spikes at 2.7 and 2.1 Ga indicate a large-scale involvement of methane in the formation of the respective kerogen precursors. Contributors to the contemporary biomass are (1) C3 plants, (2) C4 plants, (3) CAM plants, (4) eukaryotic algae, (5a,b) natural and cultured cyanobacteria, (6) groups of photosynthetic bacteria other than cyanobacteria, (7) methanogenic bacteria, (8) methanotrophic bacteria. The $\delta^{13}C_{org}$ range in recent marine sediments [45] is based on some 1600 data points (black insert covers >90% of the data base).

Isua banded iron-formation [39]. Here, carbonaceous material sealed in a mineral matrix and hosted by a carbonate-free lithology had obviously escaped high-temperature isotope exchange, thereby preserving its original biogenic $\delta^{13}C_{org}$ spread with probably little alteration. The remarkably negative $\delta^{13}C_{org}$ mean of $-37 \pm 3\text{‰}$ displayed by the apatite-hosted carbon particles from the Isua Suite certainly conjures up a methane connection for sedimentary organic carbon accumulated during the Earth's early history (see also [40]). On the other hand, an archaebacterial record confirming a major involvement of methanotrophic bacteria in the early biogeochemical carbon cycle is as yet missing, the currently available paleontological data being clearly dominated by evidence on the cyanobacterial level [23, 25, 41].

23.3 Conclusions and Outlook

With terrestrial analogues of possible relics of extraterrestrial life furnished by the Earth's oldest sediments, we have a baseline for a reasonable evaluation of remnants of fossil life if ever encountered on other planets, notably on Mars. This holds for the morphological (cellular nad biosedimentary) as well as for the biogeochemical residua of former organisms, both categories of evidence being apt to be preserved in the sedimentary records of planetary bodies. Such scenario may be specifically envisaged for the sedimentary formations generated in a water-rich period during the early history of Mars. Considering the ubiquity of fossil organic carbon in terrestrial sediments, carbonaceous (kerogenous) constituents retrieved from Martian rocks might figure as important stores of biogeochemical information. If enzymatic carboxylation reactions in exobiological systems were beset with similar kinetic fractionation effects as in Earth-hosted biochemistry, the $^{13}C/^{12}C$ signature of these carbon constituents should furnish powerful constraints for current discussions on the existence of former life on Mars [42, 43].

23.4 References

1   D.R. Hitchcock, J.E. Lovelock , Icarus 7, 149-159 (1967).

2   J.E. Lovelock, Gaia, Oxford University Press, Oxford, UK, 1979.

3   T. Owen, K. Biemann, D.R. Rushneck, J.E. Biller, D.W. Howarth, A.L. Lafleur, J. Geophys Res. 82, 4635 (1977).

4   M. Schidlowski, in: M.H. Engel, S.A. Macko (Eds.) Organic Geochemistry, Plenum Press, New York, 1993, pp. 639.

5   M. Schidlowski, in: J.M. Greenberg, C.X. Mendoza-Gomez, V. Pirronello (Eds.) The Chemistry of Life's Origin, Kluwer Academic Publ., The Netherlands, 1993, pp. 389.

6   M. Schidlowski, in: R.B. Hoover (Ed.) Instruments, Methods and Missions for Astrobiology, Proc. Int. Soc. Opt. Engin. (SPIE) 3441, Bellingham; WA, 1998, pp. 149.

7   M. Schidlowski, P. Aharon, in: M. Schidlowski, S. Golubic, M.M. Kimberley, D.M. McKirdy, P.A. Trudinger (Eds.) Early Organic Evolution: Implications for Mineral and Energy Resources, Springer Verlag, Berlin, 1992, pp. 147.

8   M.H. Carr, H. Wänke, Icarus 98, 61 (1992).

9   M.H. Carr (Ed.) *Water on Mars*, Oxford University Press, Oxford, 1996, 229 pp.

10   V.I. Moroz, L.M. Mukhin (Eds.) *About the Initial Evolution of Atmosphere and Climate of the Earth-Type Planets*. Inst. Space Res. USSR Acad. Sci., Moscow, Publication D-255, 1978, 44 pp.

11   C.P. McKay, Icarus **91**, 93 (1991).

12   C.P. McKay, R.L. Mancinelli, C.R. Stoker, R.A. Wharton, in: H. Kieffer, B.M. Jacosky, C.W. Snyder, M.S. Matthews (Eds.), *Mars*, Tuscon University of Arizona Press, 1992, pp. 1234.

13   J.W. Schopf (Ed.) *Earth's Earliest Biosphere: Its Origin and Evolution*. Princeton University Press, Princeton, NJ, XXV, 1983, 543 pp.

14   E.G. Nisbet, in: M.P. Coward, A.C. Ries (Eds.) *Early Precambian Processes*, Geol. Soc. Special Publ. 95, London, 1995, pp. 27.

15   K. Biemann, J. Oro,.P. Toulmin, L.E. Orgel, A.O. Nier, D.M. Anderson, P.G. Simmonds, D. Flory, A.V. Diaz, D.R. Rushneck, J.E. Biller, A.L. Lafleur, J. Geophys. Res. **82**, 4641 (1977).

16   C.P. McKay, Adv. Space Res. **6**(12), 269 (1986).

17   C.P. McKay, S.S. Nedell, Icarus **73**, 142 (1988).

18   R.V. Burne, L.S. Moore, Palaios **2**, 241 (1987).

19   M.R. Walter, in: J.W. Schopf (Ed.) *Earth's Earliest Biopshere: Its Origin and Evolution*, Princeton University Press, Princeton, NJ, 1983, pp. 187.

20   J.P. Grotzinger, A.H. Knoll, Annu. Rev. Earth Planet. Sci. **27**, 313 (1999).

21   H.J. Hofmann, in: R.E. Riding, S.M. Awramik (Eds.) *Microbial Sediments*, Springer Verlag, Berlin, 2000, pp. 315.

22   H.D. Pflug, Naturwissenschaften **65**, 611 (1978).

23   S.M. Awramik, J.W. Schopf, M.R. Walter, in: B. Nagy, R. Weber, J.C. Guerrero, M. Schidlowski (Eds.) *Developments and Interactions of the Precambrian Atmosphere, Lithosphere and Biosphere*, Developments in Precambrian Geology 7, Elsevier, Amsterdam, 1983, pp. 249.

24   R. Buick, Palaios **5**, 441 (1991).

25   J.W. Schopf, B.M. Packer, Science **237**, 70 (1987).

26   J.W. Schopf, Science **260**, 640 (1993).

27   B. Durand (Ed.) *Kerogen-Insoluble Organic Matter from Sedimentary Rocks*, Editions Techniq, Paris, 1980, 519 pp.

28   A. Treibs, Liebigs Ann. d. Chemie **510**, 42 (1934).

29   G. Eglinton, M. Calvin, Sci. Am. **216**, 32 (1967).

30   R.E. Summons, T.G. Powell, in: M. Schidlowski, S. Golubic, M.M. Kimberley, D.M. McKiroy, P.A. Trudinger (Eds.) *Early Organic Evolution: Implications for Mineral and Energy Resources*, Springer, Berlin, 1992, pp. 296.

31   J.J. Brocks, G.A. Logan, R. Buick, R.E. Summons, Science **285**, 1033 (1999).

32   A.B. Ronov, A.A. Yaroshevsky, A.A. Migdisov (Eds.) *Khimicheskoe Stroyenie Zemnoi Kory i Khimicheski Balans Glavnykh Elementov* (Chemical Structure of the Earth's Crust and Chemical Balance of Major Elements (in Russian)). Izdatel'stvo Nauka, Moscow, 1990, 181 pp.

33   M. Schidlowski, P.W.V. Appel, R. Eichmann, C.E. Junge, Geochim. Cosmochim. Acta **43**, 189 (1979).

34   M. Schidlowski, in: H.D. Holland, M. Schidlowski (Eds.) *Mineral Deposit and the Eolution of the Biosphere*, Springer, Berlin, 1982, pp. 103.

35 J.M. Hayes, I.R. Kaplan, K.W. Wedeking, in: J.W. Schopf (Ed.) *Earth Earliest Biosphere: Its Origin and Evolution*, Princeton University Press, Princeton, N.J, 1983, pp. 93.

36 R. Park, S. Epstein, Geochim. Cosmochim. Acta **21**, 110 (1960).

37 M.H. O'Leary, Phytochemistry **20**, 55 (1981).

38 M. Schidlowski, J.M. Hayes, I.R. Kaplan, in: J.W. Schopf (Ed.) *Earth's Earliest Biosphere: Its Origin and Evolution*, Princeton University Press, Princeton, NJ, 1983, pp. 149.

39 S.J. Mojzsis, G. Arrhenius, K.D. McKeegan, T.M. Harrison, A.P. Nutman, R.L. Friend, Nature **384**, 55 (1996).

40 J.M. Hayes, in: S. Bengtson (Ed.) *Early Life on Earth*, Nobel Symposium 84, New York Columbia Univ. Press, 1994, pp. 220.

41 J.W. Schopf (Ed.) *Cradle of Life*, Princeton Univ. Press, Princeton, N.J., 1999, 347 pp.

42 L.J. Rothschild, D. DesMarais, Adv. Space Res. **9**(6), 159 (1989).

43 M. Schidlowski, Adv. Space Res. **12**(4), 101 (1992).

44 H.D. Pflug, Topics in Current Chemistry **139**, 1 (1987).

45 P. Deines, in: P. Fritz, J.C. Fontes (Eds.) *Handbook of Environmental Isotope Geochemistry*, Elsevier, Amsterdam, (1980), Vol. 1, pp. 329.

Part VI

Forthcoming Space Missions Relevant for Astrobiology

24 Space Activities in Exo-Astrobiology

Bernard H. Foing

The origin of stars and planetary systems, life in our solar system and possibly elsewhere in the Universe are research topics, which have attracted great interest among scientists. The discoveries of proto-planetary disks around other stars and the detection of more than 50 exoplanets provides evidence that the formation of extrasolar systems may be a common process throughout the Universe. Biogenic elements such as C, H, O, N, S, and P are known to be widespread in our Galaxy and beyond. The search for organic molecules in interstellar and circumstellar environments, their incorporation into potential planet-forming disks and subsequently in solar system material has been successfully investigated within the last decade. The origin of life on planet Earth might have proceeded from simple precursor molecules to more complex self-replicating, metabolizing structures, evolving into primitive life. Extraterrestrial delivery of organic matter and water by comets and asteroids shortly after planetary formation may have triggered the emergence of life on Earth and possibly on Mars. The common interest on the origin and distribution of life in the Universe led to a new discipline, named Astrobiology, which is investigated by a great number of interdisciplinary scientists, well documented in the preceding chapters of this book. Exploration with astronomical telescopes, satellites and space missions contributes to the investigation of possible life habitats in our solar systems, the search for exoplanets and the link between infalling extraterrestrial matter and the jump-start of life on Earth. In this respect, Astrobiology will benefit from and determine a number of space exploration programs in the future. In the following a brief overview is given about astronomical and planetary space missions, which will investigate astrobiological aspects during their operation phase.

24.1 Astrobiological Potential of Space Astronomy Missions

24.1.1 Infrared Spectroscopy of Cosmic Dust and Organics

Several space missions are in progress, or are well into the planning stage, that have key objectives concerning the nature of extraterrestrial organic chemistry, the search for extrasolar systems and for traces of past or present life. Fig. 24.1 shows a number of space missions, which investigate astrobiological perspectives, in particular by providing infrared (IR) and sub-mm data. The Infrared Space Observatory ISO, in operation between 1995-1998, has revolutionized our understanding of gas and dust in interstellar and circumstellar space by monitoring the distribution of organic molecules in such regions. The exploitation of ISO data will remain a major effort of the infrared community in the following decade. In the meantime the next infrared satellite (SIRTF, US mission) is already on the start ramp. The airborne observatory SOFIA will be launched in the near future and will be able to observe parts of the near infrared spectrum.

HERSCHEL, earlier called FIRST, is one of the Cornerstone missions of ESA's Horizons 2000 program and will be launched in 2007. The Herschel Space Observatory will be the only space facility ever developed covering the far infrared to submillimeter range of the spectrum (from 80 to 670 microns). The Herschel satellite is approximately 7 meters high and 4.3 meters wide, with a launch mass of around 3.25 tons. It will carry the infrared telescope and three scientific instruments and will be located 1.5 million km away from Earth. Herschel has an operational lifetime of three years minimum. It potentially offers about 7 000 hours of science time per year. It is a multi-user observatory accessible to astronomers from all over the world. The key science goals that Herschel will achieve concern the formation of galaxies in the early universe, and how stars form. Herschel will contribute to astrobiological studies by studying the processes, by which stars, their surrounding proto-planetary disks and planets themselves are made.

Also ground-based IR facilities and (sub)millimeter interferometers, such as VLT, KECK and ALMA, will be important in the astrobiological context, since they will allow the detection of complex molecules with abundances almost a factor of a hundred below current detection limits.

24.1.2   The New Generation Space Telescope (NGST)

The new generation Space Telescope NGST will be able to penetrate the dusty envelopes around new-born stars and take a closer look at the stars themselves by using the infrared part of the spectrum. NGST will also have the sensitivity to study very small objects that are not massive enough to become stars. These objects - brown dwarfs and Jupiter-sized planets - will become targets for intensive study with NGST. The high resolution of NGST will also make it possible to see how other planetary systems form, and in this way enable us to study the origin of extrasolar systems.

24.1.3   Exoplanets from Space:
GAIA, COROT, EDDINGTON, KEPLER and DARWIN

Milestones are expected in the search for extrasolar planets. Some 50 Jupiter type planets have now been detected from ground based velocity monitoring. With the GAIA astrometric mission, we should be able to measure the stellar reflex motion to detect tens of thousands of jovian planets. The COROT mission will be able to detect the transit of giant planets, but also the presence of terrestrial planets (super Earth's) during the 150 days continuous high precision simultaneous photometry of 5000 stars.

The ESA Eddington and NASA Discovery mission KEPLER will have the capability to detect transits by Earth size planets in the habitable zones. How planetary systems form and evolve, and whether habitable or life-bearing planets exist around nearby stars are major questions to be studied with DARWIN after 2010. With the help of nulling interferometers in the thermal infrared to remove the parent star light, IRSI-DARWIN will search for the spectral signature of gases such as CH_4 and O_3 in the atmosphere of extrasolar planets in order to identify Earth-like planets capable of sustaining life. Most of the molecular oxygen in the Earth's atmosphere is thought to have been produced by bacterial activity in the last billion years. O_3 is a sensitive tracer of O_2, and its detection would give hint for formidable astrobiology developments.

24.1.4 Global Life Signatures on Earth?

In this context one should recall that the remote signature of life is common practice with Earth remote sensing of vegetation by spectral imagery (for instance with ERS or ENVISAT). However this is diluted in the global abiotic spectral signature of the planet, as witnessed with the analysis of Galileo Earth flyby observations.

24.2 Astrobiological Potential of Planetary Missions

Figure 24.2 shows the most important space missions exploring our solar system in the near future.

24.2.1 CASSINI HUYGENS

CASSINI-HUYGENS was launched on October 15, 1997 and is on its journey to explore Saturn and its moon Titan in 2004. Before the CASSINI orbiter continues to explore Saturn, its moons, magnetosphere and rings, the European HUYGENS probe will be released and parachute through the atmosphere of Titan with an entry speed of 20 000 km/h, then allowing a 2.5-hour descent of the probe. During this period six instruments will measure the properties of Titan's atmosphere, which is known to contain organic molecules and nitriles and thought to resemble that of the young Earth. Recent high resolution images through the thick haze provide evidence of hydrocarbon oceans and continents on Titan's surface. CASSINI images and spectra will also give a penetrating view until the surface of Titan. Data of the Huygens probe before its impact with an unknown surface will be of particular interest for astrobiology and deliver information on the prebiotic conditions of the second largest moon in our solar system.

392   B.H. Foing

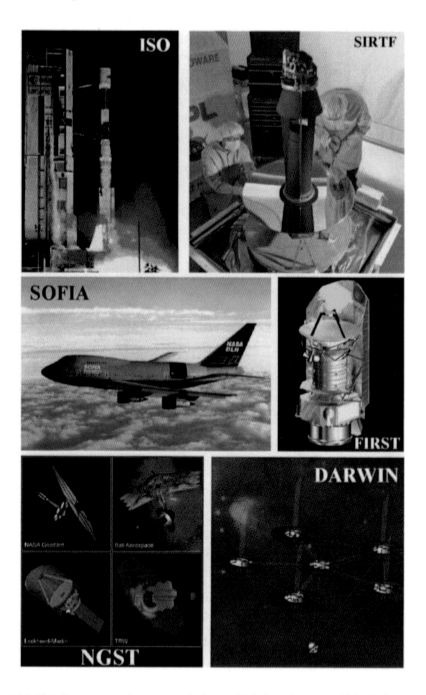

Fig. 24.1 This figure summarizes space missions, which investigate astrobiological perspectives, in particular by providing infrared (IR) and sub-mm data. Among them are the infrared satellites ISO, SIRTF, and the airborne observatory SOFIA. ESA's cornerstone mission FIRST (HERSCHEL) will observe in the wavelength range between 80-670 micron and on the NGST a mid-IR facility is planned. IRSI-DARWIN will look for Earth-like planets and monitor their atmosphere in the IR.

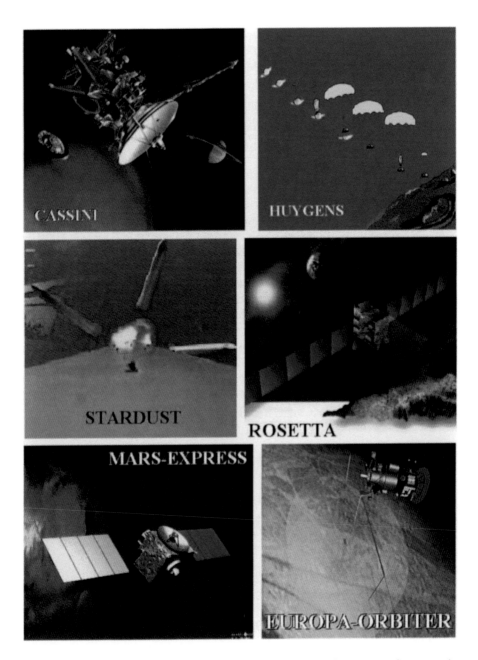

Fig. 24.2 This figure shows the most important space missions exploring our solar system in the near future with astrobiology relevance. Among them are CASSINI-HUYGENS, a mission to Saturn and Titan and the US mission STARDUST on its way to meet comet Wild-2. The ESA missions ROSETTA to rendezvous comet Wirtanen and MARS-EXPRESS/Beagle2 for Mars exploration, respectively are both launched in 2003. Planned for the future are missions to Jupiters's moon Europa to investigate its ice crust and possible subsurface oceans.

24.2.2 STARDUST

The US space mission STARDUST was launched in 1999 and is currently on the route to rendezvous comet Wild-2. One of the main tasks of this mission is to capture comet dust and volatiles by impact into an ultra-low-density aerogel at a low speed of 6.1 km/s. The encounter distance will be approximately 150 km away from the nucleus. The collected material will be dropped off in a re-entry capsule that will parachute to Earth in 2006. This will represent the first cometary material to be analyzed in Earth laboratories.

24.2.3 ROSETTA

The ROSETTA comet rendezvous mission will be launched in 2003 with the Ariane 5 rocket for a rendezvous maneuver with comet 46P/Wirtanen in 2011-2013. More than 20 instruments on the orbiter and the lander will obtain data on cometary origin and the interstellar-comet connection, which will broaden our insight into the origin of our solar system. ROSETTA will study comets Wirtanen's nucleus and its environment in great detail for nearly 2 years with far-observation activities leading ultimately to close observations (~1 km distance) and to unprecedented measurements from the ROSETTA lander.

Comets are the relics of the planet-formation process in our own solar system and are thought to contain the most pristine chemical record, since they have spent most of their life in the cold outer part far from the Sun. Knowledge on their composition is a key astrobiological objective in order to investigate what material was delivered to the early planets during cometary impacts in their early history.

24.2.4 MARS-EXPRESS and Future Mars Missions

In the search for life in our solar system the planet Mars represents the main target and will be visited by many spacecrafts in the next two decades. Also a Mars Sample Return mission is envisaged in the time frame 2009-2014, preceded by a series of automatic missions in orbit and on the surface with in situ analysis. The main goal of the European MARS-EXPRESS mission, to be launched in 2003, will be the search for water as well as for ancient and present life on Mars. The spacecraft orbiter will carry eight instruments, all of which will make a contribution to solving the mystery of the missing water. The current Mars environment is too cold (and the atmosphere is too thin) to retain liquid water on its surface. However, data from the Mars Pathfinder, which landed successfully on Mars in July 1997 suggested widespread flowing water in the previous history of Mars. Water could also be trapped as underground ice on the planet Mars. MARS-EXPRESS spectrometers will measure water in the atmosphere, while the radar will look at the surface. Just before its orbit insertion MARS-EXPRESS will jettison a lander called Beagle 2, which will head for the Martian surface where it will take in situ measurements of rocks and soil. The MARS EXPRESS Beagle 2 lander will carry a variety of scientific instruments, such as panoramic and wide field cameras and a microscope, which will look closely and investigate rocks.

The small robotic arm of Beagle 2 will carry sensors for analysis of rock fragments for the presence of organic matter, water and minerals. Beagle 2 will also deploy a mole capable of crawling short distances beneath the surface to collect soil samples for a gas analysis system. The primary aim of these experiments will be to see if any evidence of past life processes remain near the landing site. Preparation for this mission and future space missions to Mars are supported by the ground-based research of life under extreme conditions (such as permafrost, hydro-thermal vents or salt crystals).

24.2.5   Europa Missions

Jupiter's moon Europa probably hosts a subsurface water ocean beneath its outer ice crust. What geological processes create the ice rafts and other ice-tectonic processes that are at the origin of prominent surface features on Europa are currently strongly debated.

A future mission currently in the planning stage for launch by US after 2006 is to visit Jupiter's moon Europa in order to study the properties of the ice crust with radar measurements. Europa seems to have an internal energy source provided by tidal friction through its interaction with Jupiter, which could keep water in liquid state below the crust. Europa provides therefore key ingredients for life (water, energy and possibly organic molecules) and will certainly be a future target for several space missions.

24.2.6   Moon, Mercury, Formation of Planets and the Early Frustration of Life

Future Mercury missions (MESSENGER, BEPICOLOMBO) and lunar missions (SMART1, SELENE) will give information of telluric planets in the inner solar system. They will also investigate the presence of ice deposits (and eventually organics) in the permanently shadowed polar areas. They will quantify the early bombardment history in the inner solar system, with relevance to the frustration, selection and evolution of life. Comparative planetology studies will give a better understanding on geological conditions on Early Earth, when life emerged.

24.2.7   Space Exposure Experiments

The International Space Station ISS also offers facilities, which investigate issues relevant for Astrobiology. On the SEBA/EXPOSE exposure facility on the ISS EXPRESS-PALLET the radiation stability of organic molecules and primitive organisms are tested in the context of extraterrestrial delivery and panspermia. The STONE facility allows to simulate the impact of material and studies the survival and evolution of minerals, organics and spores during atmospheric entry.

Table 24.1. Space roadmap for searching for signatures of life in the solar system and beyond

	Now	2003	2010	2020
Limits of Life				
Extremophiles	In-situ			
Survival conditions on Earth		Active experiments		
Limits of life under Mars conditions		Mars missions		
Exoplanets and Habitability				
Exoplanets				
Exo-Jupiters	Ground-Based		GAIA	
Terrestrial Super Earths		COROT		
Earth-like habitable			Eddington or Kepler	
Atmospheric imprint of life			IRSI Darwin	
Habitability				
Water	MGS	Mars-Express		
Climate		Mars Climate Missions		
Frustration of life by impacts		SMART-1	Bepi Colombo	
Signatures of Life				
Biomarkers in solar systems				
Meteorites				
Instrumentation for life		Mars-Express		
characterization		Mars landers		
Biomarkers in exoplanets				
Atmospheric O_2-O_3			IRSI-Darwin	
Chlorophyll	Earth			?
Artefacts of civilisations			IR	
Radio bio-signals	SETI		lunar far-side	
Life in Solar Systems				
Mars				
Water	MGS	Mars Express		
Exobiology		Beagle 2		
Rovers		US Mars 2003		
Exobiology Mars Facility			EMF	
10 m cores			X	
In-situ			Robotic outpost	
Search for life				Sample return
Man assisted research				
Europa				
Ocean surface	Galileo			
Ocean and liquid niches			Orbiter radar	
Search for liquid life				Penetrator

Table 24.2. Space roadmap for expanding life in the solar system

Future of Life	1980	Now	2005	2010	2020	
Earth and Earth Orbit						
Ecosystems Response on human timescale			Living planet			
Earth long term habitability						
Hazards to Earth life				NEO watch		
Life in Earth orbit		ISS -------------------- ISS				
Terrestrial life in space		X				
Planetary protection						
Terrestrial life on extraterrestrial outposts				X		
Evolving extraterrestrial life				X		
Moon						
First organisms	Luna Ranger					
first humans	Apollo					
Robotic precursors		Prospector / SMART-1				
Landers			Lunar-A / SELENE-B			
Virtual telepresence			X			
Robotic Outposts				X		
Ecosystem experiments				Femme Melissa		
Vegetal, animals			X			
Resource utilisation			X			
Life support systems				X		
Evolving life on the Moon				X		
Robotic villages					X	
Humans and Lunar bases					X	
Expansion						X
Colonisation						X
Mars						
Precursor	Viking					
Orbiters		MGS	Mars express			
Landers		Pathfinder				
Virtual telepresence			X			
Robotic outposts				X		
Resource utilisation					X	
Evolving life on Mars					X	
Manned expedition						X
Live off the land						X
Colonisation						
Terraforming						

24.3 Conclusion: Roadmap for Astrobiology and Long-Term Space Exploration

24.3.1 Experimenting for Life in the Universe

As a perspective we should mention other areas where space research can contribute to Astrobiology. This is namely in the experimentation of the limits of life, the conditions of habitability on other planets, the search for signatures of life in the solar system and elsewhere in the Universe. A road map is indicated including the space missions identified (see Table 24.1.).

24.3.2 Expanding Life in the Solar System

Another aspect of Astrobiology where space research will play a key role, concerns the future of life on Earth, in Earth orbit, on the Moon, Mars, and after 2020 in the solar system and beyond. As the space missions for this exploration program are still being defined, we propose a road map for these investigations, concentrating on the next 30 years (see Table 24.2.).

Acknowledgements. The following web-sites have plenty of information and supply the reader with recent highlights on astrobiological space research: http://sci.esa.int; http://www.astrobiology.com/; and http://origins.jpl.nasa.gov/.

List of Contributors

Ralf H. Anken
Zoological Institute, University of Stuttgart-Hohenheim, Garbenstr. 30,
70593 Stuttgart, Germany,
E-mail: anken@uni-hohenheim.de

Christa Baumstark-Khan
Radiation Biology Division, Institute of Aerospace Medicine,
DLR German Aerospace Centre, Linder Höhe, 51147 Köln, Germany,
E-mail: christa.baumstark-khan@dlr.de

André Brack
Centre de Biophysique Moleculaire, CNRS, Rue Charles Sadron,
45071 Orleans Cedex 2, France
E-mail: brack@cnrs-orleans.fr

Richard Bräucker
Institute of Zoology, University of Bonn, 53115 Bonn, Germany
E-mail: richard.braeucker@planet-interkom.de

Werner von Bloh
Potsdam Institute for Climate Impact Research, PO Box 601203,
14412 Potsdam, Germany
E-mail bloh@pik-potsdam.de

Baruch S. Blumberg
NASA Astrobiology Institute, Ames Research Center, Moffett Field, CA, 94035, USA
E-mail: bblumberg@mail.arc.nasa.gov

Eberhard Bock
Institute of General Botany, Department of Microbiology, University of Hamburg,
Ohnhorststrasse 18, 22609 Hamburg, Germany
E-mail: bock@mikrobiologie.uni-hamburg.de

Christine Bounama
Potsdam Institute for Climate Impact Research, PO Box 601203,
14412 Potsdam, Germany
E-mail bounama@pik-potsdam.de

Charles S. Cockell
British Antarctic Survey, High Cross, Madingley Road, Cambridge CB3 0ET,
United Kingdom
E-mail: csco@bas.ac.uk

Augusto Cogoli
Space Biology Group and Biotechnology Space Support Center, BIOTESC,
Swiss Federal Institute of Technology, ETH Zurich, Technoparkstrasse 1,
8005 Zurich, Switzerland
E-mail cogoli@spacebiol.ethz.ch

Francis A. Cucinotta
NASA Johnson Space Center, Houston, TX 77058, USA
E-mail: fcucinot@ems.jsc.nasa.gov

H. G. M. Edwards
Department of Chemistry and Forensic Science, University of Bradford,
Bradford BD7 1DP, United Kingdom
E-mail: h.g.m.edwards@bradford.ac.uk

Pascale Ehrenfreund
Raymond and Beverly Sackler Laboratory for Astrophysics, Leiden Observatory,
PO Box 9513, 2300 RA Leiden, The Netherlands
E-mail: pascale@strwchem.strw.leidenuniv.nl

Rainer Facius
Radiation Biology Division, Institute of Aerospace Medicine,
DLR German Aerospace Center, Linder Höhe, 51147 Köln, Germany
E-mail: rainer.facius@dlr.de

Bernard H. Foing
Research Division, ESA Space Science Department, ESTEC SCI-SR,
2200 AG Noordwijk, The Netherlands
E-mail: bernard.foing@esa.int

Siegfried Franck
Potsdam Institute for Climate Impact Research, PO Box 601203,
14412 Potsdam, Germany
E-mail franck@pik-potsdam.de

Paul Geissler
Lunar and Planetary Laboratory, University of Arizona,
1629 East University Blvd. Tucson, Arizona 85721-0092, USA
E-mail: geissler@lpl.arizona.edu

David A. Gilichinsky
Soil Cryology Laboratory,
Institute of Physico-Chemical & Biological Problems in Soil Science,
Russian Academy of Sciences, Pushchino, Russia
E-mail: gilichin@issp.serpukhov.su

Brett Gladman
Observatoire de la Cote d'Azur, department Cassini, PO Box 4229,
06304 Nice, Cedex 4, France
E-mail: gladman@obs-nice.fr

Richard Greenberg
Lunar and Planetary Laboratory, University of Arizona, 1629 East University Blvd.
Tucson, Arizona 85721-0092 USA
E-mail: greenber@jupiter.lpl.arizona.edu

Ernst Hauber
Institut für Weltraumsensorik und Planetenerkundung,
DLR German Aerospace Centre, Rutherfordstr. 2, 12489 Berlin, Germany
E-mail: ernst.hauber@dlr.de

Ruth Hemmersbach,
Institute of Aerospace Medicine, DLR German Aerospace Centre, Linder Höhe,
51147 Köln, Germany
E-mail ruth.hemmersbach@dlr.de

Harald Hoffmann
Institut für Weltraumsensorik und Planetenerkundung,
DLR German Aerospace Centre, Rutherfordstr. 2, 12489 Berlin, Germany
E-mail: harald.hoffmann@dlr.de

Gregory V. Hoppa
Lunar and Planetary Laboratory, University of Arizona, 1629 East University Blvd.
Tucson, Arizona 85721-0092 USA
E-mail: hoppa@lpl.arizona.edu

Gerda Horneck
Institute of Aerospace Medicine, DLR German Aerospace Centre, Linder Höhe,
51147 Köln, Germany
E-mail: gerda.horneck@dlr.de

Ralf Jaumann
Institut für Weltraumsensorik und Planetenerkundung,
DLR German Aerospace Centre, Rutherfordstr. 2, 12489 Berlin, Germany
E-mail: ralf.jaumann@dlr.de

Hans Jörg Kunte
Institut für Mikrobiologie & Biotechnologie, University Bonn,
Meckenheimer Allee 168, 53115 Bonn, Germany
E-mail: kunte@uni-bonn.de

Helmut Lammer
Space Research Institute, Department of Extraterrestrial Physics,
Austrian Academy of Sciences, Elisabethstr. 20, 8010 Graz, Austria
E-mail: helmut.lammer@oeaw.ac.at

Julia Lanz
Institut für Weltraumsensorik und Planetenerkundung,
DLR German Aerospace Centre, Rutherfordstr. 2, 12489 Berlin, Germany
E-mail: julia.lanz@dlr.de

Michel Mayor
Département d'Astronomie, Observatoire de Genève, Université de Genève,
1290 Sauverny, Switzerland
e-mail: michel.mayor@obs.unige.ch

H. Jay Melosh
Lunar and Planetary Laboratory, University of Arizona, Tucson, Arizona, USA
E-mail: jmelosh@lpl.arizona.edu

Karl M. Menten
Max-Planck-Institut fuer Radioastronomie, Auf dem Hügel 69, 53121 Bonn,
Germany
E-mail: kmenten@mpifr-bonn.mpg.de

Curt Mileikowsky
Royal Institute of Technology, Stockholm, Sweden
E-mail: mil.behr@iprolink.ch

Gregorio J. Molina-Cuberos
Space Research Institute, Department of Experimental Space Research,
Austria Academy of Sciences, Steyrergasse 17-19, 8010 Graz, Austria
E-mail: gregorio.molina-cuberos@oeaw.ac.at

Gerhard Neukum
Institut für Weltraumsensorik und Planetenerkundung,
DLR German Aerospace Centre, Rutherfordstr. 2, 12489 Berlin, Germany
E-mail: gerhard.neukum@dlr.de

Juan Pérez-Mercader
Centro de Astrobiología (CSIC/INTA), Carretera Ajalvir, km 4,
28850 Torrejón de Ardoz, Madrid, Spain
E-mail: mercader@laeff.esa.es

Eva-Maria Pfeiffer
Alfred-Wegener-Institute for Polar and Marine Research, Columbusstrasse,
27568 Bremerhaven, Germany
E-mail: empfeiffer@awi-bremerhaven.de

Hinrich Rahmann
Zoological Institute, University of Stuttgart-Hohenheim, Garbenstr. 30,
70593 Stuttgart, Germany
E-mail: rahmann@uni-hohenheim.de

Petra Rettberg
Institute of Aerospace Medicine, DLR German Aerospace Center, Linder Höhe,
51147 Köln, Germany,
E-mail: petra.rettberg@dlr.de

Lynn J. Rothschild
Ecosystem Science and Technology, NASA Ames Research Center, Moffett Field,
CA, 94035-1000 U.S.A.
E-mail: lrothschild@mail.arc.nasa.gov

Hans-Joachim Schellnhuber
Potsdam Institute for Climate Impact Research, PO Box 601203, 14412 Potsdam,
Germany
E-mail schellnhuber@pik-potsdam.de

Manfred Schidlowski
Max-Planck-Institut für Chemie, Postfach 3060, 55020 Mainz, Germany
E-mail: paleo@mpch-mainz.mpg.de

Detlef Schönberner
Astrophysical Institute Potsdam, An der Sternwarte 16, 14482 Potsdam, Germany
E-mail deschoenberner@aip.de

Heide Schnabl
Institute of Agricultural Botany, Department of Physiology and Biotechnology of
Plants, University of Bonn, Karlrobert-Kreiten-Str. 13, 53115 Bonn, Germany
E-mail: hSchnabl@uni-bonn.de

Eva Spieck
Institute of General Botany, Department of Microbiology, University of Hamburg,
Ohnhorststrasse 18, 22609 Hamburg, Germany
E-mail: spieck@mikrobiologie.uni-hamburg.de

Helga Stan-Lotter
Institute Genetics and General Biology, University of Salzburg, Hellbrunnerstr. 34,
5020 Salzburg, Austria
E-mail: helga.stan-lotter@sbg.ac.at

Matthias Steffen
Astrophysical Institute Potsdam, An der Sternwarte 16, 14482 Potsdam
Germany
E-mail msteffen@aip.de

Karl O. Stetter
Department of Microbiology, University of Regensburg, Universitätsstr. 31,
93053 Regensburg, Germany
E-mail: karl.stetter@biologie.uni-regensburg.de

Willibald Stumptner
Space Research Institute, Department of Extraterrestrial Physics,
Austrian Academy of Sciences, Elisabethstr. 20, 8010 Graz, Austria
E-mail: stumptner.willibald@oeaw.ac.at

Hans G. Trüper
Institut für Mikrobiologie & Biotechnologie,
Rheinische Friedrich-Wilhelms-Universität, Meckenheimer Allee 168, 53115 Bonn,
Germany
E-mail: trueper@uni-bonn.de

B. Randall Tufts
Lunar and Planetary Laboratory, University of Arizona, 1629 East University Blvd.
Tucson, Arizona 85721-0092 USA
E-mail: rtufts@lpl.arizona.edu

Stéphane Udry
Département d'Astronomie, Observatoire de Genève, Université de Genève,
1290 Sauverny, Swizzerland
e-mail: stephane.udry@obs.unige.ch

Dirk Wagner
Alfred-Wegener-Institute for Polar and Marine Research, Research Department
Potsdam, Telegrafenberg A43, 14473 Potsdam, Germany
E-mail: dwagner@awi-potsdam.de

John W. Wilson
NASA Langley Research Center, Hampton, 23681 Virginia, USA
E-mail: john.w.wilson@larc.nasa.gov

David D. Wynn-Williams
British Antarctic Survey, High Cross, Madingley Road,
Cambridge CB3 0ET, United Kingdom
E-mail: ddww@bas.ac.uk

Subject Index

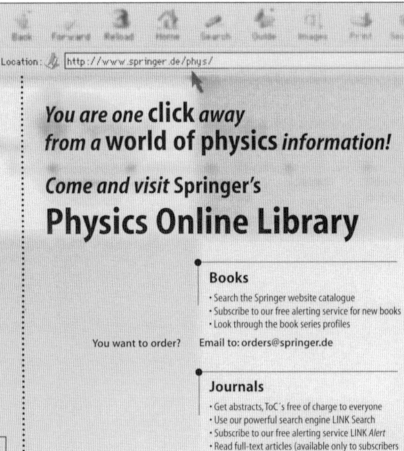